Operation Fundamentals in Bioreactor Engineering—Medium Handling

F. Xavier Malcata

Operation Fundamentals in Bioreactor Engineering—Medium Handling

Springer

F. Xavier Malcata
Department of Chemical and Biological
Engineering & LEPABE/ALiCE
University of Porto
Porto, Portugal

The author is an integrated member of LEPABE – Laboratory for Process Engineering, Environment, Biotechnology and Energy, supported by national funds through FCT/MECI: LEPABE, UID/00511/2025 (https://doi.org/10.54499/UID/00511/2025) and UID/PRR/00511/2025 (https://doi.org/10.54499/UID/PRR/00511/2025) and ALiCE, LA/P/0045/2020 (https://doi.org/10.54499/LA/P/0045/2020).

ISBN 978-3-032-03171-6 ISBN 978-3-032-03172-3 (eBook)
https://doi.org/10.1007/978-3-032-03172-3

This Springer imprint is published by the registered company Springer Nature Switzerland AG
The registered company address is: Gewerbestrasse 11, 6330 Cham, Switzerland

An expert is a person who has made all the mistakes that can be made in a very narrow field.

Niels Henrik Bohr (1885–1962)

To my family: Angela, Filipa, and Diogo.
For their everlasting understanding,
unselfish support, and endless love.

Preface

In the field of biotechnology, mastering the mathematical simulation of reactors based on living cells is crucial to optimize biomass production and metabolite generation. These two textbooks, targeted to biological and chemical engineers, constitute a set within a collection that offers a comprehensive, systematic, and integrative perspective on bioreactor engineering. *Operation Fundamentals in Bioreactor Engineering—Medium Handling* and *—Fluid Dynamics*, by complementing *Essentials of Biocatalysts: Cell Structure and Function* and *Modeling Fundamentals in Bioreactor Engineering*, aim to bridge the gap between qualitative biochemical analyses and empirical technological approaches; they indeed provide quantitative information that emphasizes dimensionless approaches and physicochemical interpretations, while offering a step-by-step scheme suitable for self-learning that highlights final formulae useful for professional practitioners.

The evolution of bioreactor engineering as a separate discipline has led to its formal integration into chemical and biochemical engineering curricula, at both undergraduate and graduate levels. However, the complex nature of the equations used in bioreactor modeling, coupled with the tendency for blind use of dedicated software to merely solve problems rather than understanding the solving strategy, does often frustrate students. The aforementioned collection fully addresses this issue by deriving analytical solutions, and maintaining simple and general mathematical forms—in line with mechanistic reasoning departing from the germane phenomena.

Operation Fundamentals in Bioreactor Engineering—Medium Handling delves specifically into the features of such subsidiary operations as medium preparation and sterilization, without which desired cells will not remain alive and metabolically active. *Operation Fundamentals in Bioreactor Engineering—Fluid Dynamics* tackles fluid flow in depth, including pumping and manipulation, without which biocatalysts will not gain access to nutrients sufficiently fast to be of commercial interest; and fluid mixing, without which concentration (and temperature) gradients will build up, thus compromising survival and metabolic activity of said biocatalysts. Despite being almost always condemned to a secondary position when describing bioreactors, subsidiary operations are a *sine qua non* for their efficient and profitable functioning; hence, they deserve fundamental reasoning and description from first principles, in much the same way as bioreactor simulation.

The first textbook begins with an introduction to bioreactor operation, followed by an overview of bioreactor performance. Formulation, inoculum preparation, buffering, and foaming are addressed under the scope of medium preparation, whereas cell death fundamentals, heating modes, choice of temperature, possibility of operating continuouswise in thermal or filter processing, selection of materials and devices, and aseptic operation are discussed for being relevant to medium sterilization.

In the second textbook, fluid flow is comprehensively explored, focusing first on homogeneous flow and then on heterogeneous one—namely involving liquid flow and gaseous bubbles, liquid flow and solid particles, and non-Newtonian behavior of liquid matrix; single bubbles and particles, and swarms and beds thereof, respectively, are sequentially tackled to some length. The subsequent chapter deals with devices that impart motion to the medium, in the form of various types of pumps, i.e., centrifugal, disk, peristaltic, and positive displacement—with accurate modeling of their performance from first principles and brief reference to cavitation, complemented by considerations on devices meant to manipulate flow, in terms of metering and control. Finally, the phenomena of vorticity and turbulence are described in detail, to set the stage for the following thorough presentation of mixers—covering both the case of mixing imparted by rotating device, viz. vertical shaft without and with impellers, horizontal radial bar, and vertical disk, and the case of mixing imparted by direct rotation of vessel, viz. vertical rotation, orbital motion with and without rotation, and horizontal rotation with liquid film or bulk liquid.

Notwithstanding the essentially chemical and mechanical nature of the topics covered, both books maintain an applied perspective—rooted in the reasoning of a bio/chemical engineer. Fundamental concepts and internal logic are thus prioritized over plain technological approaches, so as to allow students and instructors grasp the essentials behind preparation of media, and implementation of convection and mixing; rational choices of devices or operating conditions are accordingly outlined and discussed.

These textbooks have been primarily designed for advanced undergraduate and graduate students enrolled in fermentation, bioprocess, biological, or biochemical engineering courses; a background in elementary calculus and general physics is recommended to allow them to fully benefit from their content. Through carefully laid out derivations and solved illustrative problems, these books aim to equip readers with a deep understanding of the major concepts dealt with in bioreactor operation itself—taken chiefly from the point of view of culture medium and medium hydrodynamics.

Porto, Portugal F. Xavier Malcata

Contents

About the Author

F. Xavier Malcata was born in Malanje (Angola) in 1963. He obtained a B.Sc. degree in *chemical engineering* (5-year program) from the College of Engineering of the University of Porto (Portugal) in 1986, a Ph.D. degree in *chemical engineering/food science* from the University of Wisconsin (UW) (Madison, USA) in 1991, and a *habilitation*/professorship degree in *food science and engineering* from the College of Biotechnology of the Portuguese Catholic University (UCP) in 2004.

During his outstanding career in teaching and research, Prof. Malcata has held academic positions as Teaching Assistant at UCP (1985–1987) and UW (1988); Lecturer at UW (1989); Assistant Professor at UCP (1991–1998); Associate Professor at UCP (1998–2004); and Full Professor at UCP (2004–2010), Instituto Superior da Maia (2010–2012), and University of Porto (2012–). His teaching responsibilities have encompassed (among others) thermodynamics, transport phenomena, fluid mechanics, system dynamics and control, applied biochemistry, enzyme engineering, fermentation engineering, bioprocess engineering, bioreactor technology, bioresource engineering, food technology, and industrial project of B.Sc. and M.Sc. degrees in food engineering, bioengineering, and biological engineering.

Prof. Malcata has held professional positions as Dean of the College of Biotechnology of UCP (1998–2008)—with major responsibilities for the B.Sc. and M.Sc. degrees in food engineering, environmental engineering, and microbiology; Chairman of the Portuguese Society of Biotechnology (2003–2008); Coordinator of the Northern Chapter of Chemical Engineering, and Funding Member of the Specialization in Food Engineering, of the Portuguese Engineering Accreditation Board (2004–2009); Official Delegate of the Portuguese Government to the VI and VII Framework R&D Programs of the European Union, in the key areas of food quality and safety (2004–2006) and food, agriculture, and biotechnology (2006–2013); Chief Executive Officer of the nonprofit university/industry extension associations AESBUC (1998–2008) and INTERVIR^{+} (2006–2008), and the nonprofit entrepreneurial biotechnological support associations CiDEB (2005–2008) and INOVAR&CRESCER (2006–2008); and Director of the combined B.Sc./M.Sc. program in bioengineering of the University of Porto (2014–2021).

He has been the recipient of several international distinctions and awards, namely election for membership in *Phi Tau Sigma* (Honor Society of Food Science) in 1990, *Sigma Xi* (Honor Society of Scientific and Engineering Research) in 1990, *Tau Beta Pi* (Honor Society of Engineering) in 1991, and New York Academy of Sciences in 1992; *Cristiano P. Spratley Award* in 1985, *Centennial Award* in 1986, *College of Engineering Pedagogical Excellence Award* in 2021, and *College of Engineering Research Excellence Award* in 2023 by the University of Porto; *Ralph H. Potts Memorial Award* in 1991, *Young Scientist Research Award* in 2001, *Stephen S. Chang Award* in 2022, election as Fellow in 2014, and *Biotechnology Division Achievement Award* in 2025 by the American Oil Chemists' Society; *Foundation Scholar Award—dairy foods division* in 1998, *Danisco International Dairy Science Award* in 2007, *Distinguished Service Award* in 2012, election as Fellow in 2013, and *International Dairy Foods Association Teaching Award in Dairy Manufacturing* in 2020 by the American Dairy Science Association; *Canadian/International Constituency Investigator Award in Physical Sciences and Engineering* in 2002 and 2004 by Sigma Xi; *Excellence Promotion Award* in 2005 by the Portuguese Foundation for Science and Technology; *Edgar Cardoso Innovation Award* in 2007 by the Mayor of Gaia; *Scientist of the Year Award* in 2007 and election as Fellow in 2012 by the European Federation of Food Science and Technology; *Samuel C. Prescott Award* in 2008, *William V. Cruess Award* in

2014, election as Fellow in 2011, and *Distinguished Career Award in honor of Carl R. Fellers* in 2023 by the Institute of Food Technologists; *International Leadership Award* in 2008 and *Elmer Marth Educator Award* in 2011 by the International Association of Food Protection; and decoration as *Chevalier dans l'Ordre des Palmes Académiques* in 1999 by the French Government.

Among various complementary areas of research, Prof. Malcata has placed a major emphasis on microbiological and biochemical characterization, and technological improvement of traditional foods; development of nutraceutical ingredients, and nonconventional, probiotic foods and the like; design and development of enzymatic reactors for transformation of lipids; rational application of unit operations to specific agri-food processes; and enhancement of production, and eventual secretion of oil by microalgae via metabolic and genetic engineering tools. To date, he has published more than 485 papers in peer-reviewed, science journals, which have received more than 22,000 documented citations by the scientific community—supporting an *h*-index of 74; he has supervised 31 Ph.D. dissertations that were successfully concluded; he has also written 21 monographic books and edited five multiauthored books; he has further authored more than 70 chapters in edited books and 35 papers in trade journals, besides ca. 50 technical publications; he has been a member of 74 peer-reviewing committees of research projects and fellowships; he has acted as a supervisor of 106 individual fellowships, and collaborated with 61 research and development projects—of which he has served as principal investigator in 37; he has participated in 50 organizing and scientific committees of professional meetings; he has also delivered 160 invited lectures worldwide, besides more than 530 volunteer presentations in professional meetings, congresses, and workshops; he has served in the editorial board of five major journals in the food science and applied biotechnology areas; and he has reviewed more than 600 manuscripts for international journals and encyclopedia. According to the latest version of *Stanford Ranking*, Prof. Malcata ranks in the top 0.5% of the most influential scientists worldwide—among ca. seven million scientists that have published at least five papers in refereed journals.

Prof. Malcata is a long-standing professional member of the Institute of Food Technologists, as well as the American Oil Chemists' Society, the American Dairy Science Association, and the International Association of Food Protection.

Part I

Introduction to Bioreactor Operation

General Overview

1

The good thing about science is that it is true,
whether or not you believe in it.
Neil deGrasse Tyson (1958–)

Bioreactors are vessels designed to carry out reactions brought about by cells, or portions thereof (e.g., enzymes); their ultimate purpose is to generate extra biomass or specific metabolites for a number of practical applications—following harvest, separation, and purification of their contents. In either case, the biocatalyst must gain access to substrate(s)—which, in the case of living cells, also serve as nutrients for metabolism lato sensu; all such starting compounds are to be supplied in liquid form, usually dissolved in water—and even such gaseous substrates as O_2 or CO_2 can be taken up only if in solubilized form.

In the first book of this series, *Essentials of Biocatalysts: Cell Structure and Function*, extensive information was provided on biochemical reactions, from the point of view of the biocatalyst involved; it was meant to help understand how metabolic pathways lead to formation of more (or larger) cells, as well as several intermediate or final metabolites. The next book of this series, *Modeling Fundamentals in Bioreactor Engineering*, focuses on mathematical models of what happens inside the bioreactor, with a brief tour of such subsidiary operations as online control and downstream separation. However, a bioreactor will not operate unless nutrients (or reactants) are supplied via carefully designed culture media—which constitutes the emphasis of the present book on *Operation Fundamentals in Bioreactor Engineering—Medium Handling*. Therefore, preparation of an appropriate medium is the starting step and plays a critical role upon eventual success of the bioprocess.

Since nutrients in the medium are normally suitable for growth of a very many microorganisms, absence of microbial contaminants has to be guaranteed in the first

F. X. Malcata, *Operation Fundamentals in Bioreactor Engineering—Medium Handling*, https://doi.org/10.1007/978-3-032-03172-3_1

place; sterilization accordingly comes into play and classically resorts to high-temperature treatments—able to destroy not only vegetative cells, but also their spores as resistance forms. Once the medium has been sterilized and cooled down to room temperature, inoculation is in order—specifically with the microbial species designed for, or intended to operate as biocatalyst.

In order to maintain a suitable rate of growth of biomass or primary/secondary metabolism, appropriate environmental conditions have to be provided—namely of temperature and *pH*. Note that enzymes, as molecular catalysts responsible for cell metabolism, are proteins constituted by amino acid residues—many of which possess side chains that act as Bronsted and Lowry's acids, that is, they can release to or abstract protons from nearby bases or acids, respectively, depending on medium *pH*. Due to existence of hydrogen bonds involving chiefly proteinaceous oxygen and nitrogen as donor atoms, the topology of the three-dimensional network of intermolecular forces, and thus the molecular conformation developed by enzymes rely critically on their states of ionization. Maximum activity requires the correct conformation at the active site, which is in turn associated to an optimum *pH*—so media are typically buffered with strong/weak acids/bases to set the working *pH* to the desired value, while resisting change in *pH* due to consumption/formation of compounds with protolytic behavior themselves.

On the other hand, several nutrients or processing aids present in the medium decrease the surface tension of the water/air interface, thus playing the role of surfactants; this facilitates retention of bubbles of gas at the surface of the liquid medium, concomitant with dramatic increases in interfacial area materialized by the liquid envelopes of such bubbles—which are detrimental to biocatalysts, because of the preferential adsorption thereto and the stretching forces associated therewith. This is why fundamental knowledge on foaming phenomena is relevant to support precautionary measures—of either mechanical nature or produced via addition of specific chemicals, meant to avoid (excessive) foam buildup.

Problem 1.1

The concept of *pH*, originally stated by Danish chemist Søoren Sørensen in 1909 to represent the power (*p*) of hydrogen ions (*H*), entails a regular scale ranging between 0 and 14. Find examples from practical life that illustrate *pH* values along such scale.

Solution

From a brief search in literature, the following examples become readily accessible:

pH 0: Battery acid
pH 1: Stomach
pH 2: Lemon juice
pH 3: Wine, apple juice

(continued)

Problem 1.1 (continued)
pH 4: Banana, cherry, tomato juice
pH 5: Black coffee
pH 6: Milk, saliva
pH 7: Plain water
pH 8: Blood
pH 9: Egg white, baking soda
pH 10: Household bleach, hand soap
pH 11: Household ammonia
pH 12: Hair remover, strong bleach
pH 13: Oven cleaner
pH 14: Drain cleaner

In the case of the biological examples above, the stated value of *pH* is associated to the optimum performance in terms of biochemical reactions and microbial activity; when dealing with growth media, choice of *pH* is often dictated by maximum nutrient availability.

Further Reading

Atri, M. R., & Ashrafizadeh, S. N. (2010). The importance of foams and antifoaming in bioprocesses. *Pakistan Journal of Biotechnology, 7*, 19–39.

Basu, S. A., & Desai, P. K. (2005). Quality control of culture media in a microbiology laboratory. *Indian Journal of Medical Microbiology, 23*, 159–163.

Bykowski, T., Verma, A., Brissette, C. A., & Stevenson, B. (2011). Aseptic technique. In *Current protocols—Essential laboratory techniques* (pp. 4.1.1–4.1.12). Wiley.

Condayan, C. (2008). Culture media. *Nature Reviews in Microbiology, 6*, 646.

Esteban, M. A., Alcalá, M., Marcos, A., & Fernandez-Salguero, J. (2007). Water activity of culture media used in food microbiology. *International Journal of Food Science and Technology, 25*, 464–468.

Farfour, E., Limousin, L., Amandine, H., et al. (2019). Optimization of incubation duration of culture media in microbiology. *Annales de Biologie Chimique, 77*, 525–531.

Kragh, H. (2025). S. P. L. Sørensen, the pH concept and its early history. *Foundations of Chemistry*.

Popovic, M. K., & Portner, R. (2020). Bioreactors and cultivation systems for cell and tissue culture. In *Encyclopedia of life support systems*. UNESCO.

Prins, R. C., & Billerbeck, A. (2021). A buffered media system for yeast batch culture growth. *BMC Microbiology, 21*, 127.

Senz, M., Keil, C., Schmacht, M., Palinski, S., Cammerer, B., & Hagebock, M. (2019). Influence of media heat sterilization process on growth performance of representative strains of the genus *Lactobacillus*. *Fermentation, 5*, 20.

Tyagi, S., & Mani, S. (2023). Media and buffer preparation in cell culture. In S. Mani, M. Singh, & A. Kumar (Eds.), *Animal cell culture; Principles and practice* (pp. 77–88). Springer.

Zhang, S., Handa-Corrigan, A., & Spier, R. E. (1992). Foaming and media surfactant effects on the cultivation of animal cells in stirred and sparged bioreactors. *Journal of Biotechnology, 25*, 289–306.

Part II

Bioreactor Performance

Medium Preparation

2

Any sufficiently advanced technology
is equivalent to magic.
Arthur C. Clarke (1917–2008)

Bioreactors are normally used for a specific experiment or production run—which may last for hours, days, or even weeks in a row, depending on the utilized organism and intended product. Sequential steps of bioreactor preparation, medium formulation, and preculture/inoculation are required and, as a whole, they constitute upstream operations; cultivation proper comes next as onstream operation, involving physicochemical and mechanical tools; these are complemented by cell harvest, followed by extraction/separation/purification of cells or compounds—which together account for downstream operations. This book focuses on up- and onstream operations, since they are crucial to functioning of any bioreactor and ultimately leverage or constrain its performance.

In fact, the success of any biological process hinges critically upon the environmental conditions provided and designed in advance for biomass expansion or product formation; hence, criterious formulation of medium in terms of nutrients is a must. However, specific conditions of temperature, *pH*, dissolved oxygen, stirring, medium level, and foaming extent are also to be assured at each stage of operation of a bioreactor. All of them call for careful and continuous monitoring of the process, so that any significant deviation from the predefined setpoint (usually associated to optimum performance, according to some objective function) is timely corrected via an appropriate control system. Monitoring also conveys information on the progress of the process, physiological state of the cells, and contamination by spurious undesired microorganisms or degeneration of original strains. Online monitoring and control, together with downstream operations, are briefly covered in the last book of this series.

F. X. Malcata, *Operation Fundamentals in Bioreactor Engineering—Medium Handling*, https://doi.org/10.1007/978-3-032-03172-3_2

Problem 2.1

Barley malt is a crucial ingredient in typical brewing processes, where it brings along most nutrients, namely, fermentable sugars for growth and alcoholic fermentation by yeasts; however, its precursor cereal consists chiefly of starch and other complex, polymeric carbohydrates. Although wheat, spelt, rye, oats, and rice can also (or instead) be used, barley still remains the base of most beers—being subjected to essentially the same processing as 5000 years ago in Mesopotamia.

(a) Justify choice of barley as feedstock to produce microbial broth.
(b) Explain how such a solid cereal gives eventually rise to a nutritious fluid medium.
(c) Why is roasting useful in manufacture of a few specific beers?

Solution

(a) Barley is the ideal cereal grain for malting and ultimately for brewing because it is self-contained—that is, it possesses a husk that protects the germ, a high starch-to-protein ratio suitable for high yields, a balanced and complete enzyme system able to breakdown starch to small sugars, a self-adjusting pH, a light color, and a neutral flavor. Although wheat and rye have also been routinely malted in the manufacture of specialty beers, the finished malt does not perform in the brewhouse as well as malted barley.
(b) The nutritious medium required by yeasts in the brewing process must exhibit a liquid form—as obtained by an aqueous infusion of milled malt, where both simple sugars and hydrolases are released and dissolved. This requires previous malting of the feeding stock, which involves three basic steps—steeping, germinating, and drying. During the first one, and after preliminary sorting and cleaning, barley is covered with water, which, upon absorption by the kernel, triggers germination. For the next ca. 48 h, raw barley alternates between submerged and drained states until its moisture content increases from ca. 12% to ca. 44% (w/w). The absorbed water activates adventitious enzymes and stimulates the embryo to synthesize fresh enzymes; they break down the protein and carbohydrate matrix that encloses starch granules in the endosperm, thus opening the starch reserves of the seed while newly developed hormones initiate growth of the acrospire (or sprout). The chitted barley is then transferred from the steep tank to the germination compartment, and germination, already initiated in the former, continues for 4–5 days; it leads to extensive breakdown of protein and starch to simpler, and thus soluble and fermentable peptides and sugars. Germination is controlled by drawing temperature-adjusted, humidified air through the bed—and turners keep the bed from compacting and avoid rootlets from growing together (or felting).

(continued)

Problem 2.1 (continued)

Germination is halted via drying; otherwise, all starch reserves in the kernel would be eventually used up by the growing plant, and thus become unavailable to prepare the growth medium for yeasts. Base malts are kiln-dried—typically with a finish temperature of ca. 85 °C for 2–4 h; this leads also to development of specific flavors, ranging from very light malty to subtle malty.

(c) Roasting is but a more intensive heating treatment during the drying step of malting, and inherently decreases the amount of enzymes active to aid in breakdown of starch and the like (which would constrain the levels of nutrients available). However, the major purpose of roasting is to contribute unique flavors to the final beer, such as intense malty, sweet caramel, nutty, woody, coffee, or even burnt—besides conveying richer colors, ranging from golden through red and even to black.

2.1 Formulation

All growth media, manufactured whether for industrial or laboratory purposes, are expected to satisfy the needs of the microorganism in terms of carbon, nitrogen, minerals, growth factors, and (obviously) water. In addition, growth inhibitors must be absent; otherwise, biomass expansion would be compromised.

Carbon provides both energy and chemical moieties for biosynthesis; it is usually conveyed by carbohydrates, namely, simple sugars (e.g., glucose), disaccharides (e.g., sucrose), or even polysaccharides (e.g., starch, cellulose). Nitrogen is required for synthesis of amino acids, and thus proteins and enzymes therefrom, as well as enzyme cofactors; and also for synthesis of nucleotides, and thus nucleic acids therefrom. Despite its abundance in molecular form as part of the atmosphere, most cells can only uptake it in (fixed) salt form (e.g., ammonia, nitrate, nitrite). Minerals are necessary for activity of the cell, chiefly owing to their role as cofactors or constitutive part of the active sites of several crucial enzymes; this includes P, S, Mg, and Fe—together with trace amounts of Mn, B, Zn, Cu, and Mo. Growth factors are organic compounds hardly synthesized by microorganisms; this is the case of vitamins—or even some amino acids/nucleotides, in the case of fastidious microorganisms or auxotrophic strains. Finally, water serves as carrier solvent of all such ingredients—being thus of vital importance for industrial microbiology.

Problem 2.2

The biomass of *Saccharomyces cerevisiae* is approximately described by $CH_{1.703}O_{0.459}N_{0.171}$ for pseudo-chemical formula. Calculate the minimum molar concentration of ammonia in the medium that is required to fully deplete glucose

(continued)

Problem 2.2 (continued)

via aerobic growth of that yeast—knowing that the respiratory (molar) quotient, Y_{CO_2/O_2}, is 1.1.

Solution

The underlying chemical equation for aerobic growth of *S. cerevisiae* on ammonia reads

$$C_6H_{12}O_6 + aO_2 + bNH_3 \rightarrow cCH_{1.703}O_{0.459}N_{0.171} + dCO_2 + eH_2O, \qquad (2.1)$$

where a, b, c, d, and e denote (still unknown) stoichiometric coefficients; glucose, $C_6H_{12}O_6$, was deliberately set as reference compound, so it bears unity for stoichiometric coefficient. Recalling its definition, the respiratory coefficient sets

$$Y_{CO_2/O_2} = \frac{d}{a} = 1.1 \qquad (2.2)$$

when combined with Eq. (2.1) and the given datum; whereas mass balances to C, H, O, and N support

$$6 = c + d, \qquad (2.3)$$

$$12 + 3b = 1.703c + 2e, \qquad (2.4)$$

$$6 + 2a = 0.459c + 2d + e, \qquad (2.5)$$

and

$$b = 0.171c, \qquad (2.6)$$

respectively, again stemming from Eq. (2.1). The set of Eqs. (2.2)–(2.6) is represented by the following extended coefficient matrix

(continued)

Problem 2.2 (continued)

$$\begin{array}{ccccc|c} a & b & c & d & e & \\ \end{array}$$
$$\left[\begin{array}{ccccc|c} 1.1 & 0 & 0 & -1 & 0 & 0 \\ 0 & 0 & 1 & 1 & 0 & 6 \\ 0 & -3 & 1.703 & 0 & 2 & 12 \\ -2 & 0 & 0.459 & 2 & 1 & 6 \\ 0 & 1 & -0.171 & 0 & 0 & 0 \end{array}\right], \tag{2.7}$$

where exchange of the fourth and fifth rows leaves

$$\begin{array}{ccccc|c} a & b & c & d & e & \\ \end{array}$$
$$\left[\begin{array}{ccccc|c} 1.1 & 0 & 0 & -1 & 0 & 0 \\ 0 & 0 & 1 & 1 & 0 & 6 \\ 0 & -3 & 1.703 & 0 & 2 & 12 \\ 0 & 1 & -0.171 & 0 & 0 & 0 \\ -2 & 0 & 0.459 & 2 & 1 & 6 \end{array}\right]; \tag{2.8}$$

replacement of the last row by its sum with the first row multiplied by $2/1.1 = 1.818$ gives then rise to

$$\begin{array}{ccccc|c} a & b & c & d & e & \\ \end{array}$$
$$\left[\begin{array}{ccccc|c} 1.1 & 0 & 0 & -1 & 0 & 0 \\ 0 & 0 & 1 & 1 & 0 & 6 \\ 0 & -3 & 1.703 & 0 & 2 & 12 \\ 0 & 1 & -0.171 & 0 & 0 & 0 \\ 0 & 0 & 0.459 & 0.182 & 1 & 6 \end{array}\right]. \tag{2.9}$$

Swapping of the second and fourth rows transforms Eq. (2.9) to

(continued)

Problem 2.2 (continued)

$$\begin{array}{ccccc|c} a & b & c & d & e & \\ \end{array}$$
$$\left[\begin{array}{ccccc|c} 1.1 & 0 & 0 & -1 & 0 & 0 \\ 0 & 1 & -0.171 & 0 & 0 & 0 \\ 0 & -3 & 1.703 & 0 & 2 & 12 \\ 0 & 0 & 1 & 1 & 0 & 6 \\ 0 & 0 & 0.459 & 0.182 & 1 & 6 \end{array}\right], \tag{2.10}$$

where replacement of the third row by its sum with the triple of the second row produces

$$\begin{array}{ccccc|c} a & b & c & d & e & \\ \end{array}$$
$$\left[\begin{array}{ccccc|c} 1.1 & 0 & 0 & -1 & 0 & 0 \\ 0 & 1 & -0.171 & 0 & 0 & 0 \\ 0 & 0 & 1.190 & 0 & 2 & 12 \\ 0 & 0 & 1 & 1 & 0 & 6 \\ 0 & 0 & 0.459 & 0.182 & 1 & 6 \end{array}\right]; \tag{2.11}$$

exchange of third and fourth rows is now in order, namely,

$$\begin{array}{ccccc|c} a & b & c & d & e & \\ \end{array}$$
$$\left[\begin{array}{ccccc|c} 1.1 & 0 & 0 & -1 & 0 & 0 \\ 0 & 1 & -0.171 & 0 & 0 & 0 \\ 0 & 0 & 1 & 1 & 0 & 6 \\ 0 & 0 & 1.190 & 0 & 2 & 12 \\ 0 & 0 & 0.459 & 0.182 & 1 & 6 \end{array}\right]. \tag{2.12}$$

The fourth row is now to be replaced by its sum with -1.190-fold the third row, and the fifth row by its sum with -0.459-fold the third row to get

(continued)

Problem 2.2 (continued)

$$\begin{array}{c} \\ \\ \\ \\ \\ \end{array}\begin{array}{c}\begin{matrix} a & b & c & d & e & \end{matrix}\\ \left[\begin{array}{ccccc|c} 1.1 & 0 & 0 & -1 & 0 & 0 \\ 0 & 1 & -0.171 & 0 & 0 & 0 \\ 0 & 0 & 1 & 1 & 0 & 6 \\ 0 & 0 & 0 & -1.190 & 2 & 4.860 \\ 0 & 0 & 0 & -0.277 & 1 & 3.246 \end{array}\right]\end{array}, \tag{2.13}$$

while the last row will be replaced by its sum with the previous row upon multiplication by −0.277/1.190 = −0.233 to obtain

$$\begin{array}{c}\begin{matrix} a & b & c & d & e & \end{matrix}\\ \left[\begin{array}{ccccc|c} 1.1 & 0 & 0 & -1 & 0 & 0 \\ 0 & 1 & -0.171 & 0 & 0 & 0 \\ 0 & 0 & 1 & 1 & 0 & 6 \\ 0 & 0 & 0 & -1.190 & 2 & 4.860 \\ 0 & 0 & 0 & 0 & 0.534 & 2.114 \end{array}\right]\end{array}. \tag{2.14}$$

The last row in Eq. (2.14) is equivalent to

$$0.534e = 2.114 \tag{2.15}$$

that degenerates to

$$e = \frac{2.114}{0.534} = 3.959; \tag{2.16}$$

the fourth row looks like

$$-1.190d + 2e = 4.860, \tag{2.17}$$

where insertion of Eq. (2.16) unfolds

$$-1.190d + 2 \cdot 3.959 = 4.860 \tag{2.18}$$

with isolation of d leaving

(continued)

Problem 2.2 (continued)

$$d = \frac{4.860 - 7.918}{-1.190} = 2.570. \tag{2.19}$$

The condition underlying the third row of Eq. (2.14) reads

$$c + d = 6, \tag{2.20}$$

so combination with Eq. (2.19) gives

$$c + 2.570 = 6 \tag{2.21}$$

that breaks down to

$$c = 6 - 2.570 = 3.430; \tag{2.22}$$

by the same token, the second row of Eq. (2.14) supports

$$b - 0.171c = 0, \tag{2.23}$$

where insertion of Eq. (2.22) gives rise to

$$b - 0.171 \cdot 3.430 = 0 \tag{2.24}$$

equivalent to

$$b = 0.587. \tag{2.25}$$

Finally,

$$1.1a - d = 0 \tag{2.26}$$

serves as alias to the first row in Eq. (2.14), and becomes

$$1.1a - 2.570 = 0 \tag{2.27}$$

with the aid of Eq. (2.19); after solving for a, Eq. (2.27) turns to

(continued)

Problem 2.2 (continued)

$$a = \frac{2.570}{1.1} = 2.336. \tag{2.28}$$

Therefore, the liquid medium should contain (at least) 0.587 mol of NH_3 per each mole of $C_6H_{12}O_6$, as a consequence of Eqs. (2.1) and (2.25); under these conditions, 3.430 mole of biomass with chemical formula $CH_{1.703}O_{0.459}N_{0.171}$ will form, in agreement with Eqs. (2.1) and (2.22)—along with release of 2.570 mole of CO_2 and 3.959 mole of H_2O, as per Eq. (2.1) coupled with Eq. (2.19) or Eq. (2.16), respectively.

2.1.1 Laboratory Media

When microorganisms are grown at laboratory scale, as required during preliminary steps of screening for wild strains or performing recombinant DNA technology, or else in preparing inocula for larger-volume fermentation batches, media are more defined; since the amounts needed are relatively small, cost will hardly become prohibitive. Generally speaking, a growth (or culture) medium is a solid (agar) or liquid (broth) designed to support biomass expansion and metabolism of a population of microorganisms or cells; these forms of culture media convey different growth environments for microorganisms—where cells either form colonies on the agar or remain suspended in the broth. Agar agar is a natural polysaccharide produced by marine macroalgae and the most common solidifying agent; if occurrence of hydrolysis is anticipated, silica gel is used instead. Agar media are utilized to isolate colonies, whereas broth media are used for planktonic (or free-floating) cells; typically, microorganisms are first grown in broth to reach a particular growth stage—at which point cells can be plated on agar for further study or long-term storage. A major difference also exists between growth media used for (plant or animal) cell culture or microbiological culture—in that cells cannot grow in the absence of hormones or growth factors, otherwise present in the whole multicellular organisms they were derived from.

Depending on the type and combination of nutrients carried, distinct categories of media may in turn be produced—namely, culture, minimal, selective, differential, and transport. Culture media contain all ingredients necessary for growth, so they are used for general cultivation and maintenance of microorganisms in culture collections; they are labeled as either complex or defined.

Complex media entail an undefined (or complex) assortment and quantity of nutrients, in the form of water-soluble extracts of plant or animal tissues, as is the case of enzymatic digests of animal proteins, such as peptone and tryptone from beef, brain heart infusion, casein/soy digest, or yeast extract; they represent excellent sources of amino acids, peptides, and proteins and thus serve as major N source. Acid digestion is often used to bring about hydrolysis of said animal/plant tissues or microbial cultures; however, papain, pepsin, trypsin, or pancreatic juice as enzymatic cocktail may instead be utilized for the same purpose. Since they provide a variety of nutrients, they are suitable for most needs—should easy-to-grow microorganisms be under scrutiny, namely, most heterotropic microorganisms. A sugar (often glucose) is often added to serve as main carbon and energy source; this combination of extracts and sugar gives rise to a medium that ends up being rich in both minerals and organic nutrients.

Defined media are, in turn, composed of pure ingredients in carefully measured concentrations, dissolved in double-distilled water; hence, the exact qualitative and quantitative composition of the starting medium is known—as happens with nutrient agar and broth, plate count agar, or trypticase soy agar. These types of media are required for physiological and metabolic studies, encompassing organisms that require specific nutrients for growth; being suitable to grow specific heterotrophs, they are often mandatory for chemo- and photoautotrophs. Defined media classically contain a simple sugar as carbon and energy source, an inorganic nitrogen source, various mineral salts, and, if necessary, growth factors, such as pure amino acids, vitamins, purines, and pyrimidines. The most common medium ingredients/supplements are L-arabinose, Arg, Asp, biotin, citric acid, cycloheximide, Cys, dextrin, EDTA, Ehrlich's reagent, fructose, fuchsin, galactose, glucose, glycerol, glycogen, lactose, maltose, mannitol, niacin, ornithine, Phe, resazurin, ribose, sorbitol, starch, and sucrose. On the other hand, medium composition may play an important role, in which case interaction between medium components is to be taken into account. For instance, yeast extract, peptone, and amino acids bind large amounts of Hg^{2+}, Pb^{2+}, Ag^{+}, and Cu^{2+}; glucose, citrate, and Tris exhibit nonselective metal binding; EDTA chelates essential trace metals; NaCl increases Zn toxicity but decreases Cd toxicity; agar increases Sn toxicity; silica gel decreases Sn^{2+}, Cd^{2+}, Pb^{2+}, Ni^{2+}, and Zn^{2+} toxicity; phosphates precipitate Cd^{2+}, Pb^{2+}, and Fe^{2+}; and carbonates precipitate Pb^{2+}.

A defined medium that contains just enough ingredients to support growth is termed minimal medium, for example, M9 minimal medium, but the number and nature of its ingredients are heavily dependent on the organism at stake. Such type of medium usually contains a carbon and energy source, namely, glucose, or a less energy-rich source like succinate, and various salts that provide essential elements, namely, N, Mg, P, and S, so that proteins and nucleic acids can be synthesized. Supplementary minimal media also contain a single selected agent, usually an amino acid or sugar, which allows for culturing of specific lines of auxotrophic recombinants.

Selective media are used in cell culture to ensure survival or proliferation of cells bearing certain features, such as antibiotic resistance or ability to synthesize a certain metabolite; normally, the presence of a specific gene (serving as marker) confers the cell with the ability to grow in this type of medium. Such media suppress unwanted microorganisms or instead encourage desired ones (and are thus sometimes termed enrichment media); one particular case is inclusion of reducing agents when growth of obligate anaerobes is envisaged. In the case of eukaryotes, selective media often contain neomycin aimed at selecting cells that have been successfully transfected with a plasmid carrying the neomycin resistance gene for marker; conversely, ganciclovir is used as ingredient to selectively kill cells that carry the corresponding marker, the Herpes simplex virus thymidine kinase. Common examples of selective media include Simons citrate agar selective for Gram$^-$ bacteria; eosin methylene blue, triple sugar iron agar, and bromocresol purple lactose agar, all selective for Enterobacteriaceae; mannitol salt agar, selective for *Staphylococcus*; Baird-Parker agar and DNAse test agar, both selective for *Staphylococcus aureus*; YM, selective for eukaryotes; XLD agar, selective for Gram$^-$ enteric bacteria; buffered charcoal yeast extract agar, selective for (Gram$^-$) *Legionella pneumophila*; McConkey agar and xylose lysine deoxycholate, selective for growth of Gram$^-$ bacteria; HIS medium, a cell culture medium that lacks His; Sabouraud agar, selective for certain fungi; dichloran rose Bengal chloramphenicol agar, selective for enumeration of molds and yeasts; Mueller-Hinton agar, selective for *Neisseria* spp.; and Hektoen enteric agar and Salmonella-Shigella agar, both selective for *Shigella* and *Salmonella*.

A differential (or indicator) medium visibly distinguishes colonies of one microorganism type from another, by taking advantage of the biochemical characteristics of one of them when growing in the presence of specific nutrients or indicators (e.g., neutral red, phenol red, eosin y, methylene blue). They depart from either complex or defined media and include supplementation with growth-promoting or growth-inhibiting additives, and may be species- or strain-selective (e.g., cycloheximide inhibits all eukaryotic growth in mixed cultures). Illustrative examples of this type of media include blood agar with bovine heart blood that becomes transparent in the presence of such β-hemolytic organisms as *Streptococcus pyrogenes* and *Staphylococcus aureus*; Granada medium, differential for *Streptococcus agalactiae*, which grows as distinctive red colonies; eosin methylene blue and McConkey agar, both differential for lactose fermenters; mannitol salt agar, differential for mannitol fermenters; and X-gal, differential for lac operon mutants.

Transport media are designed for temporary storage of specimens, namely, during transport to the laboratory for cultivation, and to maintain viability of all microorganisms without significantly changing their concentrations. Hence, they are composed solely of buffers and salts, and lack C, N, and organic growth factors so as to prevent active multiplication; in the case of anaerobes, they should also be devoid of molecular oxygen. Examples include thioglycolate broth for strict anaerobes; Stuart transport medium, a non-nutrient, soft agar gel containing a reducing agent to prevent oxidation and charcoal for neutralization; Venkataraman-Ramakrishna medium for *Vibrio cholera*; bacterial inhibitor-containing media for *Neisseria gonorrhoeae*; and buffered glycerol saline-containing media for enteric bacilli.

The nutritional requirements posed by cyanobacteria and microalgae, as photoautotrophic organisms, vary depending on their original source (e.g., fresh water, seawater) and species, as well as production purpose (e.g., biomass, colorants, antioxidants, oils); dissolved CO_2 serves as (inorganic) source of C, but an inorganic source of N is required, besides other macro- and micronutrients, all conveyed via soluble salts.

The most popular media for general growth of freshwater specimens are Bristol medium (containing calcium and sodium chlorides, sodium nitrate, potassium phosphate, and magnesium sulfate), Waris medium (containing potassium nitrate, calcium and magnesium sulfates, and ammonium phosphate to higher concentrations, and ferric, manganese, zinc, and cobalt chlorides, sodium molybdate, and EDTA to lower concentrations), BG11 medium (containing sodium carbonate, ferric ammonium citrate, calcium chloride, sodium nitrate, potassium phosphate, magnesium sulfate, sodium thiosulfate, citric acid, and EDTA as macronutrients, as well as boric acid, magnesium chloride, sodium molybdate, cobalt nitrate, and zinc and copper sulfates as micronutrients), Bold Basal medium (containing calcium and sodium chlorides, sodium nitrate, potassium phosphate, and magnesium sulfate as macronutrients, as well as manganese chloride, cobalt nitrate, molybdenum oxide, iron, copper, zinc sulfates, boric and sulfuric acids, and EDTA as micronutrients), and WC medium (containing chiefly sodium carbonate, calcium chloride, sodium nitrate, sodium silicate, magnesium sulfate, potassium phosphate, and boric acid, besides iron, cobalt, magnesium chloride, sodium molybdate, sodium orthovanadate, copper and zinc sulfates, vitamin B_{12}, thiamine, biotin, and EDTA to much lesser levels). If growth of specific species is intended, then one may resort to Cyanophycean medium (designed for growth of cyanobacteria, and containing potassium nitrate, potassium phosphate, and magnesium sulfate as macronutrients, and cobalt, ferric, manganese, and zinc chlorides, sodium molybdate, and EDTA as micronutrients), Allen medium (designed also for cyanobacteria and containing sodium carbonate, calcium chloride, sodium nitrate, sodium silicate, potassium and magnesium sulfate, HEPES, and citric acid for macronutrients, and the same micronutrients as the previous one), DYIII medium (designed for growth of chrysophytes, and containing MES, calcium and potassium chlorides, sodium glycerophosphate, ammonium and sodium nitrate, sodium silicate, and magnesium sulfate as macronutrientes, as well as cobalt, ferric, and manganese chlorides, sodium molybdate, zinc sulfate, vitamin B_{12}, biotin, thiamine, EDTA, and boric acid as micronutrients), Botryococcus medium (designed for growth of *Botryococcus* spp., and containing calcium and sodium chloride, sodium nitrate, ammonium and potassium phosphates, and magnesium sulfate), Ochromonas medium (designed for *Ochromonas* spp., and containing glucose, tryptone, yeast extract, and liver extract infusion), and Euglena medium (designed for growth of *Euglena* and *Chlorogonium* spp., and containing sodium acetate, calcium chloride, beef and yeast extracts, and tryptone).

With regard to seawater species, the most popular media are Erdschreiber medium (designed for marine microalgae at large, and containing calcium carbonate, sodium nitrate, and sodium phosphate as macronutrients, and cobalt, ferric, manganese, and zinc chlorides and sodium molybdate, as well as vitamin B_{12} and EDTA to a lesser extent, all dissolved in pasteurized seawater containing greenhouse soil), F/2 medium (also a general purpose medium, containing sodium nitrate, sodium phosphate, and sodium silicate as macronutrients, as well as copper chloride, sodium molybdate, ammonium iron, cobalt, magnesium, and zinc sulfates, vitamin B_{12}, biotin, thiamine, and EDTA as micronutrients), enriched seawater medium (still another medium for growth of marine microalgae at large, and containing sodium nitrate and sodium glycerate as major components, as well as cobalt and ferric chlorides, ammonium ferric, magnesium, and zinc sulfates, vitamin B_{12}, biotin, thiamine, EDTA, boric acid, and HEPES as minor components), LDM medium (designed for marine diatoms, and containing calcium and sodium chlorides, sodium nitrate, potassium phosphate, magnesium sulfate, and tryptone as macronutrients, as well as cobalt, ferric, manganese, and zinc chlorides, sodium molybdate, vitamin B_{12}, and biotin as micronutrients, all dissolved also in pasteurized seawater), Porphyridium medium (designed for growth of *Porphyridium*, *Brachiomonas*, and *Cylindrotheca* spp., and containing yeast extract and tryptone as major components, as well as greenhouse soil and calcium carbonate, dissolved/suspended in nonpasteurized seawater), and artificial seawater medium (designed for growth of *Agmenellum quadruplicatum*, and containing ammonium, calcium, sodium, and potassium chlorides, sodium nitrate, potassium phosphate, magnesium sulfate, and tricine as macronutrients, as well as cobalt and ferric chlorides, magnesium and zinc sulfates, vitamin B_{12}, EDTA, and boric acid as micronutrients).

In the case of plant cell culturing, a number of carbohydrates have been used as C source (2–5%) besides sucrose—for example, lactose, galactose, maltose, and starch; myoinositol (50–5000 mg/L) is also included as growth stimulator. Although most plants can synthesize amino acids, addition thereof permits assimilation of N faster than in inorganic form; this usually takes the form of casein hydrolyzate (0.25–1 g/L), Gly (2 mg/L), Gln (8 mM), Asn (100 mg/L), Cys (10 mg/L), and Tyr (100 mg/L). Culture media should contain also such macronutrients as K (20–30 mM) and P, Ca, Mg, and S (1–3 mM) for satisfactory growth and morphogenesis; essential micronutrients encompass Fe (often the most critical one), Mn, Zn, B, Cu, and Mo. The vitamins most commonly employed are thiamine (0.1–10 mg/L), nicotinic acid (0.1–5 mg/L), and pyridoxine (0.1–10 mg/L). Some media are also supplemented with natural extracts, such as protein hydrolyzates, coconut milk, yeast extract, ground banana, and orange or tomato juice. Finally, presence of growth regulators is important for their vital role upon stem elongation, tropism, and apical dominance; the most common are auxins, cytokinins, gibberellins, and abscisic acid. The most known basic media utilized are Murashige and Skoog, Linsmaier and Skoog, Gamborg, and Nitsch and Nitsch media; they entail differing concentrations of calcium, potassium, and ammonium phosphates, calcium, ammonium, and potassium nitrates, calcium and potassium chlorides, and magnesium and potassium sulfates as macronutrients, as well as potassium iodide, boric acid, manganese, zinc, and iron sulfates, sodium molybdate, copper sulfate, cobalt and manganese chlorides, cobalt nitrate, iron tartrate, and EDTA.

Medium selection in the case of animal cells depends on the type of cells to be cultured, and the purpose of the culture, namely, growth, differentiation, or production of some desired product; the said cells are cultured using either a completely natural medium or instead an artificial medium composed of a basal synthetic medium supplemented with a few natural products. Natural media include clots (e.g., plasma separated from heparinized blood, serum, fibrinogen), tissue extracts (e.g., from chicken embryos, liver, spleen, bone marrow), and biological fluids (e.g., plasma, serum, lymph, amniotic fluid, pleural fluid). Due to their sources, these media exhibit poor reproducibility and a quite ill-defined composition.

In attempts to overcome the shortcomings of natural media, a number of artificial media have been proposed: serum-containing media (in which human, bovine, or equine serum is used as supplement), serum-free media (in which crude protein fractions, such as bovine serum albumin, or α- and β-globulin are used as supplements), xeno-free media (in which human-source components, such as serum albumin, serve as supplements, but not any other animal components), protein-free media (in which undefined components, such as peptide fractions or protein hydrolyzates, are utilized as supplements), and chemically defined media (in which highly purified components, such as recombinant proteins, function as supplements). In general, all media buffer *pH* close to physiological level and ensure isotonic conditions; if long-term survival is intended, then a balanced salt solution complemented with amino acids, oxygen, vitamins, and serum proteins (or serum itself) is necessary. In order to promote cell growth and proliferation, artificial media are normally added with natural constituents—as is the case of serum. Its role has not been fully elucidated, but is believed to arise from its content of hormones (e.g., insulin, hydrocortisone, triiodothyronine, estrogen, androgens, and growth hormone) and growth factors (e.g., epidermal, fibroblast, insulin-like, nerve, platelet-derived, and transforming growth factors), which induce proliferation, differentiation, migration, secretion, and import in cells, contribute to cell attachment via binding proteins and protein-, vitamin-, and lipid-carrying factors, minimize mechanical damage caused by viscosity, and act as buffering agent.

Unfortunately, serum (as additive) does not hold a uniform composition, as emphasized above, and may contain inadequate and uncontrollable levels of specific growth factors or even growth inhibitors, cytotoxic compounds, and virus, fungi, or mycoplasmas; furthermore, presence of serum in the medium interferes with downstream purification and isolation of cell culture products. To circumvent these difficulties, serum-free (artificial) media have been on the rise; they normally resort to deliberate addition of specific components to some basal medium of synthetic origin. One such lumped component is a protein hydrolyzate (derived from yeast, milk, or soy), which contains vitamins and lipids as inorganic salts, as well as amino acids and low-molecular weight peptides; since proteins have been retained by previous ultrafiltration, a posteriori recovery of antibodies and recombinant proteins is facilitated.

A large number of the aforementioned artificial media are indeed commercially available: Medium 199 (designed for cultivation of chicken embryonic cells, entailing sequential addition of amino acids, vitamins, and nucleic acid precursors to the aqueous matrix), CMRL 1066 (designed for culturing mouse L cells, and based on Medium 199 but with increased levels of Cys, glutathione, and ascorbic acid, elimination of fat-soluble vitamins, and addition of coenzymes), Basal Eagle Medium (BEM) (originally designed for HeLa cells but suitable for short-time survival of animal cells at large, and including 13 amino acids and 8 vitamins), Minimum Essential Medium (MEM; with twofold concentrations of most amino acids compared to BME), Dulbecco modified MEM (DMEM; designed for cultivation of polyomavirus in mouse embryonic cells, and with fourfold the concentrations of amino acids and vitamins in BME), α-MEM (designed for cultivation of mouse and hamster hybridomas, and based on MEM supplemented with nonessential amino acids, ascorbic acid, biotin, cyanocobalamin, pyruvate, lipoic acid, and nucleosides), Iscove modified DMEM (IMDM; designed for rapidly proliferating cells, and based on DMEM supplemented with cyanocobalamin, biotin, selenite, pyruvate, HEPES, transferrin, bovine serum albumin, and soybean lipids), NCTC 109 (designed for culture of L cells under protein-free conditions, and including coenzymes, nucleobases, reducing agents, and a wealth of vitamins), Ham F-10 (designed for colony formation by CHO cells, and containing serum albumin and fetuin instead of serum, as well as trace Cu and Zn), Ham F-12 (based on Ham F-10, but with the proteins replaced by linoleic acid and putrescine), Kaighn F-12 (based on Ham F-12, but with reinforced contents of amino acids, pyruvate, biotin, Ca, Mg, putrescine, and phenol red), MCDB 202, 301, 153, 110, 402, 170, and 131 (all based again on Ham F-12, but designed for culture of chicken embryo fibroblasts, CHO cells, human keratinocytes, human fibroblasts, mouse fibroblasts, mammary epithelium, and human vascular endothelial cells, respectively), RPMI 1640 (designed for long-term culture of peripheral blood lymphocytes, and containing low levels of Ca and Mg but high levels of phosphate), mixture of DMEM, Ham F-12, and RPMI 1640 (designed for culture of hybridomas, and containing extra insulin, transferrin, ethanolamine, and selenite), Waymouth MB752 (designed for culture of mouse L929 cells, and containing a total of 40 components, including glucose, inorganic salts, amino acids, purine bases, and hypoxanthine, as well as high concentrations of glucose, His, Lys, Gln, choline, and thiamine), Trowell T-8 (designed for long-term culture of adult rat liver epithelial cells and short-term culture of organ cells, and devoid of nonessential amino acids and vitamins but characterized by high glucose and insulin levels), and Leibovitz L-15 (designed for cell and tissue transport and primary cultures, with buffering capacity mediated by phosphates and free basic amino acids instead of sodium bicarbonate, and containing pyruvate and galactose to high levels in lieu of glucose to allow storage at ambient air rather than under CO_2).

Problem 2.3

When air is supplied at the bottom of a liquid medium of viscosity μ and initial height H in a bioreactor of radius R_r, the bubbles formed will rise at their terminal velocity v_∞—which, for being finite, implies development of a holdup volume of gas inside the liquid. Produce a plot allowing estimation of fraction of holdup volume as a function of bubble radius, R, assuming

$$F_d = 4\pi\mu R v_\infty \tag{2.29}$$

describes the drag force upon the bubble.

Solution

A balance of forces acting upon a gaseous bubble of mass density ρ_g, inside a liquid medium of mass density ρ_l can be set up as

(continued)

Problem 2.3 (continued)

$$\frac{4}{3}\pi R^3\left(\rho_l - \rho_g\right)g = 4\pi\mu R v_\infty, \tag{2.30}$$

on the hypothesis of a spherical bubble under gravitational field described by acceleration g—with the left-hand side pointing upward and representing the net result of buoyancy and weight of bubble, and its right-hand side pointing downward and representing the drag force upon the upward-moving bubble. After dropping $4\pi R$ from both sides, Eq. (2.30) reduces to

$$\frac{1}{3}R^2\left(\rho_l - \rho_g\right)g = \mu v_\infty, \tag{2.31}$$

where isolation of v_∞ leaves

$$v_\infty = \frac{1}{3}\frac{\rho_l - \rho_g}{\mu}gR^2; \tag{2.32}$$

the time taken for a bubble, τ, to span the whole height of liquid abides to

$$\tau = \frac{H}{v_\infty}, \tag{2.33}$$

so combination with Eq. (2.32) yields

$$\tau = \frac{H}{\frac{1}{3}\frac{\rho_l - \rho_g}{\mu}gR^2} \tag{2.34}$$

or, after reducing number of fraction levels,

$$\tau = 3\frac{\mu}{\rho_l - \rho_g}\frac{H}{g}\frac{1}{R^2}. \tag{2.35}$$

The total volume of air, V_g, retained inside the liquid matrix will accordingly be given by

$$V_g = Q_g\tau, \tag{2.36}$$

should Q_g denote volumetric flow rate of gas and assuming that the volume expansion caused by the decreasing hydrostatic pressure gradient upward essentially

(continued)

Problem 2.3 (continued)

balances the volume reduction incurred in as a consequence of mass transfer into the liquid; insertion of Eq. (2.35) leads then to

$$V_g = Q_g 3 \frac{\mu}{\rho_l - \rho_g} \frac{H}{g} \frac{1}{R^2} = 3 \frac{\mu}{\rho_l - \rho_g} \frac{HQ_g}{g} \frac{1}{R^2} \tag{2.37}$$

from Eq. (2.36). The fractional holdup volume, ε, will accordingly satisfy

$$\varepsilon \equiv \frac{V_g}{V_g + V_l}, \tag{2.38}$$

while the existing liquid volume, V_l, abides to

$$V_l = \pi R_r{}^2 H; \tag{2.39}$$

insertion of Eqs. (2.37) and (2.39) converts Eq. (2.38) to

$$\varepsilon \equiv \frac{3 \frac{\mu}{\rho_l - \rho_g} \frac{HQ_g}{g} \frac{1}{R^2}}{3 \frac{\mu}{\rho_l - \rho_g} \frac{HQ_g}{g} \frac{1}{R^2} + \pi R_r{}^2 H}, \tag{2.40}$$

while division of both numerator and denominator by $R_r{}^2 H$ produces

$$\varepsilon = \frac{3 \frac{\mu}{\rho_l - \rho_g} \frac{HQ_g}{g R_r{}^2 H} \frac{1}{R^2}}{3 \frac{\mu}{\rho_l - \rho_g} \frac{HQ_g}{g R_r{}^2 H} \frac{1}{R^2} + \pi}. \tag{2.41}$$

Cancellation of H between numerator and denominator transforms Eq. (2.41) to

$$\varepsilon = \frac{3 \frac{\mu}{\rho_l - \rho_g} \frac{Q_g}{g R_r{}^2} \frac{1}{R^2}}{3 \frac{\mu}{\rho_l - \rho_g} \frac{Q_g}{g R_r{}^2} \frac{1}{R^2} + \pi}, \tag{2.42}$$

whereas straightforward algebraic rearrangement supports

(continued)

Problem 2.3 (continued)

$$\varepsilon = \frac{\dfrac{3}{\left(R_r\sqrt{\frac{g(\rho_l-\rho_g)}{\mu Q_g}}R\right)^2}}{\dfrac{3}{\left(R_r\sqrt{\frac{g(\rho_l-\rho_g)}{\mu Q_g}}R\right)^2}+\pi} \tag{2.43}$$

or, after multiplying both numerator and denominator by $\left(R_r\sqrt{g(\rho_l-\rho_g)/\mu Q_g R}\right)^2$,

$$\varepsilon = \frac{3}{3+\pi\left(R_r\sqrt{\frac{g(\rho_l-\rho_g)}{\mu Q_g}}R\right)^2}; \tag{2.44}$$

a graphical interpretation of Eq. (2.44) is provided in Fig. 2.1—with fractional holdup volume of gas plotted as a function of dimensionless bubble radius. As expected, absence of liquid phase ($R_r \to 0$) produces a unit fractional holdup volume as per Eqs. (2.38) and (2.39), or else

$$\lim_{R_r\sqrt{\frac{g(\rho_l-\rho_g)}{\mu Q_g}}R\to 0} \varepsilon = \frac{3}{3+\pi 0^2} = \frac{3}{3} = 1 \tag{2.45}$$

(continued)

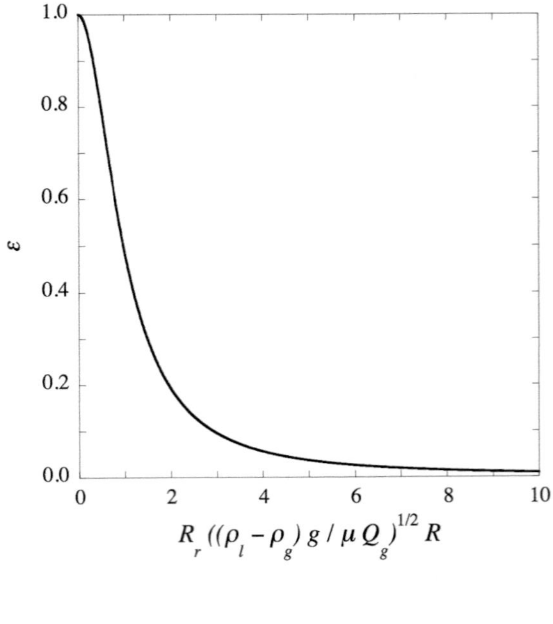

Fig. 2.1 Variation of fractional holdup volume of gas, ε, versus dimensionless bubble radius, $R_r\sqrt{g(\rho_l-\rho_g)/\mu Q_g}R$

Problem 2.3 (continued)

after Eq. (2.44), which serves as descriptor of the vertical intercept in Fig. 2.1; this would also be asymptotically attained if no driving force existed for bubble rise (either $g = 0$ or $\rho_l - \rho_g = 0$), resistance to the rising motion would be unbounded ($\mu \to \infty$), or volumetric flow rate of gas would be unbounded as well ($Q_g \to \infty$). Conversely,

$$\lim_{R_r\sqrt{\frac{g(\rho_l-\rho_g)}{\mu Q_g}R}\to\infty} \varepsilon = \frac{3}{3+\pi\,\infty^2} = \frac{3}{\infty} = 0 \tag{2.46}$$

enforces the horizontal axis as long- $R_r\sqrt{g(\rho_l-\rho_g)/\mu Q_g R}$, horizontal asymptote.

2.1.2 Industrial Media

Choice of raw materials to formulate cultivation media at industrial scale hinges on a number of factors, namely, costs of harvest, extraction, and transport; availability in terms of source and ease of disposal in terms of waste; adequate and balanced nutrient/ingredient composition; uniformity in quality and ease of standardization; and presence of relevant precursors.

The cheaper the raw material, the more competitive the selling price of the final medium, thus justifying why most raw materials are agricultural wastes. Choice of a more expensive raw material, as is the case of lactose over glucose, usually responds to some processing requirement, namely the lower rate of uptake required in penicillin manufacture. Raw materials should be readily available; otherwise, they will turn the limiting step and may even halt production; if seasonal, or imported, then storage facilities for reasonable periods will be necessary. The closer the source of a raw material to the point of use, the more suitable it becomes—should all remaining conditions prove satisfactory. Disposal of industrial wastes is subjected to stricter and stricter control in most countries, so waste materials from specific processes are preferred candidates for available downstream processing; this is the case of spent grains from breweries (or *drêche*), which are still suitable for animal feed. If no further use can be found, costs of waste treatment are to be taken into account—and disposal might actually be expensive.

A final product within the specifications set by the customer demands a reasonably constant composition of the raw material from which it is manufactured—for instance, molasses as waste product of the sugar industry; substantial variability in quality is undesirable as extra costs will arise from changes in setpoint and control actions, including analytical costs and extra nutrients to be added, so as to match the profile expected for the medium. The medium must have C, N, minerals, and vitamins to appropriate levels and proportions, as necessary for the optimum production of the commodity at stake, and in usable form by the cells; for instance, most yeasts use hexose saccharides rather than disaccharides, cellulose is not easily metabolized by most microorganisms, and molds will promptly grow in corn steep liquor while (filamentous) bacteria belonging to genus *Actinomycetes* prefer soya bean cake by far. Precursors often stimulate production of secondary metabolites—either by increasing the amount of a limiting metabolite or inducing a biosynthetic enzyme (or both); for instance, manufacture of penicillin G demands presence of a phenyl compound in the medium (as happens with corn steep liquor), whereas production of vitamin B_{12} requires cobalt, and of chlortetracycline and griseofulvin critically depends on chlorine in the medium.

Finally, recall that many microorganisms of industrial interest undergo two stages in batch cultivation—growth during the trophophase and production afterward during the idiophase. Cells multiply rapidly in the former, with little (or no) production of the desired (secondary) metabolite; while production in the idiophase follows internal synthesis of specific enzymes, so it is not accompanied by biomass expansion. Since these two phases frequently require distinct nutrients, or different proportions of the same nutrient, the medium should be complete and be able to address these requirements; for instance, high levels of glucose and phosphate inhibit the onset of the idiophase in the case of several secondary metabolites, so the levels of added components have to be such as to not adversely affect production.

The most common raw materials in compounding industrial media are corn steep liquor, pharmamedia, distillers' solubles, soya bean meal, molasses, sulfite liquor, and other substrates encompassing alcohol, acetic acid, methanol, methane, and fractions of crude petroleum; these are illustrated in Fig. 2.2. Corn steep liquor is a byproduct of starch manufacture from maize; for being highly acidic, it requires preliminary neutralization via $CaCO_3$. Although its low *pH* inhibits growth of most microorganisms, it permits development of adventitious lactic acid bacteria (especially *Lactobacillus* spp.). Despite some

Fig. 2.2 Common raw materials used for medium formulation in industrial bioprocesses: (**a**) corn steep liquor, (**b**) pharmamedia, (**c**) distillers' solubles, (**d**) soya bean meal, (**e**) molasses, (**f**) sulfite liquor, and (**g**) fraction of crude petroleum

natural variability, a typical composition of 77% (w/w) solids is found—of which 45% protein, 21% lactic acid, 10.5% ash, 3% reducing sugars, and 0.1% lipids, on a dry weight basis, besides 16,000 ppm of P, 29500 ppm of K, 5800 ppm of Mg, 810 ppm of Ca, and 210 ppm of Fe. Pharmamedia is a fine, yellow powder made from cottonseed embryo; it has been employed in manufacture of tetracycline and several semisynthetic penicillins. It contains ca. 56% (w/w) protein, 24% carbohydrates, 5% oil, and 4% ash on a weight dry basis, and is also rich in Ca, Fe, Cl, P, and SO_4^{2-}. Distillers' solubles is a byproduct of distillation of alcohol from fermented grain; it is prepared by filtering away the solids left in the kettle upon distillation and concentrating afterward to syrup form—which is finally drum-dried to facilitate storage. This raw material is rich in N, minerals, and growth factors. Soya bean meal has ca. 30% (w/w) carbohydrates and 11% nitrogen-containing material; the latter is more complex than that found in corn steep liquor and not readily available, except to actinomycetes. Molasses are obtained as a byproduct of the sugar industry—from either beetroot or sugarcane; they are used mainly to produce industrial alcohol via fermentation, as well as acetone, citric acid, and glycerol—further to yeast biomass. Sulfite liquor (also known as waste/spent sulfite liquor) is the aqueous effluent from the sulfite process in the manufacture of cellulose or pulp from wood; its composition depends on the severity of processing and type of source wood, and becomes suitable to grow microorganisms following neutralization with $CaCO_3$, complemented by enrichment afterward with ammonium salts or urea. Hardwoods not only yield a higher amount of sugar (up to 3% dry weight), but also convey most sugars in the form of pentoses, for example, xylose; their hydrolyzates contain a significant amount of acetic acid as well. Soft woods are instead richer in hexoses (up to 75% dry weight), chiefly mannose.

Growth factors are usually required in small amounts and obtained from a number of sources; the most common examples are vitamin B at large—obtained from rice polishing, wheat germ, or waste yeast. More specifically, cereals and corn steep liquor are good sources of vitamin B_2, liver or penicillin spent liquor of vitamin B_3 (alias nicotinamide), corn steep liquor and waste yeast of vitamin B_5 (alias pantothenic acid), and liver, silage, and meat of vitamin B_{12}.

For being the dominant component of industrial media, water is used to large quantities; in several bioprocesses, for example, brewing, the quality of the final product depends heavily on the quality of water utilized in the first place. In order to ensure constancy of product quality, water must be regularly analyzed for minerals and color, and *pH* adjusted as deemed necessary.

Although a number of carbohydrates suitable for medium formulation can be retrieved from agricultural products, they often require previous hydrolysis to release simple and thus fermentable sugars; the most common examples, that is, cassava, sweet potato, yam, cocoyam, millets, rice, Jerusalem artichoke, and sorghum, are illustrated in Fig. 2.3. Cassava (*Manihot esculenta*), common in the tropical world, requires little attention when cultivated, and its roots can remain in the ground for long periods without deterioration before harvest; their inner, fleshy portion is a rich source of starch and has served (upon hydrolysis) as carbon source for single cell protein, ethanol, and even beer. Despite being a

Fig. 2.3 Common sources of carbohydrates used for medium formulation in industrial bioprocesses: (**a**) cassava, (**b**) sweet potato, (**c**) yam, (**d**) cocoyam, millets in (**e**) plant and (**f**) grain forms, (**g**) rice, (**h**) Jerusalem artichoke, and (**i**) sorghum

minor source of carbohydrates when compared to cassava, sweet potato (*Ipomoea batatas*) is a warm-climate crop and requires as well little agronomical attention, while attaining yields up to 40 ton ha^{-1}; it has been used as a source of sugar on a semicommercial basis because their fleshy roots contain saccharolytic enzymes. The sugar-rich syrup obtained from boiling the tubers has been used in compounding industrial media, namely, to produce butyl alcohol, acetone, and ethanol via fermentation. Yams (*Dioscorea* spp.) are common in the tropics, yet their cultivation is tedious; when used in flour or flake form, the residues left behind by peeling hold substantial amounts of carbohydrates suitable for microbial media. Cocoyam is a blanket name for several edible members of a monocotyledonous plant aroid (*Araceae* spp.), common in tropical areas; its starch has met with acceptable quality for pharmaceutical fermentations, pending hydrolysis, despite their being laborious to cultivate, requiring large quantities of moisture, and exhibiting a short shelf life. Millets is a collective name for tropical and subtropical cereals (*Pennisetum americanum*, *Setaria italica*, *Panicum miliaceum*, *Echinochloa frumentacea*, or *Eleusine coracana*)—with seeds small compared to those of regular cereals; considered as minor cereals for not constituting a regular component of human diet, they are quite hardy and thus tolerate drought and persistent heat, grow on poor soil, and mature quickly. Millet starch has been hydrolyzed by malting for alcohol production. Rice (*Oryza sativa*) is a leading food crop worldwide, especially in tropical areas; although being a high-cost commodity, it holds the advantages of ease of mechanization, storability, and availability of improved seeds, and an increasing plant productivity is likely to bring its cost sufficiently down. Rice is currently used as brewing adjunct and has been successfully malted for brewing itself. Jerusalem artichoke (*Helianthus tuberosus*) is a root crop of the Compositae family that grows in temperate and semitropical climates, and resorts to inulin rather than starch as storage carbohydrate; this polymer accordingly releases fructose following hydrolysis. Finally, sorghum (*Sorghum bicolor*), the fourth cereal in worldwide production after wheat, rice, and corn, is currently employed in the manufacture of special lager beers; upon successful malting, the resulting beer compares favorably with most barley malt-based beers.

In addition, plant agricultural wastes are plentiful and renewable, and these traits make them sources of carbohydrates suitable to produce industrial media; its major constituents are starch, on the one hand, and cellulose, hemicelluloses and lignin, on the other.

Remember that starch is a mixture of two polymers, amylose (20–30%) and amylopectin (remainder); the exact proportion depends on its source. The former is an insoluble polysaccharide made of 200–1000 D-glucose units, bound to each other through $\alpha_{1\rightarrow4}$ glycosidic bonds; because of its tightly packed helical structure, amylose is relatively resistant to digestion. Amylopectin is a water-soluble polysaccharide, exhibiting a linear chain with 2000–200,000 glucose units, along with side chains every ca. 25 residues, which may further branch; glucose residues are bound by $\alpha_{1\rightarrow4}$ glycosidic bonds throughout the linear chain, whereas branches are supported by $\alpha_{1\rightarrow6}$ glycosidic bonds. Starch occurs as discrete crystalline granules in plants and is accordingly resistant to enzyme action; however, it gelatinizes and dissolves in water upon heating to 55–82 °C, thus facilitating hydrolysis (termed saccharification) to sequentially smaller carbohydrates, driven by either (dilute) acids or enzymes generated during cereal pregermination (or malting). The most common enzymes for this purpose are enzymes able to cleave only the $\alpha_{1\rightarrow4}$ bond and bypass the $\alpha_{1\rightarrow6}$ bond, for example, α-amylases found in virtually any living cell; these enzymes are able to cleave anywhere, provided that there are at least six glucose residues on one side and at least three on the other side of the target glycosidic bond. Conversely, β-amylases act differently upon amylose and amylopectin because they can cleave $\alpha_{1\rightarrow4}$ bonds but cannot bypass their $\alpha_{1\rightarrow6}$ counterparts, while amyloglucosidase and glucoamylase cleave either type of bond. Finally, pullulanase and isoamylase act only upon $\alpha_{1\rightarrow6}$ bonds, and are thus termed debranching enzymes. In industry, conversion of starch to sugar is measured in terms of dextrose equivalents (DE) as percent of total solids, and a measure of reducing sugar content takes advantage of Fehling's solution; the practical upper limit for acid saccharification is 55 DE—a threshold that may be overcome should enzymes be utilized. Barley malt is widely used for saccharification of starch, owing to its high content in amylase and amyloglucosidase, able to further split saccharides to glucose; however, saccharifying enzymes from *Bacillus*, *Streptomyces*, *Aspergillus*, *Penicillium*, *Mucor*, and *Rhizopus* genera have also been in common use.

Cellulose is the most abundant organic polymer on the planet, and a polysaccharide consisting of a linear chain of several hundred to many thousands of $\beta_{1\rightarrow4}$-linked D-glucose units; a major structural component of the primary cell wall of green plants

(40–50%) and cotton fiber (ca. 90%), it cannot be found pure in nature. Cellulose usually appears combined with hemicellulose and lignin, at typical ratios 4:3:3 in wood; while the former is crystalline, strong, and resistant to hydrolysis, hemicelluloses exhibit random, amorphous structure bearing little strength and susceptible to chemical and enzymatic attack. In cotton, hemicelluloses are pectic substances, that is, polymers of galactose; in wood, they instead consist of short-branched heteropolymers of glucose, xylose, galactose, mannose, and arabinose—further to uronic acids of glucose and galactose, connected by $\beta_{1\rightarrow3}$, $\beta_{1\rightarrow4}$, and $\beta_{1\rightarrow6}$ glycosidic bonds. Lignins are a class of complex organic polymers that perform key structural roles in the supporting tissues of most plants, since they lend rigidity and do not rot easily; they result from cross-linking in diverse ways of phenolic precursors, especially coniferyl, sinapyl, and *p*-coumaryl alcohols. In view of the above structure, physical or chemical methods are a must to expose lignocellulosics to hydrolytic attack; the former encompass grinding, irradiation, or plain heating of wood, whereas the latter resort to such swelling agents as steam, alkalis as sodium or ammonium hydroxide, acids as concentrated sulfuric or hydrochloric acid, oxidizing agents as sodium hypochlorite or hydrogen peroxide, and organic solvents as butanol or ethylamine to introduce water between the cellulose crystals. After pretreatment, wood is exposed for a long period to dilute hydrochloric or sulfuric acid, or calcium, magnesium, or sodium sulfites under high temperature and pressure; in alternative, fungal cellulolytic enzymes can be employed—usually obtained from *Trichoderma viride* or *T. koningii*, and *Penicillium funiculosum* or *Fusarium solani* to a lesser extent. Three major components have been identified in such enzyme cocktails: C_1-enzymes that attack crystalline cellulose, thus loosening the cellulose chains; C_x-enzymes or $\beta_{1\rightarrow4}$ glucanases, able to hydrolyze soluble derivatives of cellulose and swollen or partially degraded cellulose; and β-glucosidases, which hydrolyze cellobiose and short-chain oligosaccharides from cellulose down to glucose form, yet unable to act upon plain cellulose.

Various leguminous seeds are available for protein source/supplementation of industrial fermentation media; one of the most popular is ground peanuts (*Arachis hypogea*) (see Fig. 2.4a). The (pressed) groundnut cake left after defatting is indeed used as a rich source of protein, while the oil expressed may be used as antifoam agent. Blood consists of ca. 82% (w/w) water, 16.4% nitrogen, 0.7% ash, 0.6% fat, and 0.1% carbohydrate; it constitutes a waste product in abattoirs. Drying thereof is achieved by passing live steam through the blood until temperature reaches ca. 100 °C, a treatment meant not only for sterilization but also for clotting; it is then drained, pressed to remove serum, further dried, and ground (see Fig. 2.4b). The resulting product is chocolate-colored and contains ca. 80% protein—along with small amounts of ash and lipids. Fish meal is rich in protein (ca. 65%) and minerals (ca. 6% Ca and 3.5% P); it is obtained by drying leftover fish with steam under a partial vacuum, or plain atmospheric drying under sunlight. In alternative, hot air may be passed over the fish when placed in revolving drums prior to grinding into a fine powder as per Fig. 2.4c.

Fig. 2.4 Common sources of proteins used for medium formulation in industrial bioprocesses: (**a**) peanut meal, (**b**) blood meal, and (**c**) fish meal

Problem 2.4

Aerobic metabolism in aqueous media is crucial for growth of many microorganisms, in view of the energy load made available thereby; however, molecular oxygen, serving as final electron acceptor, is poorly soluble in water, a situation that worsens as temperature is increased. Experimental data on oxygen solubility in water are provided below, at various temperatures.

i	Temperature (°C) [x_i]	Oxygen solubility (mg L^{-1}) [y_i]
1	0	14.8
2	10	11.5
3	15	10.4
4	20	9.45
5	25	8.69
6	26	8.55
7	27	8.42
8	28	8.29
9	29	8.17
10	30	8.05
11	35	7.52
12	40	7.07

Assume the data to be independently, identically, and normally distributed with constant variance.

(a) Fit a polynomial of increasing degree, up to quadratic, and find the best estimates for the corresponding parameters.
(b) Plot the three fitted models superimposed on the experimental data and decide, on statistical grounds, whether a linear model is preferable to a zeroth-order one, and a quadratic model is preferable to its first-order counterpart.

(continued)

Problem 2.4 (continued)

Solution

(a) The basic requirements for validity of the classical regression analysis are, per hypothesis, satisfied by the set of 12 data points provided; hence, one can safely resort to the least squares estimator as best unbiased estimator. The basic zeroth order model reads merely

$$y_i = a_0, \tag{2.47}$$

where y denotes oxygen solubility and a_0 denotes zeroth-order coefficient; the sum of squares of residuals, S_{sqr}, then reads

$$S_{sqr,0} \equiv \sum_{i=1}^{12} (y_i - a_0)^2, \tag{2.48}$$

and the best estimate of a_0 ensues from

$$\begin{aligned} \frac{dS_{sqr,0}}{da_0} &= \frac{d}{da_0}\left(\sum_{i=1}^{12} (y_i - a_0)^2\right) = \sum_{i=1}^{12} \frac{d}{da_0}\left((y_i - a_0)^2\right) \\ &= \sum_{i=1}^{12} 2(y_i - a_0)(-1) = -2\sum_{i=1}^{12} (y_i - a_0) = 0 \end{aligned} \tag{2.49}$$

where the classical rules of differentiation were meanwhile used. Upon dividing both sides by -2 and splitting the summation afterward, Eq. (2.49) produces

$$\sum_{i=1}^{12} y_i - \sum_{i=1}^{12} a_0 = \sum_{i=1}^{12} y_i - a_0 \sum_{i=1}^{12} 1 = 0 \tag{2.50}$$

along with factoring out of (constant) a_0, which will be solved for a_0 as

$$a_0 = \frac{\sum\limits_{i=1}^{12} y_i}{\sum\limits_{i=1}^{12} 1} \tag{2.51}$$

(continued)

Problem 2.4 (continued)

with $\sum_{i=1}^{12} 1$ and $\sum_{i=1}^{12} y_i$ given below.

i	1	y_i
1	1	14.8
2	1	11.5
3	1	10.4
4	1	9.45
5	1	8.69
6	1	8.55
7	1	8.42
8	1	8.29
9	1	8.17
10	1	8.05
11	1	7.52
12	1	7.07
$\sum_{i=1}^{12}$	12	110.9

In view of Eq. (2.51), the best estimate for the zeroth-order parameter is the average of all data, which boils down to

$$a_0 = \frac{110.9}{12} = 9.242 \tag{2.52}$$

pertaining to the zeroth-order model conveyed by Eq. (2.47). If a linear model is postulated of the form

$$y_i = a_0 + a_1 x_i, \tag{2.53}$$

then the corresponding sum of squares of residuals looks like

$$S_{sqr,1} \equiv \sum_{i=1}^{12} \left(y_i - (a_0 + a_1 x_i)\right)^2 = \sum_{i=1}^{12} (y_i - a_0 - a_1 x_i)^2; \tag{2.54}$$

the best estimates must satisfy

(continued)

Problem 2.4 (continued)

$$\begin{aligned}\frac{\partial S_{sqr,1}}{\partial a_0} &= \frac{\partial}{\partial a_0}\left(\sum_{i=1}^{12}(y_i - a_0 - a_1x_i)^2\right) = \sum_{i=1}^{12}\frac{\partial}{\partial a_0}\left((y_i - a_0 - a_1x_i)^2\right) \\ &= \sum_{i=1}^{12}2(y_i - a_0 - a_1x_i)(-1) = -2\sum_{i=1}^{12}(y_i - a_0 - a_1x_i) = 0\end{aligned} \tag{2.55}$$

referring to parameter a_0, coupled with

$$\begin{aligned}\frac{\partial S_{sqr,1}}{\partial a_1} &= \frac{\partial}{\partial a_1}\left(\sum_{i=1}^{12}(y_i - a_0 - a_1x_i)^2\right) = \sum_{i=1}^{12}\frac{\partial}{\partial a_1}\left((y_i - a_0 - a_1x_i)^2\right) \\ &= \sum_{i=1}^{12}2(y_i - a_0 - a_1x_i)(-x_i) = -2\sum_{i=1}^{12}x_i(y_i - a_0 - a_1x_i) = 0\end{aligned} \tag{2.56}$$

pertaining to parameter a_1. Division of both sides of Eq. (2.55) by -2, followed by splitting of summation and factoring out of (constant) parameters give rise to

$$\sum_{i=1}^{12}(y_i - a_0 - a_1x_i) = \sum_{i=1}^{12}y_i - \sum_{i=1}^{12}a_0 - \sum_{i=1}^{12}a_1x_i = \sum_{i=1}^{12}y_i - a_0\sum_{i=1}^{12}1 - a_1\sum_{i=1}^{12}x_i = 0, \tag{2.57}$$

whereas Eq. (2.56) likewise generates

$$\begin{aligned}\sum_{i=1}^{12}x_i(y_i - a_0 - a_1x_i) &= \sum_{i=1}^{12}\left(x_iy_i - a_0x_i - a_1x_i^2\right) \\ &= \sum_{i=1}^{12}x_iy_i - \sum_{i=1}^{12}a_0x_i - \sum_{i=1}^{12}a_1x_i^2 = \sum_{i=1}^{12}x_iy_i - a_0\sum_{i=1}^{12}x_i - a_1\sum_{i=1}^{12}x_i^2 = 0\end{aligned} \tag{2.58}$$

once x_i is factored in; after realizing that Es. (2.57) and (2.58) are equivalent to

$$a_0\sum_{i=1}^{12}1 + a_1\sum_{i=1}^{12}x_i = \sum_{i=1}^{12}y_i \tag{2.59}$$

and

$$a_0\sum_{i=1}^{12}x_i + a_1\sum_{i=1}^{12}x_i^2 = \sum_{i=1}^{12}x_iy_i, \tag{2.60}$$

(continued)

Problem 2.4 (continued)

respectively, Cramer's rule will be invoked to access a_0 and a_1, namely,

$$a_0 = \frac{\begin{vmatrix} \sum_{i=1}^{12} y_i & \sum_{i=1}^{12} x_i \\ \sum_{i=1}^{12} x_i y_i & \sum_{i=1}^{12} x_i^2 \end{vmatrix}}{\begin{vmatrix} \sum_{i=1}^{12} 1 & \sum_{i=1}^{12} x_i \\ \sum_{i=1}^{12} x_i & \sum_{i=1}^{12} x_i^2 \end{vmatrix}} \tag{2.61}$$

and

$$a_1 = \frac{\begin{vmatrix} \sum_{i=1}^{12} 1 & \sum_{i=1}^{12} y_i \\ \sum_{i=1}^{12} x_i & \sum_{i=1}^{12} x_i y_i \end{vmatrix}}{\begin{vmatrix} \sum_{i=1}^{12} 1 & \sum_{i=1}^{12} x_i \\ \sum_{i=1}^{12} x_i & \sum_{i=1}^{12} x_i^2 \end{vmatrix}}. \tag{2.62}$$

Definition of second-order determinant justifies replacement of Eqs. (2.61) and (2.62) by

$$a_0 = \frac{\sum_{i=1}^{12} y_i \sum_{i=1}^{12} x_i^2 - \sum_{i=1}^{12} x_i y_i \sum_{i=1}^{12} x_i}{\sum_{i=1}^{12} 1 \sum_{i=1}^{12} x_i^2 - \sum_{i=1}^{12} x_i \sum_{i=1}^{12} x_i} = \frac{\sum_{i=1}^{12} x_i^2 \sum_{i=1}^{12} y_i - \sum_{i=1}^{12} x_i \sum_{i=1}^{12} x_i y_i}{\sum_{i=1}^{12} 1 \sum_{i=1}^{12} x_i^2 - \left(\sum_{i=1}^{12} x_i\right)^2} \tag{2.63}$$

and

$$a_1 = \frac{\sum_{i=1}^{12} 1 \sum_{i=1}^{12} x_i y_i - \sum_{i=1}^{12} x_i \sum_{i=1}^{12} y_i}{\sum_{i=1}^{12} 1 \sum_{i=1}^{12} x_i^2 - \sum_{i=1}^{12} x_i \sum_{i=1}^{12} x_i} = \frac{\sum_{i=1}^{12} 1 \sum_{i=1}^{12} x_i y_i - \sum_{i=1}^{12} x_i \sum_{i=1}^{12} y_i}{\sum_{i=1}^{12} 1 \sum_{i=1}^{12} x_i^2 - \left(\sum_{i=1}^{12} x_i\right)^2}, \tag{2.64}$$

(continued)

Problem 2.4 (continued)

respectively; the values for the outstanding summations are tabulated below.

i	1	x_i	x_i^2	y_i	x_iy_i
1	1	0	0	14.8	0
2	1	10	100	11.5	115
3	1	15	225	10.4	156
4	1	20	400	9.45	189
5	1	25	625	8.69	217.3
6	1	26	676	8.55	222.3
7	1	27	729	8.42	227.3
8	1	28	784	8.29	232.1
9	1	29	841	8.17	236.9
10	1	30	900	8.05	241.5
11	1	35	1225	7.52	263.2
12	1	40	1600	7.07	282.8
$\sum_{i=1}^{12}$	12	285	8105	110.9	2383

Insertion of the summations above in Eqs. (2.63) and (2.64) leads to

$$a_0 = \frac{8105 \cdot 110.9 - 285 \cdot 2383}{12 \cdot 8105 - 285^2} = \frac{219690}{16035} = 13.7007 \tag{2.65}$$

and

$$a_1 = \frac{12 \cdot 2383 - 285 \cdot 110.9}{12 \cdot 8105 - 285^2} = \frac{-3010.5}{16035} = -0.18775; \tag{2.66}$$

Equations (2.65) and (2.66) convey the best estimates for the parameters in the linear model conveyed by Eq. (2.53). Consider finally a quadratic model, that is,

$$y_i = a_0 + a_1 x_i + a_2 x_i^2, \tag{2.67}$$

where a_2 denotes second-order coefficient and for which the sum of squares of residuals takes the form

$$S_{sqr,2} \equiv \sum_{i=1}^{12} \left(y_i - \left(a_0 + a_1 x_i + a_2 x_i^2 \right) \right)^2 = \sum_{i=1}^{12} \left(y_i - a_0 - a_1 x_i - a_2 x_i^2 \right)^2; \tag{2.68}$$

the necessary condition for a minimum sum of squares of residuals now reads

(continued)

Problem 2.4 (continued)

$$\frac{\partial S_{sqr,2}}{\partial a_0}=\frac{\partial}{\partial a_0}\left(\sum_{i=1}^{12}\left(y_i-a_0-a_1x_i-a_2x_i^2\right)^2\right)=\sum_{i=1}^{12}\frac{\partial}{\partial a_0}\left(\left(y_i-a_0-a_1x_i-a_2x_i^2\right)^2\right)$$
$$=\sum_{i=1}^{12}2\left(y_i-a_0-a_1x_i-a_2x_i^2\right)(-1)=-2\sum_{i=1}^{12}\left(y_i-a_0-a_1x_i-a_2x_i^2\right)=0, \tag{2.69}$$

$$\frac{\partial S_{sqr,2}}{\partial a_1}=\frac{\partial}{\partial a_1}\left(\sum_{i=1}^{12}\left(y_i-a_0-a_1x_i-a_2x_i^2\right)^2\right)=\sum_{i=1}^{12}\frac{\partial}{\partial a_1}\left(\left(y_i-a_0-a_1x_i-a_2x_i^2\right)^2\right)$$
$$=\sum_{i=1}^{12}2\left(y_i-a_0-a_1x_i-a_2x_i^2\right)(-x_i)=-2\sum_{i=1}^{12}x_i\left(y_i-a_0-a_1x_i-a_2x_i^2\right)=0, \tag{2.70}$$

and

$$\frac{\partial S_{sqr,2}}{\partial a_2}=\frac{\partial}{\partial a_2}\left(\sum_{i=1}^{12}\left(y_i-a_0-a_1x_i-a_2x_i^2\right)^2\right)=\sum_{i=1}^{12}\frac{\partial}{\partial a_2}\left(\left(y_i-a_0-a_1x_i-a_2x_i^2\right)^2\right)$$
$$=\sum_{i=1}^{12}2\left(y_i-a_0-a_1x_i-a_2x_i^2\right)\left(-x_i^2\right)=-2\sum_{i=1}^{12}x_i^2\left(y_i-a_0-a_1x_i-a_2x_i^2\right)=0, \tag{2.71}$$

pertaining to the first, second, and third parameter, respectively. When both sides are divided by −2, the summation split afterward, and constant factors taken off the summations, Eq. (2.69) produces

$$\sum_{i=1}^{12}\left(y_i-a_0-a_1x_i-a_2x_i^2\right)=\sum_{i=1}^{12}y_i-\sum_{i=1}^{12}a_0-\sum_{i=1}^{12}a_1x_i-\sum_{i=1}^{12}a_2x_i^2$$
$$=\sum_{i=1}^{12}y_i-a_0\sum_{i=1}^{12}1-a_1\sum_{i=1}^{12}x_i-a_2\sum_{i=1}^{12}x_i^2=0, \tag{2.72}$$

while Eq. (2.70) gives similarly rise to

(continued)

Problem 2.4 (continued)

$$\sum_{i=1}^{12} x_i\left(y_i - a_0 - a_1x_i - a_2x_i^2\right) = \sum_{i=1}^{12}\left(x_iy_i - a_0x_i - a_1x_i^2 - a_2x_i^3\right)$$

$$= \sum_{i=1}^{12} x_iy_i - \sum_{i=1}^{12} a_0x_i - \sum_{i=1}^{12} a_1x_i^2 - \sum_{i=1}^{12} a_2x_i^3 = \sum_{i=1}^{12} x_iy_i - a_0\sum_{i=1}^{12} x_i - a_1\sum_{i=1}^{12} x_i^2 - a_2\sum_{i=1}^{12} x_i^3 = 0 \tag{2.73}$$

after initial elimination of parenthesis; Eq. (2.71) generates

$$\sum_{i=1}^{12} x_i^2\left(y_i - a_0 - a_1x_i - a_2x_i^2\right) = \sum_{i=1}^{12}\left(x_i^2y_i - a_0x_i^2 - a_1x_i^3 - a_2x_i^4\right)$$

$$= \sum_{i=1}^{12} x_i^2y_i - \sum_{i=1}^{12} a_0x_i^2 - \sum_{i=1}^{12} a_1x_i^3 - \sum_{i=1}^{12} a_2x_i^4$$

$$= \sum_{i=1}^{12} x_i^2y_i - a_0\sum_{i=1}^{12} x_i^2 - a_1\sum_{i=1}^{12} x_i^3 - a_2\sum_{i=1}^{12} x_i^4 = 0, \tag{2.74}$$

upon factoring in of x_i^2. Equations (2.72)–(2.74) can be redone as

$$a_0\sum_{i=1}^{12} 1 + a_1\sum_{i=1}^{12} x_i + a_2\sum_{i=1}^{12} x_i^2 = \sum_{i=1}^{12} y_i, \tag{2.75}$$

$$a_0\sum_{i=1}^{12} x_i + a_1\sum_{i=1}^{12} x_i^2 + a_2\sum_{i=1}^{12} x_i^3 = \sum_{i=1}^{12} x_iy_i, \tag{2.76}$$

and

$$a_0\sum_{i=1}^{12} x_i^2 + a_1\sum_{i=1}^{12} x_i^3 + a_2\sum_{i=1}^{12} x_i^4 = \sum_{i=1}^{12} x_i^2y_i, \tag{2.77}$$

respectively; Cramer's rule will again prove helpful, namely,

(continued)

Problem 2.4 (continued)

$$a_0 = \frac{\begin{vmatrix} \sum_{i=1}^{12} y_i & \sum_{i=1}^{12} x_i & \sum_{i=1}^{12} x_i^2 \\ \sum_{i=1}^{12} x_i y_i & \sum_{i=1}^{12} x_i^2 & \sum_{i=1}^{12} x_i^3 \\ \sum_{i=1}^{12} x_i^2 y_i & \sum_{i=1}^{12} x_i^3 & \sum_{i=1}^{12} x_i^4 \end{vmatrix}}{\begin{vmatrix} \sum_{i=1}^{12} 1 & \sum_{i=1}^{12} x_i & \sum_{i=1}^{12} x_i^2 \\ \sum_{i=1}^{12} x_i & \sum_{i=1}^{12} x_i^2 & \sum_{i=1}^{12} x_i^3 \\ \sum_{i=1}^{12} x_i^2 & \sum_{i=1}^{12} x_i^3 & \sum_{i=1}^{12} x_i^4 \end{vmatrix}}, \tag{2.78}$$

$$a_1 = \frac{\begin{vmatrix} \sum_{i=1}^{12} 1 & \sum_{i=1}^{12} y_i & \sum_{i=1}^{12} x_i^2 \\ \sum_{i=1}^{12} x_i & \sum_{i=1}^{12} x_i y_i & \sum_{i=1}^{12} x_i^3 \\ \sum_{i=1}^{12} x_i^2 & \sum_{i=1}^{12} x_i^2 y_i & \sum_{i=1}^{12} x_i^4 \end{vmatrix}}{\begin{vmatrix} \sum_{i=1}^{12} 1 & \sum_{i=1}^{12} x_i & \sum_{i=1}^{12} x_i^2 \\ \sum_{i=1}^{12} x_i & \sum_{i=1}^{12} x_i^2 & \sum_{i=1}^{12} x_i^3 \\ \sum_{i=1}^{12} x_i^2 & \sum_{i=1}^{12} x_i^3 & \sum_{i=1}^{12} x_i^4 \end{vmatrix}}, \tag{2.79}$$

and

(continued)

Problem 2.4 (continued)

$$a_2 = \frac{\begin{vmatrix} \sum_{i=1}^{12} 1 & \sum_{i=1}^{12} x_i & \sum_{i=1}^{12} y_i \\ \sum_{i=1}^{12} x_i & \sum_{i=1}^{12} x_i^2 & \sum_{i=1}^{12} x_i y_i \\ \sum_{i=1}^{12} x_i^2 & \sum_{i=1}^{12} x_i^3 & \sum_{i=1}^{12} x_i^2 y_i \end{vmatrix}}{\begin{vmatrix} \sum_{i=1}^{12} 1 & \sum_{i=1}^{12} x_i & \sum_{i=1}^{12} x_i^2 \\ \sum_{i=1}^{12} x_i & \sum_{i=1}^{12} x_i^2 & \sum_{i=1}^{12} x_i^3 \\ \sum_{i=1}^{12} x_i^2 & \sum_{i=1}^{12} x_i^3 & \sum_{i=1}^{12} x_i^4 \end{vmatrix}} \tag{2.80}$$

as formal estimates of a_0, a_1, and a_2, respectively. Laplace's theorem supports transformation of Eq. (2.78) to

$$a_0 = \frac{(-1)^{1+1} \sum_{i=1}^{12} y_i \begin{vmatrix} \sum_{i=1}^{12} x_i^2 & \sum_{i=1}^{12} x_i^3 \\ \sum_{i=1}^{12} x_i^3 & \sum_{i=1}^{12} x_i^4 \end{vmatrix} + (-1)^{2+1} \sum_{i=1}^{12} x_i y_i \begin{vmatrix} \sum_{i=1}^{12} x_i & \sum_{i=1}^{12} x_i^2 \\ \sum_{i=1}^{12} x_i^3 & \sum_{i=1}^{12} x_i^4 \end{vmatrix} + (-1)^{3+1} \sum_{i=1}^{12} x_i^2 y_i \begin{vmatrix} \sum_{i=1}^{12} x_i & \sum_{i=1}^{12} x_i^2 \\ \sum_{i=1}^{12} x_i^2 & \sum_{i=1}^{12} x_i^3 \end{vmatrix}}{(-1)^{1+1} \sum_{i=1}^{12} 1 \begin{vmatrix} \sum_{i=1}^{12} x_i^2 & \sum_{i=1}^{12} x_i^3 \\ \sum_{i=1}^{12} x_i^3 & \sum_{i=1}^{12} x_i^4 \end{vmatrix} + (-1)^{1+2} \sum_{i=1}^{12} x_i \begin{vmatrix} \sum_{i=1}^{12} x_i & \sum_{i=1}^{12} x_i^3 \\ \sum_{i=1}^{12} x_i^2 & \sum_{i=1}^{12} x_i^4 \end{vmatrix} + (-1)^{1+3} \sum_{i=1}^{12} x_i^2 \begin{vmatrix} \sum_{i=1}^{12} x_i & \sum_{i=1}^{12} x_i^2 \\ \sum_{i=1}^{12} x_i^2 & \sum_{i=1}^{12} x_i^3 \end{vmatrix}} \tag{2.81}$$

using the first column and the first row for pivotal decomposition of the third-order determinants in numerator and denominator, respectively, where definition of second-order determinant leads, in turn, to

$$a_0 = \frac{\sum_{i=1}^{12} y_i \left(\sum_{i=1}^{12} x_i^2 \sum_{i=1}^{12} x_i^4 - \left(\sum_{i=1}^{12} x_i^3 \right)^2 \right) - \sum_{i=1}^{12} x_i y_i \left(\sum_{i=1}^{12} x_i \sum_{i=1}^{12} x_i^4 - \sum_{i=1}^{12} x_i^2 \sum_{i=1}^{12} x_i^3 \right) + \sum_{i=1}^{12} x_i^2 y_i \left(\sum_{i=1}^{12} x_i \sum_{i=1}^{12} x_i^3 - \left(\sum_{i=1}^{12} x_i^2 \right)^2 \right)}{\sum_{i=1}^{12} 1 \left(\sum_{i=1}^{12} x_i^2 \sum_{i=1}^{12} x_i^4 - \left(\sum_{i=1}^{12} x_i^3 \right)^2 \right) - \sum_{i=1}^{12} x_i \left(\sum_{i=1}^{12} x_i \sum_{i=1}^{12} x_i^4 - \sum_{i=1}^{12} x_i^2 \sum_{i=1}^{12} x_i^3 \right) + \sum_{i=1}^{12} x_i^2 \left(\sum_{i=1}^{12} x_i \sum_{i=1}^{12} x_i^3 - \left(\sum_{i=1}^{12} x_i^2 \right)^2 \right)}; \tag{2.82}$$

Equation (2.79) similarly yields

(continued)

Problem 2.4 (continued)

$$a_1=\frac{(-1)^{1+2}\sum_{i=1}^{12}y_i\begin{vmatrix}\sum_{i=1}^{12}x_i & \sum_{i=1}^{12}x_i^3\\ \sum_{i=1}^{12}x_i^2 & \sum_{i=1}^{12}x_i^4\end{vmatrix}+(-1)^{2+2}\sum_{i=1}^{12}x_iy_i\begin{vmatrix}\sum_{i=1}^{12}1 & \sum_{i=1}^{12}x_i^2\\ \sum_{i=1}^{12}x_i^2 & \sum_{i=1}^{12}x_i^4\end{vmatrix}+(-1)^{3+2}\sum_{i=1}^{12}x_i^2y_i\begin{vmatrix}\sum_{i=1}^{12}1 & \sum_{i=1}^{12}x_i^2\\ \sum_{i=1}^{12}x_i & \sum_{i=1}^{12}x_i^3\end{vmatrix}}{(-1)^{2+1}\sum_{i=1}^{12}x_i\begin{vmatrix}\sum_{i=1}^{12}x_i & \sum_{i=1}^{12}x_i^2\\ \sum_{i=1}^{12}x_i^3 & \sum_{i=1}^{12}x_i^4\end{vmatrix}+(-1)^{2+2}\sum_{i=1}^{12}x_i^2\begin{vmatrix}\sum_{i=1}^{12}1 & \sum_{i=1}^{12}x_i^2\\ \sum_{i=1}^{12}x_i^2 & \sum_{i=1}^{12}x_i^4\end{vmatrix}+(-1)^{2+3}\sum_{i=1}^{12}x_i^3\begin{vmatrix}\sum_{i=1}^{12}1 & \sum_{i=1}^{12}x_i\\ \sum_{i=1}^{12}x_i^2 & \sum_{i=1}^{12}x_i^3\end{vmatrix}},\tag{2.83}$$

where the second column and the second row were instead used to expand the third-order determinants in numerator and denominator, respectively, with algebraic rearrangement based on the definition of second-order determinant leading to

$$a_1=\frac{-\sum_{i=1}^{12}y_i\left(\sum_{i=1}^{12}x_i\sum_{i=1}^{12}x_i^4-\sum_{i=1}^{12}x_i^2\sum_{i=1}^{12}x_i^3\right)+\sum_{i=1}^{12}x_iy_i\left(\sum_{i=1}^{12}1\sum_{i=1}^{12}x_i^4-\left(\sum_{i=1}^{12}x_i^2\right)^2\right)-\sum_{i=1}^{12}x_i^2y_i\left(\sum_{i=1}^{12}1\sum_{i=1}^{12}x_i^3-\sum_{i=1}^{12}x_i\sum_{i=1}^{12}x_i^2\right)}{-\sum_{i=1}^{12}x_i\left(\sum_{i=1}^{12}x_i\sum_{i=1}^{12}x_i^4-\sum_{i=1}^{12}x_i^2\sum_{i=1}^{12}x_i^3\right)+\sum_{i=1}^{12}x_i^2\left(\sum_{i=1}^{12}1\sum_{i=1}^{12}x_i^4-\left(\sum_{i=1}^{12}x_i^2\right)^2\right)-\sum_{i=1}^{12}x_i^3\left(\sum_{i=1}^{12}1\sum_{i=1}^{12}x_i^3-\sum_{i=1}^{12}x_i\sum_{i=1}^{12}x_i^2\right)}.\tag{2.84}$$

By the same token, Eq. (2.80) is equivalent to

$$a_2=\frac{(-1)^{1+3}\sum_{i=1}^{12}y_i\begin{vmatrix}\sum_{i=1}^{12}x_i & \sum_{i=1}^{12}x_i^2\\ \sum_{i=1}^{12}x_i^2 & \sum_{i=1}^{12}x_i^3\end{vmatrix}+(-1)^{2+3}\sum_{i=1}^{12}x_iy_i\begin{vmatrix}\sum_{i=1}^{12}1 & \sum_{i=1}^{12}x_i\\ \sum_{i=1}^{12}x_i^2 & \sum_{i=1}^{12}x_i^3\end{vmatrix}+(-1)^{3+3}\sum_{i=1}^{12}x_i^2y_i\begin{vmatrix}\sum_{i=1}^{12}1 & \sum_{i=1}^{12}x_i\\ \sum_{i=1}^{12}x_i & \sum_{i=1}^{12}x_i^2\end{vmatrix}}{(-1)^{3+1}\sum_{i=1}^{12}x_i^2\begin{vmatrix}\sum_{i=1}^{12}x_i & \sum_{i=1}^{12}x_i^2\\ \sum_{i=1}^{12}x_i^2 & \sum_{i=1}^{12}x_i^3\end{vmatrix}+(-1)^{3+2}\sum_{i=1}^{12}x_i^3\begin{vmatrix}\sum_{i=1}^{12}1 & \sum_{i=1}^{12}x_i^2\\ \sum_{i=1}^{12}x_i & \sum_{i=1}^{12}x_i^3\end{vmatrix}+(-1)^{3+3}\sum_{i=1}^{12}x_i^4\begin{vmatrix}\sum_{i=1}^{12}1 & \sum_{i=1}^{12}x_i\\ \sum_{i=1}^{12}x_i & \sum_{i=1}^{12}x_i^2\end{vmatrix}}\tag{2.85}$$

once the third-order determinants in numerator and denominator have been expanded through their third column and third row, respectively, which degenerates to

(continued)

Problem 2.4 (continued)

$$a_2=\frac{\sum_{i=1}^{12}y_i\left(\sum_{i=1}^{12}x_i\sum_{i=1}^{12}x_i^3-\left(\sum_{i=1}^{12}x_i^2\right)^2\right)-\sum_{i=1}^{12}x_iy_i\left(\sum_{i=1}^{12}1\sum_{i=1}^{12}x_i^3-\sum_{i=1}^{12}x_i\sum_{i=1}^{12}x_i^2\right)+\sum_{i=1}^{12}x_i^2y_i\left(\sum_{i=1}^{12}1\sum_{i=1}^{12}x_i^2-\left(\sum_{i=1}^{12}x_i\right)^2\right)}{\sum_{i=1}^{12}x_i^2\left(\sum_{i=1}^{12}x_i\sum_{i=1}^{12}x_i^3-\left(\sum_{i=1}^{12}x_i^2\right)^2\right)-\sum_{i=1}^{12}x_i^3\left(\sum_{i=1}^{12}1\sum_{i=1}^{12}x_i^3-\sum_{i=1}^{12}x_i\sum_{i=1}^{12}x_i^2\right)+\sum_{i=1}^{12}x_i^4\left(\sum_{i=1}^{12}1\sum_{i=1}^{12}x_i^2-\left(\sum_{i=1}^{12}x_i\right)^2\right)} \tag{2.86}$$

in view again of the definition of second-order determinant; the values for the summations in Eqs. (2.82), (2.84), and (2.86) are provided below.

i	1	x_i	x_i^2	x_i^3	x_i^4	y_i	x_iy_i	$x_i^2y_i$
1	1	0	0	0	0	14.8	0	0
2	1	10	100	1000	10,000	11.5	115	1150
3	1	15	225	3375	50,625	10.4	156	2340
4	1	20	400	8000	160,000	9.45	189	3780
5	1	25	625	15,625	390,625	8.69	217.3	5431.3
6	1	26	676	17,576	456,976	8.55	222.3	5779.8
7	1	27	729	19,683	531,441	8.42	227.3	6138.2
8	1	28	784	21,952	614,656	8.29	232.1	6499.4
9	1	29	841	24,389	707,281	8.17	236.9	6871.0
10	1	30	900	27,000	810,000	8.05	241.5	7245.0
11	1	35	1225	42,875	1,500,625	7.52	263.2	9212.0
12	1	40	1600	64,000	2,560,000	7.07	282.8	11312.0
$\sum_{i=1}^{12}$	12	285	8105	245,475	7,792,229	110.9	2383	65,759

Insertion of the above summations in Eqs. (2.82), (2.84), and (2.86) leads to

$$a_0=\frac{110.9\left(8105\cdot7792229-245475^2\right)-2383(285\cdot7792229-8105\cdot245475)+65759\left(285\cdot245475-8105^2\right)}{12\left(8105\cdot7792229-245475^2\right)-285(285\cdot7792229-8105\cdot245475)+8105\left(285\cdot245475-8105^2\right)}$$
$$=\frac{5.11665\times10^{10}}{0.348461\times10^{10}}=14.6836, \tag{2.87}$$

$$a_1=\frac{-110.9(285\cdot7792229-8105\cdot245475)+2383\left(12\cdot7792229-8105^2\right)-65759(12\cdot245475-285\cdot8105)}{-285(285\cdot7792229-8105\cdot245475)+8105\left(12\cdot7792229-8105^2\right)-245475(12\cdot245475-285\cdot8105)}$$
$$=\frac{-1.16429\times10^9}{3.48461\times10^9}=-0.334123, \tag{2.88}$$

(continued)

Problem 2.4 (continued)

and

$$a_2 = \frac{110.9\left(285 \cdot 245475 - 8105^2\right) - 2383(12 \cdot 245475 - 285 \cdot 8105) + 65759\left(12 \cdot 8105 - 285^2\right)}{8105\left(285 \cdot 245475 - 8105^2\right) - 245475(12 \cdot 245475 - 285 \cdot 8105) + 7792229\left(12 \cdot 8105 - 285^2\right)}$$
$$= \frac{0.0128647 \times 10^9}{3.48461 \times 10^9} = 0.00369186, \tag{2.89}$$

respectively, for best estimates of the three parameters germane in the second-order model consubstantiated in Eq. (2.67).

(b) The accuracy in prediction of the three models fitted in *a*) can be graphically grasped in Fig. 2.5. Although improvement in fit is dramatic when going from the zeroth- to the first-order model, the conclusion is not so sharp when going from the first- to the second-order model. An incremental sum of squares analysis is then in order, in attempts to validate the need for a more accurate model relative to another nested therein; hence, one should calculate the sum of squares of residuals, *SSR*, of each of the aforementioned models, as tabulated below, with indication at the bottom of the sum of the entries in the same column and corresponding number of degrees of

(continued)

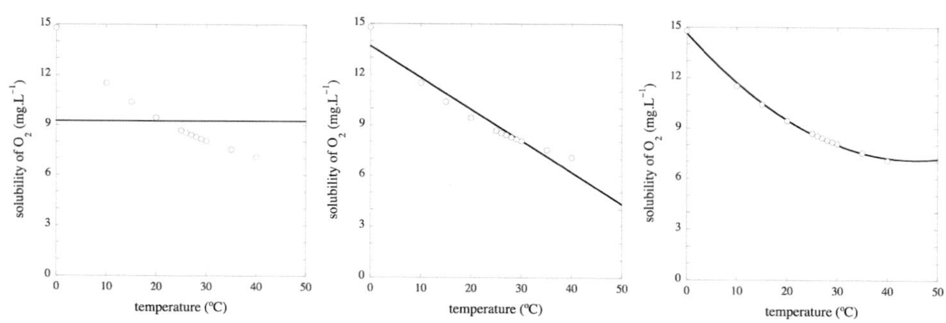

Fig. 2.5 Variation of solubility of molecular oxygen in plain water versus temperature, as experimental data (○) and best fit (—) via zeroth-order model (left), first-order model (center), and second-order model (right)

Problem 2.4 (continued)

freedom, ν, given by number of data points minus number of model parameters.

i	y_i	a_0	$(y_i - a_0)^2$	$a_0 + a_1x_i$	$(y_i - (a_0 + a_1x_i))^2$	$a_0 + a_1x_i + a_2x_i^2$	$(y_i - (a_0 + a_1x_i + a_2x_i^2))^2$
1	14.8	9.242	30.8914	13.700	1.2100	14.684	0.0135
2	11.5	9.242	5.0986	11.823	0.1043	11.379	0.0146
3	10.4	9.242	1.3410	10.888	0.2381	10.502	0.0104
4	9.45	9.242	0.0433	9.946	0.2460	9.478	0.0008
5	8.69	9.242	0.3047	9.007	0.1005	8.638	0.0027
6	8.55	9.242	0.4789	8.819	0.0724	8.492	0.0034
7	8.42	9.242	0.6757	8.631	0.0445	8.354	0.0044
8	8.29	9.242	0.9063	8.444	0.0237	8.223	0.0045
9	8.17	9.242	1.1492	8.256	0.0074	8.099	0.0050
10	8.05	9.242	1.4209	8.068	0.0003	7.983	0.0045
11	7.52	9.242	2.9653	7.129	0.1529	7.512	0.0001
12	7.07	9.242	4.7176	6.191	0.7726	7.226	0.0243
SSR	–	–	49.9929	–	2.9727	–	0.0882
ν	–	–	11	–	10	–	9

The change in sum of squares of residuals when going from zeroth- to first-order model, $\Delta SSR_{0\to1}$, normalized by the corresponding change in number of degrees of freedom, $\Delta\nu_{0\to1} = \nu_0 - \nu_1 = 11 - 10 = 1$, should then be compared to an F-distribution with $\Delta\nu_{0\to1} = 1$ and $\nu_1 = 10$ degrees of freedom, that is, $F(1,10)$, and likewise when going from first- to second-order model, $\Delta SSR_{1\to2}$, to be normalized by the associated change in number of degrees of freedom, $\Delta\nu_{1\to2} = \nu_1 - \nu_2 = 10 - 9 = 1$, and then compared to an F distribution with $\Delta\nu_{1\to2} = 1$ and $\nu_2 = 9$ degrees of freedom, that is, $F(1,9)$, as done below.

	Zeroth-order model	First-order model	Second-order model	Zeroth- to first-order model	First- to second-order model
SSR	49.9929	2.9727	0.0882	47.02	2.885
ν	11	10	9	1	1
$\frac{SSR}{\nu}$	–	0.2972	0.0098	47.02	2.885
$\frac{\frac{\Delta SSR_{i\to j}}{\Delta\nu_{i\to j}}}{\frac{SSR_j}{\nu_j}}$	–	–	–	$\frac{47.02}{0.2972} = 158.2$	$\frac{2.885}{0.0098} = 294.4$
F-statistic	–	–	–	$F(1,9;1\%) = 10.56$	$F(1,8;1\%) = 11.26$
P (ratio $< F$)	–	–	–	0.0000	0.0000

One therefore concludes that the fit provided by the first-order model is statistically better than that provided by the zeroth-order model or, in other

(continued)

Problem 2.4 (continued)

words, the probability that such a large value for the F-ratio, when going from the zeroth- to the first-order model, is due to pure chance is far less than 1%; and an identical conclusion applies when the second-order model is used relative to the first-order model.

2.2 Inoculum Preparation

The cultivation medium to be used in a regular bioreactor is first inoculated with a preculture—oftentimes previously grown in a shaker or incubator; smaller bioreactors may also be utilized for this purpose. This process of inoculum development from a preculture is intended to generate cellular biomass rather than fermentation products; the result should be microorganisms in optimal physiological condition, compatible with inoculation afterward of the production bioreactor, which normally requires a sufficient level of viable and competent biomass. In addition, the inoculum should be free from contamination and hold product-forming ability; its potential may be further enhanced by strain improvement techniques or cell immobilization technology. During this preparation process, monitorization of biomass is in order, which may involve electrochemical sensors, exhaust gas analysis, and even mass spectrometers.

Microorganisms for industrial production are usually sourced from their natural habitat or purchased in pure form from culture collection organizations (e.g., American Type Culture Collection, ATCC). When at laboratory scale, the various forms taken by the medium undergoing inoculation are agar plates, slant cultures, and stab cultures, all based on a solid matrix, or else broth cultures should the medium be kept in liquid form. In the first three cases, agar is originally poured onto a Petri dish or a test tube, and let to solidify therein; if the test tube is kept slanted, then a greater surface area is made available for the microorganism to grow on, while minimizing water loss by capping said tube. For Petri- and slanted tube-contained agar, an aliquot is then introduced onto the agar surface using a streaking loop or needle, dipped in advance into the original cell-containing medium; incubation at appropriate temperature and time ensues. If a vertical test tube is utilized, then an inoculation needle is used to introduce the microorganisms, and the punctured area allows inner growth of the inoculum; this is a particularly useful strategy for shipping or short-term storage.

Upon isolation, or purchase and retrieval from their transport/preservation culture, microorganisms are to be activated; preparation of a proper inoculum is indeed crucial for success of any bioprocess with regard to lag time, growth kinetics, productivity, and product specification. More specifically, the inoculum must be in a healthy and active state, be available in sufficiently large volume by the time of (planned) inoculation, be in a suitable morphological state, be devoid of contamination, and retain its product-forming capabilities. Remember that cost minimization and profit maximization are major drivers in industrial processes, so an efficient inoculum should exhibit high biomass productivity and/or high product yield.

The first-generation culture is accordingly prepared from the original culture stored on agar slants; sterile saline water or liquid nutrient medium containing suspended saline beads is then added so that a microbial suspension results, which is transferred to a flat bed bottle containing sterile agar medium. Incubation at a convenient temperature, under gentle mixing, will then support expansion of biomass. In small-scale operation, this working culture may serve to directly inoculate the actual bioreactor.

In large-scale processes, inoculum preparation requires additional steps that go beyond the working culture. The cells from the aforementioned flat bed bottles are further transferred to shaker flasks containing sterile liquid nutrient medium, placed on a rotary shaker, and duly incubated; microbial growth is enhanced due to aeration. This preparation is then used as seed culture in a small fermenter and allowed to further grow for 1–2 days, under environmental conditions that mimic the target bioprocess. The resulting microorganisms play the role of actual inoculum and are finally transferred to the industrial scale bioreactor containing a larger volume of medium, which is then allowed to run for the stipulated time, and under the processing conditions leading to optimal biomass formation and/or product generation.

In parallel to inoculum preparation, the bioreactor and its feed lines and sensors need to be sterilized, and medium is to be prepared to sufficient amounts, either to fill the bioreactor at startup if operated batchwise or to be gradually pumped into the bioreactor if operated continuous or fedbatchwise, respectively.

Problem 2.5

Consider that rate of biomass growth, r_X, is described by Monod's equation, namely,

$$r_X = \frac{\mu_{max} C_S}{K_M + C_S} C_X, \tag{2.90}$$

with kinetic parameters denoted by μ_{max} and K_M, and biomass and (limiting) substrate concentrations denoted by C_X and C_S, respectively. The starting inoculum, with biomass $n_{X,0}$, may either be inoculated once into a bioreactor of volume V at concentration $C_{X,0} \equiv n_{X,0}/V$, containing substrate at initial concentration $C_{S,0}$, and to be operated for batch time τ, or be inoculated in a first bioreactor, of volume $V/2$ at concentration $2C_{X,0} \equiv 2(n_{X,0}/V) = n_{X,0}/(V/2)$, containing substrate also at initial concentration $C_{S,0}$, and to be operated for batch time $\tau/2$, before reinoculating said batch into a second bioreactor of volume V, already containing a volume $V/2$ of sterile, plain fresh medium, with substrate at initial concentration $C_{S,0}$, and to be operated for another batch time $\tau/2$. Assume a constant yield of biomass on substrate, $Y_{X/S}$.

(a) Compare the two inoculation strategies, in terms of final biomass obtained, and illustrate with preselected, convenient values for the germane (dimensionless) parameters.
(b) Describe the asymptotic behavior at long batch times.

(continued)

Problem 2.5 (continued)

(c) Prove that the specific growth rate of the culture is always enhanced at the moment of reinoculation.

(d) Derive the condition that guarantees a higher growth rate at the moment of reinoculation.

Solution

(a) A balance to biomass, in either the single or the first batch bioreactor of the cascade, reads

$$\frac{dC_X}{dt} = \frac{\mu_{max} C_S}{K_M + C_S} C_X, \tag{2.91}$$

whereas definition of $Y_{X/S}$ has it that

$$Y_{X/S} \equiv \frac{\frac{dC_X}{dt}}{-\frac{dC_S}{dt}}, \tag{2.92}$$

which will be taken as constant by hypothesis. After getting rid of denominators, Eq. (2.92) unfolds

$$\frac{dC_X}{dt} = -Y_{X/S} \frac{dC_S}{dt}, \tag{2.93}$$

where multiplication of both sides by dt leaves

$$dC_X = -Y_{X/S}\, dC_S; \tag{2.94}$$

the initial conditions for the single bioreactor read

$$C_X|_{t=0} = C_{X,0} \tag{2.95}$$

and

$$C_S|_{t=0} = C_{S,0}, \tag{2.96}$$

with known $C_{X,0}$ and $C_{S,0}$, with the latter susceptible of replacement by a combination of Eqs. (2.95) and (2.96), that is,

(continued)

Problem 2.5 (continued)

$$C_S\big|_{C_X=C_{X,0}}=C_{S,0}. \tag{2.97}$$

Equation (2.94) will be integrated, with the aid of Eq. (2.97), to generate

$$\int_{C_{X,0}}^{C_X} d\tilde{C}_X=-Y_{X/S}\int_{C_{S,0}}^{C_S} d\tilde{C}_S; \tag{2.98}$$

trivial application of the fundamental theorem of integral calculus leads then to

$$\tilde{C}_X\bigg|_{C_{X,0}}^{C_X}=-Y_{X/S}\ \tilde{C}_S\bigg|_{C_{S,0}}^{C_S} \tag{2.99}$$

or, equivalently,

$$C_X-C_{X,0}=-Y_{X/S}(C_S-C_{S,0}), \tag{2.100}$$

where isolation of C_S unfolds

$$C_S=C_{S,0}-\frac{C_X-C_{X,0}}{Y_{X/S}} \tag{2.101}$$

while isolation of C_X instead produces

$$C_X=C_{X,0}+Y_{X/S}(C_{S,0}-C_S). \tag{2.102}$$

Insertion of Eq. (2.101) converts Eq. (2.91) to

$$\frac{dC_X}{dt}=\frac{\mu_{max}\left(C_{S,0}-\frac{C_X-C_{X,0}}{Y_{X/S}}\right)}{K_M+C_{S,0}-\frac{C_X-C_{X,0}}{Y_{X/S}}}C_X, \tag{2.103}$$

with elimination of parenthesis leaving

$$\frac{dC_X}{dt}=\frac{\mu_{max}C_{S,0}+\frac{\mu_{max}C_{X,0}}{Y_{X/S}}-\frac{\mu_{max}}{Y_{X/S}}C_X}{K_M+C_{S,0}+\frac{C_{X,0}}{Y_{X/S}}-\frac{C_X}{Y_{X/S}}}C_X. \tag{2.104}$$

(continued)

Problem 2.5 (continued)

Multiplication of both numerator and denominator of Eq. (2.104) by $Y_{X/S}$ supports, in turn,

$$\frac{dC_X}{dt} = \frac{\mu_{max} Y_{X/S} C_{S,0} + \mu_{max} C_{X,0} - \mu_{max} C_X}{Y_{X/S}(K_M + C_{S,0}) + C_{X,0} - C_X} C_X, \tag{2.105}$$

so integration via separation of variables ensues as

$$\int_{C_{X,0}}^{C_X} \frac{Y_{X/S}(K_M + C_{S,0}) + C_{X,0} - \widetilde{C}_X}{\widetilde{C}_X \left(\mu_{max} \left(Y_{X/S} C_{S,0} + C_{X,0} \right) - \mu_{max} \widetilde{C}_X \right)} d\widetilde{C}_X = \int_0^t d\widetilde{t} \tag{2.106}$$

with the aid of Eq. (2.95). Heaviside's expansion of the kernel in the left-hand side of Eq. (2.106) takes the form

$$\begin{aligned} &\frac{Y_{X/S}(K_M + C_{S,0}) + C_{X,0} - C_X}{C_X \left(\mu_{max} \left(Y_{X/S} C_{S,0} + C_{X,0} \right) - \mu_{max} C_X \right)} \\ &= \frac{A}{C_X} + \frac{B}{\mu_{max} \left(Y_{X/S} C_{S,0} + C_{X,0} \right) - \mu_{max} C_X}, \end{aligned} \tag{2.107}$$

where A and B denote constants still to be found; elimination of denominators permits simplification to

$$Y_{X/S}(K_M + C_{S,0}) + C_{X,0} - C_X = A \left(\mu_{max} \left(Y_{X/S} C_{S,0} + C_{X,0} \right) - \mu_{max} C_X \right) + B C_X. \tag{2.108}$$

Equation (2.108) must be satisfied irrespective of the values attributed to C_X, and, in particular, should be valid when $C_X = 0$ and $C_X = Y_{X/S} C_{S,0} + C_{X,0}$; in the former case, one gets

$$Y_{X/S}(K_M + C_{S,0}) + C_{X,0} = A \mu_{max} \left(Y_{X/S} C_{S,0} + C_{X,0} \right) \tag{2.109}$$

and the latter case enforces

$$Y_{X/S}(K_M + C_{S,0}) + C_{X,0} - \left(Y_{X/S} C_{S,0} + C_{X,0} \right) = B \left(Y_{X/S} C_{S,0} + C_{X,0} \right), \tag{2.110}$$

(continued)

Problem 2.5 (continued)

both readily obtained from Eq. (2.108). Isolation of A from Eq. (2.109) prompts

$$A = \frac{Y_{X/S}(K_M + C_{S,0}) + C_{X,0}}{\mu_{max}\left(Y_{X/S}C_{S,0} + C_{X,0}\right)} \tag{2.111}$$

and likewise with regard to B based on Eq. (2.110), that is,

$$B = \frac{Y_{X/S}(K_M + C_{S,0}) + C_{X,0} - Y_{X/S}C_{S,0} - C_{X,0}}{Y_{X/S}C_{S,0} + C_{X,0}}, \tag{2.112}$$

along with selective elimination of parenthesis; upon cancellation of symmetrical terms, Eq. (2.112) becomes

$$B = \frac{Y_{X/S}(K_M + C_{S,0}) - Y_{X/S}C_{S,0}}{Y_{X/S}C_{S,0} + C_{X,0}}, \tag{2.113}$$

which justifies factoring out of $Y_{X/S}$ as

$$B = \frac{Y_{X/S}(K_M + C_{S,0} - C_{S,0})}{Y_{X/S}C_{S,0} + C_{X,0}}, \tag{2.114}$$

or merely

$$B = \frac{Y_{X/S}K_M}{Y_{X/S}C_{S,0} + C_{X,0}}, \tag{2.115}$$

since $C_{S,0}$ cancels out with its negative. Equation (2.106) can now be integrated with the aid of Eq. (2.107), namely,

$$\int_{C_{X,0}}^{C_X} \left(\frac{A}{\widetilde{C}_X} + \frac{B}{\mu_{max}\left(Y_{X/S}C_{S,0} + C_{X,0}\right) - \mu_{max}\widetilde{C}_X} \right) d\widetilde{C}_X = \int_0^t d\widetilde{t}, \tag{2.116}$$

where the integral in the left-hand side will be split to get

(continued)

Problem 2.5 (continued)

$$\int_{C_{X,0}}^{C_X} \frac{A}{\widetilde{C}_X} d\widetilde{C}_X + \int_{C_{X,0}}^{C_X} \frac{\frac{B}{\mu_{max}}}{Y_{X/S}C_{S,0} + C_{X,0} - \widetilde{C}_X} d\widetilde{C}_X = \int_0^t d\widetilde{t} \tag{2.117}$$

together with factoring out of μ_{max} in denominator and lumping it with numerator afterward; the fundamental theorem of integral calculus is then to be invoked to write

$$A \ln \widetilde{C}_X \Bigg|_{C_{X,0}}^{C_X} - \frac{B}{\mu_{max}} \ln\left\{ Y_{X/S}C_{S,0} + C_{X,0} - \widetilde{C}_X \right\} \Bigg|_{C_{X,0}}^{C_X} = \widetilde{t} \Bigg|_0^t, \tag{2.118}$$

with replacement by upper and lower limits of integration leaving

$$A \ln \frac{C_X}{C_{X,0}} - \frac{B}{\mu_{max}} \ln \frac{Y_{X/S}C_{S,0} + C_{X,0} - C_X}{Y_{X/S}C_{S,0} + C_{X,0} - C_{X,0}} = t - 0 \tag{2.119}$$

or, after dropping symmetrical terms in denominator and simplifying the right-hand side,

$$A \ln \frac{C_X}{C_{X,0}} - \frac{B}{\mu_{max}} \ln \frac{Y_{X/S}C_{S,0} + C_{X,0} - C_X}{Y_{X/S}C_{S,0}} = t. \tag{2.120}$$

Replacement of A and B by Eqs. (2.111) and (2.115), respectively, converts Eq. (2.120) to

$$\frac{Y_{X/S}(K_M + C_{S,0}) + C_{X,0}}{\mu_{max}\left(Y_{X/S}C_{S,0} + C_{X,0}\right)} \ln \frac{C_X}{C_{X,0}} - \frac{Y_{X/S}K_M}{\mu_{max}\left(Y_{X/S}C_{S,0} + C_{X,0}\right)} \ln \frac{Y_{X/S}C_{S,0} + C_{X,0} - C_X}{Y_{X/S}C_{S,0}} = t, \tag{2.121}$$

while elimination of denominators between sides generates

$$\left(Y_{X/S}(K_M + C_{S,0}) + C_{X,0}\right) \ln \frac{C_X}{C_{X,0}} - Y_{X/S}K_M \ln \frac{Y_{X/S}C_{S,0} + C_{X,0} - C_X}{Y_{X/S}C_{S,0}} = \mu_{max}\left(Y_{X/S}C_{S,0} + C_{X,0}\right)t; \tag{2.122}$$

(continued)

Problem 2.5 (continued)

factoring in $Y_{X/S}$ and dividing by $C_{X,0}$ whenever possible finally support conversion to

$$\left(\frac{Y_{X/S}K_M}{C_{X,0}}+\frac{Y_{X/S}C_{S,0}}{C_{X,0}}+1\right)\ln\frac{C_X}{C_{X,0}}-\frac{Y_{X/S}K_M}{C_{X,0}}\ln\frac{\frac{Y_{X/S}C_{S,0}}{C_{X,0}}+1-\frac{C_X}{C_{X,0}}}{\frac{Y_{X/S}C_{S,0}}{C_{X,0}}}$$
$$=\left(\frac{Y_{X/S}C_{S,0}}{C_{X,0}}+1\right)\mu_{max}t. \tag{2.123}$$

Equation (2.123) allows calculation (in implicit form) of dimensionless biomass concentration, $C_{X,I}/C_{X,0}$, attained by dimensionless time, $\mu_{max}t$, when a single batch bioreactor is employed, namely,

$$\left(\frac{Y_{X/S}K_M}{C_{X,0}}+\frac{Y_{X/S}C_{S,0}}{C_{X,0}}+1\right)\ln\frac{C_{X,I}}{C_{X,0}}-\frac{Y_{X/S}K_M}{C_{X,0}}\ln\frac{\frac{Y_{X/S}C_{S,0}}{C_{X,0}}+1-\frac{C_{X,I}}{C_{X,0}}}{\frac{Y_{X/S}C_{S,0}}{C_{X,0}}}$$
$$=\left(\frac{Y_{X/S}C_{S,0}}{C_{X,0}}+1\right)\mu_{max}t; \tag{2.124}$$

consequently,

$$\left(\frac{Y_{X/S}K_M}{C_{X,0}}+\frac{Y_{X/S}C_{S,0}}{C_{X,0}}+1\right)\ln\frac{C_{X,I,\tau}}{C_{X,0}}-\frac{Y_{X/S}K_M}{C_{X,0}}\ln\frac{\frac{Y_{X/S}C_{S,0}}{C_{X,0}}+1-\frac{C_{X,I,\tau}}{C_{X,0}}}{\frac{Y_{X/S}C_{S,0}}{C_{X,0}}}$$
$$=\left(\frac{Y_{X/S}C_{S,0}}{C_{X,0}}+1\right)\mu_{max}\tau \tag{2.125}$$

ensues after setting $t=\tau$, and thus $C_{X,I,\tau}\equiv C_{X,I}|_{t=\tau}$. On the other hand, Eq. (2.95) should be replaced by

$$C_X|_{t=0}=2C_{X,0} \tag{2.126}$$

to serve as initial condition, while Eq. (2.96) remains valid in what concerns the first unit in the series of two batch bioreactors under scrutiny; Eq. (2.123) should accordingly take the alternative form

(continued)

Problem 2.5 (continued)

$$\left(\frac{Y_{X/S}K_M}{2C_{X,0}}+\frac{Y_{X/S}C_{S,0}}{2C_{X,0}}+1\right)\ln\frac{C_{X,1,II}}{2C_{X,0}}-\frac{Y_{X/S}K_M}{2C_{X,0}}\ln\frac{\frac{Y_{X/S}C_{S,0}}{2C_{X,0}}+1-\frac{C_{X,1,II}}{2C_{X,0}}}{\frac{Y_{X/S}C_{S,0}}{2C_{X,0}}}$$
$$=\left(\frac{Y_{X/S}C_{S,0}}{2C_{X,0}}+1\right)\mu_{max}t, \tag{2.127}$$

in attempts to calculate the dimensionless biomass concentration, $C_{X,1,II}/C_{X,0}$, prevailing in the first bioreactor of the series at generic dimensionless time, $\mu_{max}t$, and likewise

$$\left(\frac{Y_{X/S}K_M}{2C_{X,0}}+\frac{Y_{X/S}C_{S,0}}{2C_{X,0}}+1\right)\ln\frac{C_{X,1,II,\tau/2}}{2C_{X,0}}-\frac{Y_{X/S}K_M}{2C_{X,0}}\ln\frac{\frac{Y_{X/S}C_{S,0}}{2C_{X,0}}+1-\frac{C_{X,1,II,\tau/2}}{2C_{X,0}}}{\frac{Y_{X/S}C_{S,0}}{2C_{X,0}}}$$
$$=\left(\frac{Y_{X/S}C_{S,0}}{2C_{X,0}}+1\right)\mu_{max}\frac{\tau}{2}, \tag{2.128}$$

pertaining to $C_{X,1,II,\tau/2}\equiv C_{X,1,II}\big|_{t=(\tau/2)^-}$ at batch time $t=(\tau/2)^-$. In this train of two bioreactors, the batch at $t=(\tau/2)^-$ will be mixed with a batch of plain fresh substrate at concentration $C_{S,0}$ but without fresh biomass, so one initial condition for the second bioreactor should look like

$$C_{S,2,II}\Big|_{t=\left(\frac{\tau}{2}\right)^+}=\frac{C_{S,1,II}\Big|_{t=\left(\frac{\tau}{2}\right)^-}+C_{S,0}}{2}=\frac{C_{S,0}-\frac{C_{X,1,II,\tau/2}-2C_{X,0}}{Y_{X/S}}+C_{S,0}}{2}$$
$$=\frac{C_{S,0}}{2}-\frac{C_{X,1,II,\tau/2}-2C_{X,0}}{2Y_{X/S}}+\frac{C_{S,0}}{2} \tag{2.129}$$

in agreement with Eq. (2.101); one may then proceed to

$$C_{S,2,II}\Big|_{t=\left(\frac{\tau}{2}\right)^+}=C_{S,0}-\frac{C_{X,1,II,\tau/2}-2C_{X,0}}{2Y_{X/S}}=C_{S,0}\left(1-\frac{1}{2}\,\frac{\frac{C_{X,1,II,\tau/2}}{C_{X,0}}-2}{\frac{Y_{X/S}C_{S,0}}{C_{X,0}}}\right) \tag{2.130}$$

after lumping similar terms, factoring out $C_{S,0}$, and dividing both numerator and denominator by $C_{X,0}$. Equation (2.130) is to be coupled with

(continued)

Problem 2.5 (continued)

$$C_{X,2,II}\Big|_{t=\left(\frac{\tau}{2}\right)^+}=\frac{C_{X,1,II,\tau/2}+0}{2}=C_{X,0}\left(\frac{1}{2}\frac{C_{X,1,II,\tau/2}}{C_{X,0}}\right) \tag{2.131}$$

as the other initial condition—corresponding to mere dilution of volume $V/2$ of biomass generated in the first bioreactor up to volume V in the second bioreactor, without addition of extra inoculum. Replacement of $C_{S,0}$ and $C_{X,0}$ by $C_{S,2,II}\Big|_{t=(\tau/2)^+}$ as per Eq. (2.130) and $C_{X,2,II}\Big|_{t=(\tau/2)^+}$ as per Eq. (2.131), respectively, and of t by $t-\tau/2$ converts Eq. (2.124) to

$$\left(\frac{Y_{X/S}K_M}{C_{X,0}\frac{1}{2}\frac{C_{X,1,II,\tau/2}}{C_{X,0}}}+\frac{Y_{X/S}C_{S,0}\left(1-\frac{1}{2}\frac{\frac{C_{X,1,II,\tau/2}}{C_{X,0}}-2}{\frac{Y_{X/S}C_{S,0}}{C_{X,0}}}\right)}{C_{X,0}\frac{1}{2}\frac{C_{X,1,II,\tau/2}}{C_{X,0}}}+1\right)\ln\frac{C_{X,2,II}}{C_{X,0}\frac{1}{2}\frac{C_{X,1,II,\tau/2}}{C_{X,0}}}$$

$$-\frac{Y_{X/S}K_M}{C_{X,0}\frac{1}{2}\frac{C_{X,1,II,\tau/2}}{C_{X,0}}}\ln\frac{\frac{Y_{X/S}C_{S,0}\left(1-\frac{1}{2}\frac{\frac{C_{X,1,II,\tau/2}}{C_{X,0}}-2}{\frac{Y_{X/S}C_{S,0}}{C_{X,0}}}\right)}{C_{X,0}\frac{1}{2}\frac{C_{X,1,II,\tau/2}}{C_{X,0}}}+1-\frac{C_{X,2,II}}{C_{X,0}\frac{1}{2}\frac{C_{X,1,II,\tau/2}}{C_{X,0}}}}{\frac{Y_{X/S}C_{S,0}\left(1-\frac{1}{2}\frac{\frac{C_{X,1,II,\tau/2}}{C_{X,0}}-2}{\frac{Y_{X/S}C_{S,0}}{C_{X,0}}}\right)}{C_{X,0}\frac{1}{2}\frac{C_{X,1,II,\tau/2}}{C_{X,0}}}}$$

$$=\left(\frac{Y_{X/S}C_{S,0}\left(1-\frac{1}{2}\frac{\frac{C_{X,1,II,\tau/2}}{C_{X,0}}-2}{\frac{Y_{X/S}C_{S,0}}{C_{X,0}}}\right)}{C_{X,0}\frac{1}{2}\frac{C_{X,1,II,\tau/2}}{C_{X,0}}}+1\right)\mu_{max}\left(t-\frac{\tau}{2}\right), \tag{2.132}$$

which yields $C_{X,2,II}$ as biomass concentration in the second batch bioreactor; Eq. (2.132) may be algebraically rearranged to read

(continued)

Problem 2.5 (continued)

$$
\left(\frac{\frac{Y_{X/S}K_M}{C_{X,0}}}{\frac{1}{2}\frac{C_{X,1,II,\tau/2}}{C_{X,0}}}+\frac{Y_{X/S}C_{S,0}}{C_{X,0}}\frac{1-\frac{1}{2}\frac{\frac{C_{X,1,II,\tau/2}}{C_{X,0}}-2}{\frac{Y_{X/S}C_{S,0}}{C_{X,0}}}}{\frac{1}{2}\frac{C_{X,1,II,\tau/2}}{C_{X,0}}}+1\right)\ln\frac{\frac{C_{X,2,II}}{C_{X,0}}}{\frac{1}{2}\frac{C_{X,1,II,\tau/2}}{C_{X,0}}}
$$

$$
-\frac{\frac{Y_{X/S}K_M}{C_{X,0}}}{\frac{1}{2}\frac{C_{X,1,II,\tau/2}}{C_{X,0}}}\ln\frac{\frac{Y_{X/S}C_{S,0}}{C_{X,0}}\frac{1-\frac{1}{2}\frac{\frac{C_{X,1,II,\tau/2}}{C_{X,0}}-2}{\frac{Y_{X/S}C_{S,0}}{C_{X,0}}}}{\frac{1}{2}\frac{C_{X,1,II,\tau/2}}{C_{X,0}}}+1-\frac{\frac{C_{X,2,II}}{C_{X,0}}}{\frac{1}{2}\frac{C_{X,1,II,\tau/2}}{C_{X,0}}}}{\frac{Y_{X/S}C_{S,0}}{C_{X,0}}\frac{1-\frac{1}{2}\frac{\frac{C_{X,1,II,\tau/2}}{C_{X,0}}-2}{\frac{Y_{X/S}C_{S,0}}{C_{X,0}}}}{\frac{1}{2}\frac{C_{X,1,II,\tau/2}}{C_{X,0}}}}
$$

$$
=\left(\frac{Y_{X/S}C_{S,0}}{C_{X,0}}\frac{1-\frac{1}{2}\frac{\frac{C_{X,1,II,\tau/2}}{C_{X,0}}-2}{\frac{Y_{X/S}C_{S,0}}{C_{X,0}}}}{\frac{1}{2}\frac{C_{X,1,II,\tau/2}}{C_{X,0}}}+1\right)\mu_{max}\left(t-\frac{\tau}{2}\right); \tag{2.133}
$$

specifically for $t=\tau$ and $C_{X,2,II,\tau}\equiv C_{X,2,II}|_{t=\tau}$, Eq. (2.133) turns to

(continued)

Problem 2.5 (continued)

$$\left(\frac{\frac{Y_{X/S}K_M}{C_{X,0}}}{\frac{1}{2}\frac{C_{X,1,II,\tau/2}}{C_{X,0}}} + \frac{Y_{X/S}C_{S,0}}{C_{X,0}} \frac{1-\frac{1}{2}\frac{\frac{C_{X,1,II,\tau/2}}{C_{X,0}}-2}{\frac{Y_{X/S}C_{S,0}}{C_{X,0}}}}{\frac{1}{2}\frac{C_{X,1,II,\tau/2}}{C_{X,0}}} + 1 \right) \ln \frac{\frac{C_{X,2,II,\tau}}{C_{X,0}}}{\frac{1}{2}\frac{C_{X,1,II,\tau/2}}{C_{X,0}}}$$

$$- \frac{\frac{Y_{X/S}K_M}{C_{X,0}}}{\frac{1}{2}\frac{C_{X,1,II,\tau/2}}{C_{X,0}}} \ln \frac{\frac{Y_{X/S}C_{S,0}}{C_{X,0}} \frac{1-\frac{1}{2}\frac{\frac{C_{X,1,II,\tau/2}}{C_{X,0}}-2}{\frac{Y_{X/S}C_{S,0}}{C_{X,0}}}}{\frac{1}{2}\frac{C_{X,1,II,\tau/2}}{C_{X,0}}} + 1 - \frac{\frac{C_{X,2,II,\tau}}{C_{X,0}}}{\frac{1}{2}\frac{C_{X,1,II,\tau/2}}{C_{X,0}}}}{\frac{Y_{X/S}C_{S,0}}{C_{X,0}} \frac{1-\frac{1}{2}\frac{\frac{C_{X,1,II,\tau/2}}{C_{X,0}}-2}{\frac{Y_{X/S}C_{S,0}}{C_{X,0}}}}{\frac{1}{2}\frac{C_{X,1,II,\tau/2}}{C_{X,0}}}}$$

$$= \left(\frac{Y_{X/S}C_{S,0}}{C_{X,0}} \frac{1-\frac{1}{2}\frac{\frac{C_{X,1,II,\tau/2}}{C_{X,0}}-2}{\frac{Y_{X/S}C_{S,0}}{C_{X,0}}}}{\frac{1}{2}\frac{C_{X,1,II,\tau/2}}{C_{X,0}}} + 1 \right) \frac{\mu_{max}\tau}{2}. \tag{2.134}$$

Comparison between the two inoculation strategies is facilitated upon setting both dimensionless parameters $Y_{X/S}K_M/C_{X,0}$ and $Y_{X/S}C_{S,0}/C_{X,0}$ equal to unity, that is,

$$\frac{Y_{X/S}K_M}{C_{X,0}} = \frac{Y_{X/S}C_{S,0}}{C_{X,0}} = 1, \tag{2.135}$$

and defining dimensionless biomass concentration by

$$\chi \equiv \frac{C_X}{C_{X,0}} \tag{2.136}$$

and dimensionless time by

$$T \equiv \mu_{max}t; \tag{2.137}$$

this leads to

(continued)

Problem 2.5 (continued)

$$(1+1+1)\ln \chi_I - 1\ln \frac{1+1-\chi_I}{1} = (1+1)T, \tag{2.138}$$

or else

$$3\ln \chi_I - \ln\{2-\chi_I\} = 2T \tag{2.139}$$

in lieu of Eq. (2.124). A similar reasoning will now depart from Eq. (2.127), revisited as

$$\left(\frac{1}{2}+\frac{1}{2}+1\right)\ln \frac{\frac{C_{X,1,II}}{C_{X,0}}}{2} - \frac{1}{2}\ln \frac{\frac{1}{2}+1-\frac{\frac{C_{X,1,II}}{C_{X,0}}}{2}}{\frac{1}{2}} = \left(\frac{1}{2}+1\right)\mu_{max}t \tag{2.140}$$

with the aid of Eq. (2.135), or else

$$2\ln \frac{\chi_{1,II}}{2} - \frac{1}{2}\ln \frac{\frac{3}{2}-\frac{\chi_{1,II}}{2}}{\frac{1}{2}} = \frac{3}{2}T \tag{2.141}$$

in view of Eqs. (2.136) and (2.137); this further reduces to

$$2\ln \frac{\chi_{1,II}}{2} - \frac{1}{2}\ln\{3-\chi_{1,II}\} = \frac{3}{2}T, \tag{2.142}$$

after getting rid of denominators. By the same token, definition of auxiliary variables as

$$\alpha \equiv 1 - \frac{1}{2}\,\frac{\frac{C_{X,1,II,\tau/2}}{C_{X,0}} - 2}{\frac{Y_{X/S}C_{S,0}}{C_{X,0}}} \tag{2.143}$$

and

$$\beta \equiv \frac{1}{2}\,\frac{C_{X,1,II,\tau/2}}{C_{X,0}} \tag{2.144}$$

permits conversion of Eq. (2.133) to

(continued)

Problem 2.5 (continued)

$$\left(\frac{1}{\beta}+1\frac{\alpha}{\beta}+1\right)\ln\frac{\frac{c_{X,2,II}}{c_{X,0}}}{\beta}-\frac{1}{\beta}\ln\frac{1\frac{\alpha}{\beta}+1-\frac{\frac{c_{X,2,II}}{c_{X,0}}}{\beta}}{1\frac{\alpha}{\beta}}=\left(1\frac{\alpha}{\beta}+1\right)\left(\mu_{max}t-\mu_{max}\frac{\tau}{2}\right) \tag{2.145}$$

or merely

$$\left(\frac{1}{\beta}+\frac{\alpha}{\beta}+1\right)\ln\frac{\chi_{2,II}}{\beta}-\frac{1}{\beta}\ln\frac{\frac{\alpha}{\beta}+1-\frac{\chi_{2,II}}{\beta}}{\frac{\alpha}{\beta}}=\left(\frac{\alpha}{\beta}+1\right)\left(T-\frac{\Theta}{2}\right) \tag{2.146}$$

once Eqs. (2.136) and (2.137) are also brought on board—complemented by

$$\Theta \equiv \mu_{max}\tau. \tag{2.147}$$

Note that Eqs. (2.143) and (2.144) can be rewritten as

$$\alpha=1-\frac{1}{2}\frac{\chi_{1,II,\tau/2}-2}{1}=1-\frac{1}{2}\chi_{1,II,\tau/2}+1=2-\frac{1}{2}\chi_{1,II,\tau/2} \tag{2.148}$$

and

$$\beta=\frac{1}{2}\chi_{1,II,\tau/2}, \tag{2.149}$$

respectively, upon insertion of Eqs. (2.135) and (2.136), which also supports

$$\frac{\alpha}{\beta}=\frac{2-\frac{1}{2}\chi_{1,II,\tau/2}}{\frac{1}{2}\chi_{1,II,\tau/2}}=\frac{4-\chi_{1,II,\tau/2}}{\chi_{1,II,\tau/2}}=\frac{4}{\chi_{1,II,\tau/2}}-1 \tag{2.150}$$

following ordered division of Eq. (2.148) by Eq. (2.149), and multiplication of both numerator and denominator by 2 afterward. In view of Eqs. (2.149) and (2.150), it becomes possible to rewrite Eq. (2.146) as

(continued)

Problem 2.5 (continued)

$$\left(\frac{1}{\frac{1}{2}\chi_{1,II,\tau/2}}+\frac{4}{\chi_{1,II,\tau/2}}-1+1\right)\ln\frac{\chi_{2,II}}{\frac{1}{2}\chi_{1,II,\tau/2}}$$

$$-\frac{1}{\frac{1}{2}\chi_{1,II,\tau/2}}\ln\frac{\frac{4}{\chi_{1,II,\tau/2}}-1+1-\frac{\chi_{2,II}}{\frac{1}{2}\chi_{1,II,\tau/2}}}{\frac{4}{\chi_{1,II,\tau/2}}-1}=\left(\frac{4}{\chi_{1,II,\tau/2}}-1+1\right)\left(T-\frac{\Theta}{2}\right), \tag{2.151}$$

which degenerates to

$$\left(\frac{2}{\chi_{1,II,\tau/2}}+\frac{4}{\chi_{1,II,\tau/2}}\right)\ln\frac{2\chi_{2,II}}{\chi_{1,II,\tau/2}}-\frac{2}{\chi_{1,II,\tau/2}}\ln\frac{\frac{4}{\chi_{1,II,\tau/2}}-2\frac{\chi_{2,II}}{\chi_{1,II,\tau/2}}}{\frac{4}{\chi_{1,II,\tau/2}}-1}=\frac{4}{\chi_{1,II,\tau/2}}\left(T-\frac{\Theta}{2}\right) \tag{2.152}$$

after cancellation of unity with its negative and reduction in number of fraction levels; lumping of terms alike, coupled with multiplication of both sides by $\chi_{1,II,\tau/2}/2$ unfold

$$3\ln\frac{2\chi_{2,II}}{\chi_{1,II,\tau/2}}-\ln\frac{\frac{4}{\chi_{1,II,\tau/2}}-2\frac{\chi_{2,II}}{\chi_{1,II,\tau/2}}}{\frac{4}{\chi_{1,II,\tau/2}}-1}=2\left(T-\frac{\Theta}{2}\right), \tag{2.153}$$

which finally gives rise to

$$3\ln\frac{2\chi_{2,II}}{\chi_{1,II,\tau/2}}-\ln\frac{4-2\chi_{2,II}}{4-\chi_{1,II,\tau/2}}=2\left(T-\frac{\Theta}{2}\right) \tag{2.154}$$

upon multiplication of both numerator and denominator of the argument of the second logarithmic function by $\chi_{1,II,\tau/2}$, with $\chi_{1,II,\tau/2}$ obtainable, in turn, from solution of Eq. (2.142) for $\chi_{1,II}$ after setting $T = \Theta/2$. A graphical interpretation of the single- and double-inoculation approaches is provided in Fig. 2.6. As expected, growth is faster in the first bioreactor of the cascade because its initial concentration is twice that in the single bioreactor, and also in the second bioreactor because fresh substrate is made available; hence, the culture is kept in a more intensive metabolic state at all times. Although the cascade approach takes longer, the same final biomass concentration will eventually be attained by both configurations because depletion of the same total amount of substrate would occur.

(continued)

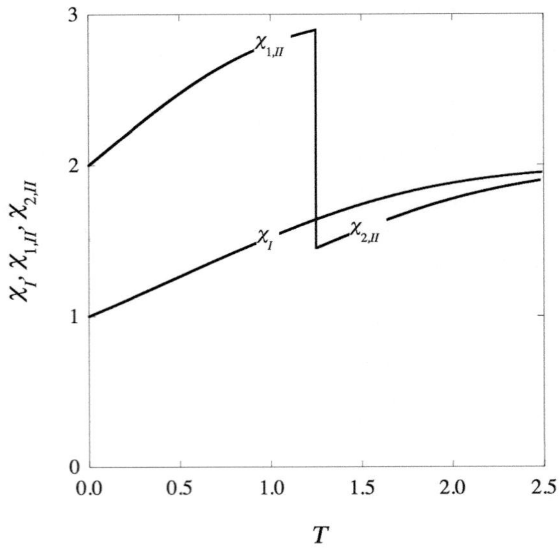

Fig. 2.6 Evolution of dimensionless biomass concentration in single bioreactor, χ_I, and first and second batch bioreactors in cascade, $\chi_{1,II}$ and $\chi_{2,II}$, respectively, versus dimensionless time, T

Problem 2.5 (continued)

(b) The asymptotic behavior of the single inoculation approach takes the form

$$\lim_{T \to \infty} \chi_I = \lim_{\frac{3 \ln \chi_I - \ln\{2 - \chi_I\}}{2} \to \infty} \chi_I \tag{2.155}$$

stemming from Eq. (2.139), which, in turn, gives rise to

$$\lim_{T \to \infty} \chi_I = \lim_{3 \ln \chi_I - \ln\{2 - \chi_I\} \to \infty} \chi_I = \lim_{-\ln\{2 - \chi_I\} \to \infty} \chi_I, \tag{2.156}$$

because $\chi_I > 1$ as per Eq. (2.136), with $C_X > C_{X,0}$ coupled with $C_{S,0} > 0$; further algebraic manipulation leads then to

$$\lim_{T \to \infty} \chi_I = \lim_{\ln\{2 - \chi_I\} \to -\infty} \chi_I = \lim_{2 - \chi_I \to 0^+} \chi_I = \lim_{\chi_I \to 2 - 0^+} \chi_I = \lim_{\chi_I \to 2} \chi_I = 2. \tag{2.157}$$

The result conveyed by Eq. (2.157) is anticipated in view of Eq. (2.102), reformulated here to advantage as

(continued)

Problem 2.5 (continued)

$$\lim_{T\to\infty}\frac{C_{X,I}}{C_{X,0}}=\lim_{\frac{C_{S,I}}{C_{S,0}}\to 0}\left(1+\frac{Y_{X/S}C_{S,0}}{C_{X,0}}\left(1-\frac{C_{S,I}}{C_{S,0}}\right)\right); \tag{2.158}$$

combination with Eqs. (2.135) and (2.136) leaves indeed

$$\lim_{T\to\infty}\chi_I=\lim_{\frac{C_{S,I}}{C_{S,0}}\to 0}\left(1+1\left(1-\frac{C_{S,I}}{C_{S,0}}\right)\right)=1+1(1-0)=1+1=2 \tag{2.159}$$

as vertical intercept of the horizontal asymptote in the plot labeled as Fig. 2.6, and pertaining to a single bioreactor. With regard to the second mode of inoculation, one should follow a similar reasoning, but start with Eq. (2.154) according to

$$\lim_{T\to\infty}\chi_{2,II}=\lim_{T-\frac{\Theta}{2}\to\infty}\chi_{2,II}=\lim_{\frac{3\ln\frac{2\chi_{2,II}}{\chi_{1,II,\tau/2}}-\ln\frac{4-\chi_{2,II}}{4-\chi_{1,II,\tau/2}}}{2}\to\infty}\chi_{2,II}, \tag{2.160}$$

as per the finite nature of $\Theta/2$ versus the infinite nature of T; elementary algebraic manipulation produces

$$\lim_{T\to\infty}\chi_{2,II}=\lim_{3\ln\frac{2\chi_{2,II}}{\chi_{1,II,\tau/2}}-\ln\frac{4-2\chi_{2,II}}{4-\chi_{1,II,\tau/2}}\to\infty}\chi_{2,II}=\lim_{-\ln\frac{4-2\chi_{2,II}}{4-\chi_{1,II,\tau/2}}\to\infty}\chi_{2,II}, \tag{2.161}$$

since $\chi_{2,II}>\chi_{1,II,\tau/2}/2$ in agreement with Eqs. (2.131) and (2.136)—further to realization that $C_{S,0}>0$ in the second bioreactor. Algebraic rearrangement of Eq. (2.161) generates

$$\begin{aligned}\lim_{T\to\infty}\chi_{2,II}&=\lim_{\ln\frac{4-2\chi_{2,II}}{4-\chi_{1,II,\tau/2}}\to-\infty}\chi_{2,II}=\lim_{\frac{4-2\chi_{2,II}}{4-\chi_{1,II,\tau/2}}\to 0^+}\chi_{2,II}\\&=\lim_{4-2\chi_{2,II}\to 0^+\,\vee\,4-\chi_{1,II,\tau/2}\to+\infty}\chi_{2,II}=\lim_{4-2\chi_{2,II}\to 0^+}\chi_{2,II},\end{aligned} \tag{2.162}$$

based on the operational features of logarithms; along with realization that $4-\chi_{1,II,\tau/2}$ cannot approach $+\infty$, in view of $\chi_{1,II,\tau/2}>0$ as per Eq. (2.136), coupled with $C_X>0$ implying $4-\chi_{1,II,\tau/2}<4$. Equation (2.162) readily degenerates to

(continued)

Problem 2.5 (continued)

$$\lim_{T\to\infty}\chi_{2,II}=\lim_{2\chi_{2,II}\to 4-0^+}\chi_{2,II}=\lim_{2\chi_{2,II}\to 4}\chi_{2,II}=\lim_{\chi_{2,II}\to 2}\chi_{2,II}=2, \tag{2.163}$$

thus confirming that the biomass concentration in the second unit of the cascade of bioreactors shares the long-T horizontal asymptote with its single unit counterpart (see Eq. (2.159)). Conversely, $T\to\infty$ prompts

$$\lim_{T\to\infty}\chi_{1,II}=\lim_{\frac{2\ln\frac{\chi_{1,II}}{2}-\frac{1}{2}\ln\{3-\chi_{1,II}\}}{\frac{3}{2}}\to\infty}\chi_{1,II} \tag{2.164}$$

stemming from Eq. (2.142); one may proceed to

$$\lim_{T\to\infty}\chi_{1,II}=\lim_{2\ln\frac{\chi_{1,II}}{2}-\frac{1}{2}\ln\{3-\chi_{1,II}\}\to\infty}\chi_{1,II}=\lim_{-\frac{1}{2}\ln\{3-\chi_{1,II}\}\to\infty}\chi_{1,II}=\lim_{\ln\{3-\chi_{1,II}\}\to-\infty}\chi_{1,II} \tag{2.165}$$

since $\chi_{1,II}>0$ is necessarily finite, which reduces to

$$\lim_{T\to\infty}\chi_{1,II}=\lim_{3-\chi_{1,II}\to 0^+}\chi_{1,II}=\lim_{\chi_{1,II}\to 3-0^+}\chi_{1,II}=\lim_{\chi_{1,II}\to 3}\chi_{1,II}=3 \tag{2.166}$$

after realizing that a logarithm tends to $-\infty$ when its argument approaches 0^+. Equation (2.166) is consistent with the graphical pattern apparent in Fig. 2.6, pertaining to the first bioreactor in the cascade. As discussed above, the same final concentration of biomass will be eventually attained for a given initial load of substrate and biomass, irrespective of whether a single batch bioreactor or a series of two equal-sized batch bioreactors is utilized. This conclusion may be drawn following a distinct path by first revisiting Eq. (2.158) as

$$\lim_{T\to\infty}\frac{C_{X,1,II}}{2C_{X,0}}=\lim_{\frac{C_{S,1,II}}{C_{S,0}}\to 0}\left(1+\frac{Y_{X/S}C_{S,0}}{2C_{X,0}}\left(1-\frac{C_{S,1,II}}{C_{S,0}}\right)\right), \tag{2.167}$$

pertaining to the first unit in the train of two equal-sized bioreactors and reflecting Eq. (2.126); insertion of Eqs. (2.135) and (2.136) in Eq. (2.167) produces

(continued)

Problem 2.5 (continued)

$$\lim_{T\to\infty}\frac{\chi_{1,II}}{2}=\lim_{\frac{C_{S,1,II}}{C_{S,0}}\to 0}\left(1+\frac{1}{2}\left(1-\frac{C_{S,1,II}}{C_{S,0}}\right)\right)=1+\frac{1}{2}(1-0)=1+\frac{1}{2}=\frac{3}{2} \quad (2.168)$$

that prompts

$$\lim_{T\to\infty}\chi_{1,II}=3 \quad (2.169)$$

upon multiplication of both sides by 2. The second unit in series would then justify

$$\lim_{T\to\infty}\frac{C_{X,2,II}}{\frac{3}{2}C_{X,0}}=\lim_{\frac{C_{S,1,II}}{C_{S,0}}\to 0}\left(1+\frac{Y_{X/S}\frac{1}{2}C_{S,0}}{\frac{3}{2}C_{X,0}}\left(1-\frac{C_{S,1,II}}{\frac{1}{2}C_{S,0}}\right)\right), \quad (2.170)$$

as alias to Eq. (2.102) obtained after dividing both sides by the initial biomass concentration $3C_{X,0}/2$ and factoring out the initial substrate concentration $C_{S,0}/2$, in the present case characterized by $C_{S,1,II}\to 0$ on account of $T\to\infty$. In fact, Eq. (2.129) has it that

$$\lim_{\frac{C_{S,1,II}}{C_{S,0}}\to 0} C_{S,2,II}\Big|_{t=\left(\frac{\tau}{2}\right)^+}=\lim_{\frac{C_{S,1,II}\big|_{t=\left(\frac{\tau}{2}\right)^-}}{C_{S,0}}\to 0}\frac{C_{S,1,II}\big|_{t=\left(\frac{\tau}{2}\right)^-}+C_{S,0}}{2}=\frac{0+C_{S,0}}{2}=\frac{1}{2}C_{S,0}, \quad (2.171)$$

and Eq. (2.131) likewise justifies

$$\lim_{\frac{C_{S,1,II}}{C_{S,0}}\to 0} C_{X,2,II}\Big|_{t=\left(\frac{\tau}{2}\right)^+}=\lim_{\frac{C_{S,1,II}}{C_{S,0}}\to 0}\frac{C_{X,1,II,\tau/2}+0}{2}=\lim_{\frac{C_{S,1,II}}{C_{S,0}}\to 0}\frac{C_{X,1,II,\tau/2}}{2}$$

$$=\lim_{T\to\infty}\frac{\chi_{1,II,\tau/2}C_{X,0}}{2}=\frac{\lim_{T\to\infty}\chi_{1,II,\tau/2}}{2}C_{X,0}=\frac{3}{2}C_{X,0}, \quad (2.172)$$

when complemented by Eqs. (2.136) and (2.168). Notice that Eq. (2.170) reduces to

(continued)

Problem 2.5 (continued)

$$\lim_{T\to\infty}\frac{\chi_{2,II}}{\frac{3}{2}}=\lim_{\frac{C_{S,1,II}}{C_{S,0}}\to 0}\left(1+\frac{\frac{1}{2}}{\frac{3}{2}}\frac{Y_{X/S}C_{S,0}}{C_{X,0}}\left(1-\frac{1}{\frac{1}{2}}\frac{C_{S1,II}}{C_{S,0}}\right)\right) \tag{2.173}$$

at the expense again of Eq. (2.136) while Eq. (2.135) justifies

$$\lim_{T\to\infty}\frac{\chi_{2,II}}{\frac{3}{2}}=\lim_{\frac{C_{S,1,II}}{C_{S,0}}\to 0}\left(1+\frac{1}{3}1\left(1-2\frac{C_{S,1,II}}{C_{S,0}}\right)\right)$$
$$=\lim_{\frac{C_{S,1,II}}{C_{S,0}}\to 0}\left(1+\frac{1}{3}\left(1-2\frac{C_{S,1,II}}{C_{S,0}}\right)\right)=1+\frac{1}{3}(1-2\cdot 0)=1+\frac{1}{3}=\frac{4}{3}; \tag{2.174}$$

multiplication of both sides by 3/2 finally yields

$$\lim_{T\to\infty}\chi_{2,II}=\frac{3}{2}\frac{4}{3}=\frac{4}{2}=2 \tag{2.175}$$

that retrieves the result conveyed by Eq. (2.163), as expected.

(c) The specific growth rate of the culture satisfies

$$\mu_X\equiv\frac{r_X}{C_X}=\frac{\mu_{max}C_S}{K_M+C_S}, \tag{2.176}$$

obtained from Eq. (2.90) upon division of both sides by C_X; at the moment of completion of the first batch in the double-bioreactor train, that is, by $t=(\tau/2)^-$, Eq. (2.176) becomes

$$\mu_X\Big|_{t=\left(\frac{\tau}{2}\right)^-}=\frac{\mu_{max}C_{S,1,II}\Big|_{t=\left(\frac{\tau}{2}\right)^-}}{K_M+C_{S,1,II}\Big|_{t=\left(\frac{\tau}{2}\right)^-}}\equiv\frac{\mu_{max}C_{S,1,II,\tau/2}}{K_M+C_{S,1,II,\tau/2}}, \tag{2.177}$$

whereas the moment of departure of the second batch bioreactor is characterized by

(continued)

Problem 2.5 (continued)

$$\mu_X\Big|_{t=\left(\frac{\tau}{2}\right)^+} = \frac{\mu_{max} C_{S,2,II}\Big|_{t=\left(\frac{\tau}{2}\right)^+}}{K_M + C_{S,2,II}\Big|_{t=\left(\frac{\tau}{2}\right)^+}} = \frac{\mu_{max} \frac{C_{S,1,II}\big|_{t=\left(\frac{\tau}{2}\right)^-} + C_{S,0}}{2}}{K_M + \frac{C_{S,1,II}\big|_{t=\left(\frac{\tau}{2}\right)^-} + C_{S,0}}{2}} = \frac{\mu_{max} \frac{C_{S,1,II,\tau/2} + C_{S,0}}{2}}{K_M + \frac{C_{S,1,II,\tau/2} + C_{S,0}}{2}} \tag{2.178}$$

consistent with Eq. (2.129). On the other hand, differentiation of Eq. (2.176) with regard to C_S unfolds

$$\frac{d\mu_X}{dC_S} = \frac{\mu_{max} 1 (K_M + C_S) - \mu_{max} C_S 1}{(K_M + C_S)^2} = \frac{\mu_{max} K_M + \mu_{max} C_S - \mu_{max} C_S}{(K_M + C_S)^2} \tag{2.179}$$

along with elimination of parenthesis, or else

$$\frac{d\mu_X}{dC_S} = \frac{\mu_{max} K_M}{(K_M + C_S)^2} > 0 \tag{2.180}$$

following condensation of terms alike; the positive sign of $d\mu_X/dC_S$ guarantees that μ_X is a monotonically increasing function of C_S. Therefore, Eqs. (2.176) and (2.178), along with Eq. (2.180) support

$$\frac{\mu_{max} \frac{C_{S,1,II,\tau/2} + C_{S,0}}{2}}{K_M + \frac{C_{S,1,II,\tau/2} + C_{S,0}}{2}} = \mu_X\Bigg|_{C_S = \frac{C_{S,1,II,\tau/2} + C_{S,0}}{2}} > \mu_X\Big|_{C_S = C_{S,1,II,\tau/2}} = \frac{\mu_{max} C_{S,1,II,\tau/2}}{K_M + C_{S,1,II,\tau/2}} = \frac{\mu_{max} C_S}{K_M + C_S}\Bigg|_{C_S = C_{S,1,II,\tau/2}} \tag{2.181}$$

due to

$$\frac{C_{S,1,II,\tau/2} + C_{S,0}}{2} > \frac{C_{S,1,II,\tau/2} + C_{S,1,II,\tau/2}}{2} = C_{S,1,II,\tau/2}, \tag{2.182}$$

coupled with

$$0 < C_{S,1,II,\tau/2} < C_{S,0} \tag{2.183}$$

as physical constraint. Therefore, Eq. (2.181) confirms the jump up in specific rate of growth, from $\mu_X\big|_{t=(\tau/2)^-}$ to $\mu_X\big|_{t=(\tau/2)^+}$, at the moment of reinoculation.

(continued)

Problem 2.5 (continued)

(d) Equations (2.177) and (2.178) can be rewritten as

$$r_X\Big|_{t=\left(\frac{\tau}{2}\right)^-} = \frac{\mu_{max} C_{S,1,II}\Big|_{t=\left(\frac{\tau}{2}\right)^-}}{K_M + C_{S,1,II}\Big|_{t=\left(\frac{\tau}{2}\right)^-}} C_{X,1,II}\Big|_{t=\left(\frac{\tau}{2}\right)^-} = \frac{\mu_{max} C_{S,1,II,\tau/2}}{K_M + C_{S,1,II,\tau/2}} C_{X,1,II,\tau/2} \tag{2.184}$$

and

$$r_X\Big|_{t=\left(\frac{\tau}{2}\right)^+} = \frac{\mu_{max} C_{S,2,II}\Big|_{t=\left(\frac{\tau}{2}\right)^+}}{K_M + C_{S,2,II}\Big|_{t=\left(\frac{\tau}{2}\right)^+}} C_{X,2,II}\Big|_{t=\left(\frac{\tau}{2}\right)^+} = \frac{\mu_{max} \frac{C_{S,1,II,\tau/2}+C_{S,0}}{2}}{K_M + \frac{C_{S,1,II,\tau/2}+C_{S,0}}{2}} \frac{C_{X,1,II,\tau/2}}{2}, \tag{2.185}$$

respectively, after having taken Eqs. (2.129), (2.131), and (2.176) on board; a higher growth rate at the moment of reinoculation means that

$$r_X\Big|_{t=\left(\frac{\tau}{2}\right)^+} > r_X\Big|_{t=\left(\frac{\tau}{2}\right)^-}, \tag{2.186}$$

which is equivalent to writing

$$\frac{\mu_{max} \frac{C_{S,1,II,\tau/2}+C_{S,0}}{2}}{K_M + \frac{C_{S,1,II,\tau/2}+C_{S,0}}{2}} \frac{C_{X,1,II,\tau/2}}{2} > \frac{\mu_{max} C_{S,1,II,\tau/2}}{K_M + C_{S,1,II,\tau/2}} C_{X,1,II,\tau/2} \tag{2.187}$$

as per Eqs. (2.184) and (2.185). Elimination of denominators converts Eq. (2.187) to

$$\mu_{max} \frac{C_{S,1,II,\tau/2} + C_{S,0}}{2} \frac{C_{X,1,II,\tau/2}}{2} \left(K_M + C_{S,1,II,\tau/2}\right)$$
$$> \mu_{max} C_{S,1,II,\tau/2} C_{X,1,II,\tau/2} \left(K_M + \frac{C_{S,1,II,\tau/2} + C_{S,0}}{2}\right), \tag{2.188}$$

while multiplication of both sides by 4 leaves

(continued)

Problem 2.5 (continued)

$$\begin{aligned} &\mu_{max} C_{X,1,II,\tau/2}\left(K_M + C_{S,1,II,\tau/2}\right)\left(C_{S,0} + C_{S,1,II,\tau/2}\right) \\ &> 2\mu_{max} C_{X,1,II,\tau/2} C_{S,1,II,\tau/2}\left(2K_M + C_{S,0} + C_{S,1,II,\tau/2}\right); \end{aligned} \quad (2.189)$$

a further cancelling of $\mu_{max} C_{X,1,II,\tau/2}$ between sides produces

$$\begin{aligned} &\left(K_M + C_{S,1,II,\tau/2}\right)\left(C_{S,0} + C_{S,1,II,\tau/2}\right) \\ &> 2C_{S,1,II,\tau/2}\left(2K_M + C_{S,0} + C_{S,1,II,\tau/2}\right). \end{aligned} \quad (2.190)$$

Selective elimination of parentheses transforms Eq. (2.190) to

$$\begin{aligned} &K_M C_{S,0} + K_M C_{S,1,II,\tau/2} + C_{S,0} C_{S,1,II,\tau/2} + C_{S,1,II,\tau/2}{}^2 \\ &> 2(2K_M + C_{S,0}) C_{S,1,II,\tau/2} + 2C_{S,1,II,\tau/2}{}^2 \end{aligned} \quad (2.191)$$

while lumping of quadratic terms leaves

$$K_M C_{S,0} + K_M C_{S,1,II,\tau/2} + C_{S,0} C_{S,1,II,\tau/2} > 2(2K_M + C_{S,0}) C_{S,1,II,\tau/2} + C_{S,1,II,\tau/2}{}^2; \quad (2.192)$$

$C_{S,1,II,\tau/2}$ will now be factored out as

$$0 > C_{S,1,II,\tau/2}{}^2 + (4K_M + 2C_{S,0} - K_M - C_{S,0}) C_{S,1,II,\tau/2} - K_M C_{S,0}, \quad (2.193)$$

and condensation of similar terms permits simplification to

$$C_{S,1,II,\tau/2}{}^2 + (3K_M + C_{S,0}) C_{S,1,II,\tau/2} - K_M C_{S,0} < 0. \quad (2.194)$$

Since the coefficient of its quadratic term is positive, the inequality labeled as Eq. (2.194) is mathematically equivalent to

$$\begin{aligned} &\frac{-(3K_M + C_{S,0}) - \sqrt{(3K_M + C_{S,0})^2 + 4K_M C_{S,0}}}{2} < C_{S,1,II,\tau/2} \\ &< \frac{-(3K_M + C_{S,0}) + \sqrt{(3K_M + C_{S,0})^2 + 4K_M C_{S,0}}}{2}, \end{aligned} \quad (2.195)$$

(continued)

Problem 2.5 (continued)

with the two roots obtained via straight application of the classical solving formula for a quadratic equation in canonical form; given the obvious physical constraint $C_{S,1,II,\tau/2} > 0$, Eq. (2.195) reduces to just

$$0 < C_{S,1,II,\tau/2} < \frac{\sqrt{(3K_M + C_{S,0})^2 + 4K_M C_{S,0}} - (3K_M + C_{S,0})}{2}, \tag{2.196}$$

so the range for $C_{S,1,II,\tau/2}$ depends on just two parameters, K_M and C_S, with $\sqrt{(3K_M + C_{S,0})^2 + 4K_M C_{S,0}} > \sqrt{(3K_M + C_{S,0})^2} = 3K_M + C_{S,0}$ assuring that the right-hand side of Eq. (2.196) is indeed positive, with the left-hand side in Eq. (2.195) being a fortiori negative. If Eq. (2.196) is satisfied, then the biomass will grow faster right after reinoculation than right before; a pattern distinct from the (universal) trend encompassing the specific growth rate is thus found. If Eq. (2.102) were reformulated to

$$\frac{C_X}{2C_{X,0}} = 1 + \frac{Y_{X/S}\ C_{S,0}}{2C_{X,0}}\left(1 - \frac{C_S}{C_{S,0}}\right), \tag{2.197}$$

following replacement of $C_{X,0}$ by $2C_{X,0}$ so as to reflect Eq. (2.126), coupled with division of both sides by $2C_{X,0}$ and factoring out of $C_{S,0}$ in the right-hand side afterward, then Eq. (2.135) might be retrieved to obtain

$$\frac{C_X}{C_{X,0}} = 2\left(1 + \frac{1}{2}\left(1 - \frac{C_S}{C_{S,0}}\right)\right) = 2 + 1 - \frac{C_S}{C_{S,0}} = 3 - \frac{C_S}{C_{S,0}}; \tag{2.198}$$

Equation (2.196) may, in turn, be rewritten as

$$0 < \frac{C_{S,1,II,\tau/2}}{C_{S,0}} < \frac{\sqrt{\left(3\frac{K_M}{C_{S,0}} + 1\right)^2 + 4\frac{K_M}{C_{S,0}}} - \left(3\frac{K_M}{C_{S,0}} + 1\right)}{2} \tag{2.199}$$

upon division of all sides by $C_{S,0}$. Meanwhile, the first equality of Eq. (2.135) implies

$$K_M = C_{S,0}, \tag{2.200}$$

in which case Eq. (2.199) turns to

(continued)

Problem 2.5 (continued)

$$0 < \frac{C_{S,1,II,\tau/2}}{C_{S,0}} < \frac{\sqrt{\left(3\frac{C_{S,0}}{C_{S,0}}+1\right)^2 + 4\frac{C_{S,0}}{C_{S,0}}} - \left(3\frac{C_{S,0}}{C_{S,0}}+1\right)}{2} = \frac{\sqrt{(3+1)^2+4}-(3+1)}{2} \tag{2.201}$$

or, upon simplification,

$$0 < \frac{C_{S,1,II,\tau/2}}{C_{S,0}} < \frac{\sqrt{4^2+4}-4}{2} = \frac{\sqrt{20}-4}{2} = \sqrt{\frac{20}{4}} - 2 = \sqrt{5} - 2; \tag{2.202}$$

Equation (2.202) proves equivalent to

$$0 > -\frac{C_{S,1,II,\tau/2}}{C_{S,0}} > 2 - \sqrt{5} \tag{2.203}$$

after taking negatives of all sides, or else

$$3 > 3 - \frac{C_{S,1,II,\tau/2}}{C_{S,0}} > 3 + 2 - \sqrt{5} = 5 - \sqrt{5} \tag{2.204}$$

upon addition of 3 to all sides. In view of Eq. (2.198), one may revisit Eq. (2.204) as

$$3 > \frac{C_{X,1,II,\tau/2}}{C_{X,0}} > 5 - \sqrt{5}, \tag{2.205}$$

which implies

$$5 - \sqrt{5} < \chi_{1,II,\tau/2} < 3 \tag{2.206}$$

as per Eq. (2.136). Inspection of Fig. 2.6 indicates that

$$2.764 = 5 - \sqrt{5} < \chi_{1,II}\Big|_{t=\left(\frac{\tau}{2}\right)^-} < 3, \tag{2.207}$$

so the slope of the geometric tangent to the $\chi_{1,II}$-curve at $(\tau/2)^+$ is higher that its counterpart at $(\tau/2)^-$; consequently,

(continued)

Problem 2.5 (continued)

$$r_X\Big|_{t=\left(\frac{\tau}{2}\right)^+} > r_X\Big|_{t=\left(\frac{\tau}{2}\right)^-} \tag{2.208}$$

since the definition of rate has it that

$$r_X \equiv \frac{dC_X}{dt} = \frac{d\left(C_{X,0}\frac{C_X}{C_{X,0}}\right)}{d\left(\frac{1}{\mu_{max}}\mu_{max}t\right)} = \mu_{max}C_{X,0}\frac{d\left(\frac{C_X}{C_{X,0}}\right)}{d(\mu_{max}t)} \equiv \mu_{max}C_{X,0}\frac{d\chi}{dT}, \tag{2.209}$$

following multiplication and division by $C_{X,0}$ or μ_{max} (as appropriate), coupled with Eqs. (2.136) and (2.137).

2.3 Buffering

Carboxylic acids are a common nutrient/metabolite in the microbial world; these molecules possess the ability to release H^+, which will compete with similar hydrogen ions released by water molecules and will eventually fix *pH* of the medium. On the other hand, the catalytic activity of cellular enzymes is strongly dependent on (local) environmental *pH* because their amino acid residues, for being Bronsted and Lowry's acid/bases, ionize as a result of release/abstraction of H^+; this will, in turn, affect electrostatic forces and hydrogen bonding with those amino acid residues, responsible for the conformation taken by the peptide backbone. Protolysis features of amino acids in free form, directly related to their α-carboxy or α-amine moieties, are presented in Table 2.1. The dissociation constant (usually reported as pK_a^*; see definition below) of the carboxy or the amine group attached to the central carbon, $^\alpha$C, of an amino acid is strictly constant, at a given temperature, if in free form. However, when incorporated in a protein as residue, it may change within a relatively narrow range due to interaction with other amino acid residues, neighbors in space but not necessarily along the peptide sequence because of protein folding upon itself, while a free amino acid is surrounded by water molecules only that form a standard solvation shell. The effect of the molecular environment accordingly decreases the dissociation constant of $-^\alpha$C–C(=O)–OH, but increases its $-^\alpha$C–NH_3^+ counterpart; therefore, release of H^+ by said –C(=O)–OH group is hampered, while abstraction of H^+ by said $-NH_2$ is facilitated when the corresponding amino acid is part of a protein as compared to its original (free) form.

Since the peptide bond, –C(=O)–NH–, in enzymes (and proteins at large) arises from condensation of the α-carboxylic group, –C(=O)–OH, of an amino acid with the α-amine group, H–NH–, from another amino acid, and concomitant release of a water molecule, H_2O, via condensation of –OH with H–, respectively, only the α-carboxy moiety of the

Table 2.1 Protolysis constants, in free and residue forms, of α-carboxy and α-amine moieties of protein terminal amino acids, with indication of structural formulae

			pK_a^*			
			free form		residue form	
Designation	Acronym	Structural formula	α-carboxy group	α-amine group	α-carboxy group	α-amine group
Cysteine	Cys	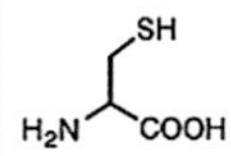	1.71	10.78	3.0–3.2	7.5–8.5
Histidine	His	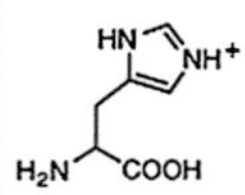	1.82	9.17		
Phenylalanine	Phe	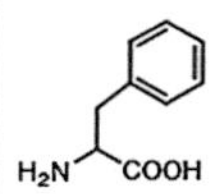	1.83	9.13		
Proline	Pro	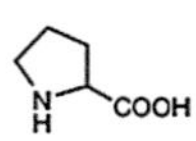	1.99	10.60		
Arginine	Arg	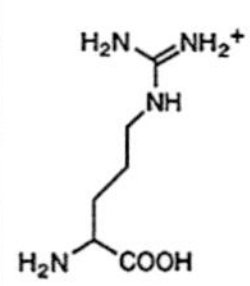	2.01	9.04		
Asparagine	Asn	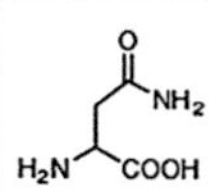	2.02	8.84		
Aspartic acid	Asp	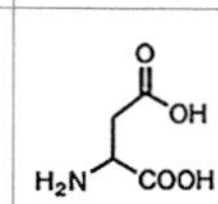	2.09	9.82		
Glutamine	Gln	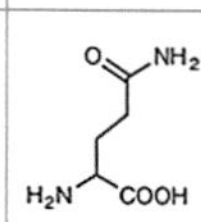	2.17	9.13		
Lysine	Lys	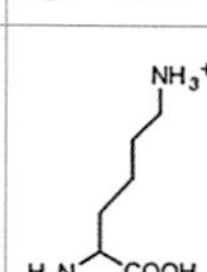	2.18	8.95		
Glutamic acid	Glu	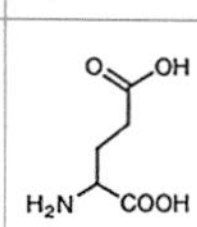	2.19	9.67		

(continued)

Table 2.1 (continued)

Tyrosine	Tyr	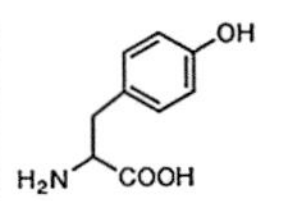	2.20	9.11		
Serine	Ser	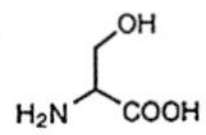	2.21	9.15		
Methionine	Met	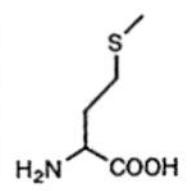	2.28	9.21		
Valine	Val	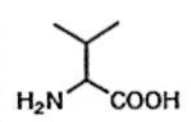	2.32	9.62		
Alanine	Ala	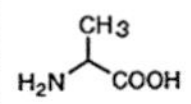	2.34	9.69		
Glycine	Gly	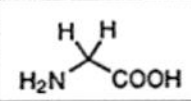		9.60		
Isoleucine	Ile	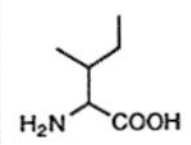	2.36	9.68		
Leucine	Leu	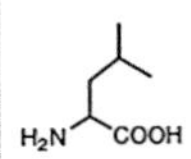		9.60		
Tryptophan	Try	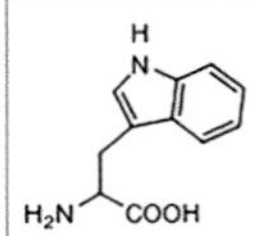	2.38	9.39		
Threonine	Thr	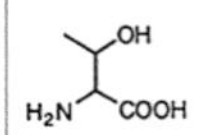	2.63	9.10		

first residue and the α-amine moiety of the last residue can actually undergo protolysis. Therefore, release/capture of H^+ by the ionizable side chains of amino acid residues ends up being more relevant than their molecular termini; the available possibilities are depicted in Table 2.2. Some amino acid residues exhibit an acid behavior per se as is the case of Asp and Glu, and His to a lesser extent, whereas others show a preference for abstraction of H^+ from solution, namely, Cys, Lys, and Arg, as well as Ser and Thr to a lesser degree. Note again the range of variation of protolysis constants when amino acids serve as residues in enzymes—centered at their value in free form, with an amplitude of 1–2 (decimal) logarithmic cycles.

Due to existence of *pH* optima for metabolically relevant enzymes, and consequently for cells as a whole, the *pH* of the medium should somehow be kept constant and as close as possible to the cell metabolic optimum, even knowing that cytoplasmic *pH* does not necessarily match outside *pH* because the endocellular composition typically differs from its environmental counterpart. Extreme *pH* values also prove detrimental to the enzyme

Table 2.2 Protolysis constants, in free and residue forms, of side chains of protein amino acids with indication of structural formulae

Nature	Designation	Acronym	Structural formula	pK_a^* of side chain	
				free	residue
Acid	Aspartic acid	Asp	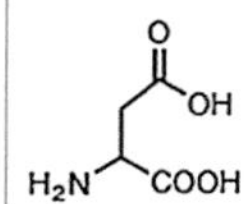	3.90	3.0–5.0
	Glutamic acid	Glu	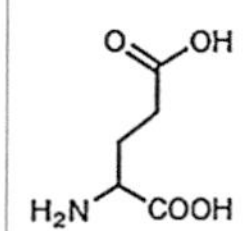	4.07	
	Histidine	His	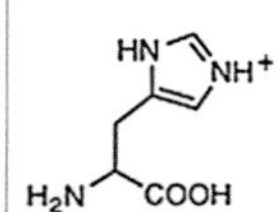	6.04	5.5–7.0
Base	Cysteine	Cys	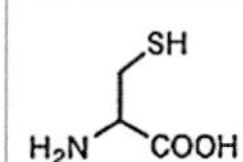	8.35	8.0–8.5
	Lysine	Lys	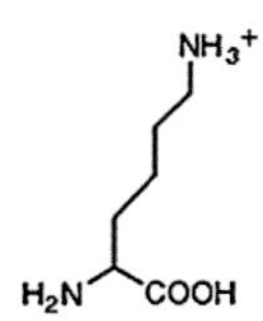	10.79	9.5–10.6
	Arginine	Arg	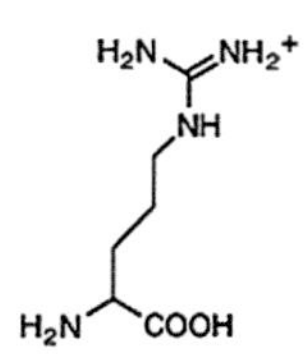	12.48	11.6–12.6
	Serine	Ser	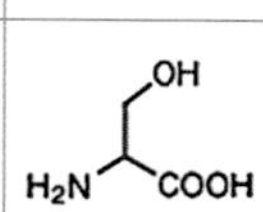	~16	–
	Threonine	Thr	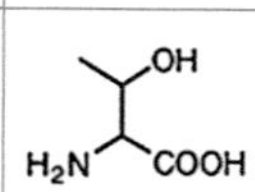		

molecules themselves because irreversible denaturation of their peptide backbones will compromise ability to bring about catalysis; note that the whole cell proves normally more sensitive to *pH* changes than its constitutive enzymes per se, owing to metabolic synergisms and pathway integration. A constant *pH* can be achieved via continuous control thereof, which requires measurement using a suitable electrode submerged in solution, coupled with addition of a volume of (strong) acid or base, as deemed necessary by the control mechanism, to restore the intended *pH* should significant deviations be detected. Devices for control of *pH* in bioreactors are illustrated in Fig. 2.7.

A balanced and steady *pH* is therefore germane for optimal cultures, in terms of productivity and stability; it lies in the vicinity of 7, for most cell types. Note that atmospheric CO_2 dissolves spontaneously in water, thus producing carbonic acid, H_2CO_3, characterized by $pK_a^* = 6.35$, which undergoes protolysis to generate HCO_3^-; hence, the pair H_2CO_3/HCO_3^- tends to bring *pH* down to the acidic side, should bubbling of atmospheric (or even CO_2-enriched) air be employed.

In alternative, the medium can be chemically buffered by including and acid/base conjugate pair in its formulation; this allows the medium to effectively resist changes in

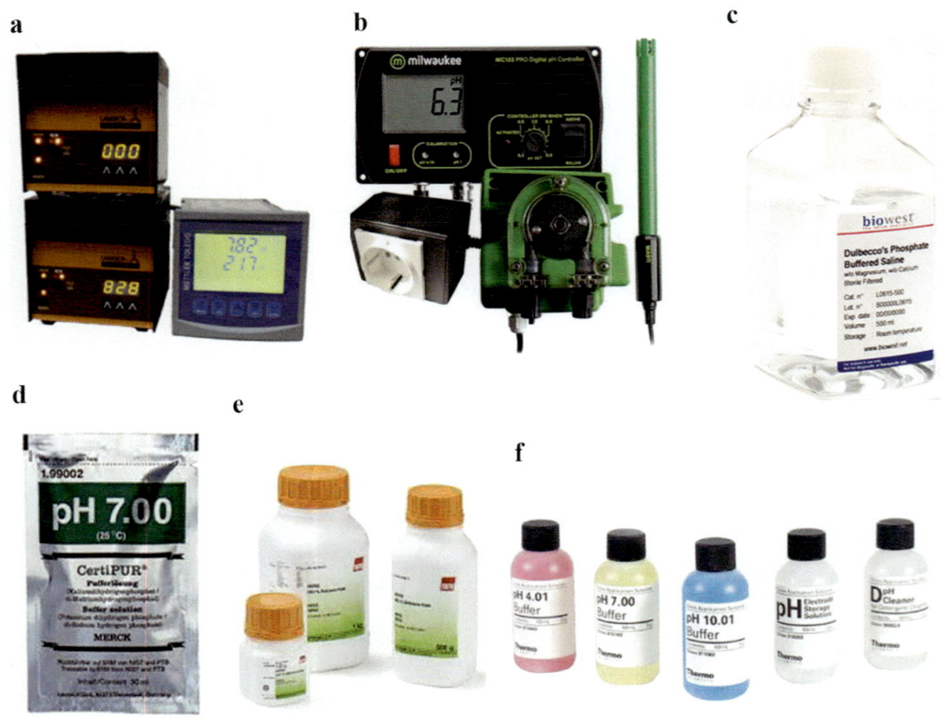

Fig. 2.7 Examples of automatic *pH* control in bioreactors, as system of peristaltic pumps with *pH* meter and controller (**a**) at bench scale (courtesy of Lambda Instruments, Brno, Czech Republic) or (**b**) portable nature (courtesy of Milwaukee Instruments, Milwaukee WI, USA); physiological buffers containing potassium dihydrogenophosphate and disodium hydrogenophosphate phosphate (**c**) in liquid form (courtesy of Biowest, Nuaillé, France) or (**d**) solid form (courtesy of Merck, Kenilworth NJ, USA), and (**e**) Good's buffers (courtesy of Roth Industries, Dautphetal, Germany); and (**f**) kit for calibration, storage, and cleaning of *pH* meters (courtesy of Orion Thermo, Chelmsford MA, USA)

pH, despite protolysis undergone by compounds generated or uptaken by living biomass. This approach is also recommended for bioreactors operated under chemostat conditions, so as to prevent spurious variations in *pH*—which last for some time before the feedback control system is able to respond and compensate for the disturbance, besides avoiding unwanted *pH* changes associated with addition of too much (strong) acid or base by the actuator device, which often operates on an on/off mode. Examples of classical buffers are tabulated in Table 2.3. The dependence of *pH* on (absolute) temperature T follows, in general, van't Hoff's law, yet a correlation of the form

$$pH\{T\} = \frac{a_1}{T} + a_2 + a_3 T + a_4 T^2 \tag{2.210}$$

has gained wide acceptance in practice; estimates of parameters a_1, a_2, a_3, and a_4 are also provided (whenever available) as entries to Table 2.3.

Classical buffers are, however, susceptible of affecting biological systems, namely, cell growth and lipid interaction with cell membranes; Sorensen's (physiological) buffer exhibits indeed a few shortcomings since its components inhibit carboxypeptidase, fumarase, urease, many kinases, and dehydrogenases, while serving as precursors for phosphate esters taken up by a number of cellular enzymes. Furthermore, phosphates possess complexing capabilities with polyvalent cations so they will inhibit metal ion-dependent biochemical reactions, a feature shared by citrate, succinate, and imidazole buffers; whereas borate buffer complexes with a wide variety of important respiratory metabolites. On the other hand, TRIS possesses a potentially reactive primary amine that often acts as metabolic inhibitor; its appreciable solubility in organic solvents favors penetration of cell membranes, thus accounting for its toxicity in mammalian cell culturing. Finally, glycylglycine is an expensive buffer, also susceptible of complexing cations.

Ideal buffers for use with biological systems should cumulatively meet the following criteria: (1) coverage of *pH* values between 6 and 8, typical of most biological reactions; (2) maximal water solubility to allow use of concentrated stock solutions, but minimal lipid solubility to prevent permeation of the cell membrane; (3) consistent protolysis constant, and thus slight effect of temperature, besides ionic strength, upon pK_a^*; (4) inability to form complexes with cations, unless such complexes are soluble; (5) stability and inertness, to avoid acting as enzyme inhibitors or substrate analogues; (6) inability to absorb light above 240 nm, to prevent interference in spectrophotometric assays within the UV and visible regions; and (7) ease of preparation from inexpensive chemicals. Zwitterionic *N*-substituted aminosulfonic acids seem to meet most of such criteria, and constitute the

Table 2.3 Classical buffers, with indication of qualitative composition, useful pH range, protolysis constant, and coefficients of variation of pH with temperature (using pH^{θ} for reference value)

Designation	Components	Useful pH range	pK_a^* (25 ° C)	Temperature variation coefficients	pH^{θ}
Oxalate	Oxalic acid Sodium hydrogenooxalate	1.1–1.8	1.30	$a_1 = -3.628 \times 10^2$ $a_2 = 6.177$ $a_3 = -1.871 \times 10^{-2}$ $a_4 = 2.585 \times 10^{-5}$	1.679
Gly-HCl buffer	Hydrochloric acid Glycine	2.2–3.6	2.35	–	–
Phthalate	Hydrochloric acid Potassium hydrogenophthalate	2.2–4.1	2.90	$a_2 = 6.615$ $a_3 = -1.851 \times 10^{-2}$ $a_4 = 3.272 \times 10^{-5}$	4.005
Citric acid/ phosphate buffer (McIlvane's buffer)	Citric acid Disodium monohydrogenophosphate	2.6–7.6	6.40 7.20	–	–
Citrate buffer	Citric acid Sodium citrate	3.0–6.2	6.40	$a_1 = 1.280 \times 10^3$ $a_2 = -4.165$ $a_3 = 1.223 \times 10^{-2}$	3.776
Succinate	Succinic acid	3.2–5.2	4.21	–	–
Acetate buffer	Acetic acid Sodium acetate	3.7–5.6	4.76	$a_2 = 7.425$ $a_3 = -1.875 \times 10^{-2}$ $a_4 = 3.167 \times 10^{-5}$	4.650
Phosphate buffer (Sorensen's buffer)	Sodium monohydrogenophosphate Sodium dihydrogenophosphate	5.8–8.0	7.20	$a_1 = 1.723 \times 10^3$ $a_2 = -3.679$ $a_3 = 1.644 \times 10^{-2}$	7.000
Glyoxaline	Imidazole Hydrochloric acid	6.2–7.8	7.10	–	–
TRIS	2-amino-2-(hydroxymethyl)-1,3-propanediol	7.0–9.0	8.20	$a_1 = 3.879 \times 10^3$ $a_2 = -1.298 \times 10$ $a_3 = 3.554 \times 10^{-2}$ $a_4 = -3.289 \times 10^{-5}$	7.699
Glycylglycine	2-[(2-aminoacetyl)amino] acetic acid	7.5–8.9	8.25	–	–
Borate	Boric acid Sodium borate	8.2–10.2	9.20	$a_1 = 5.259 \times 10^3$ $a_2 = -3.311 \times 10$ $a_3 = 1.148 \times 10^{-1}$ $a_4 = -1.079 \times 10^{-4}$	9.180
Gly-NaOH	Glycine Sodium hydroxide	8.6–10.6	9.78	–	–
Carbonate	Sodium carbonate Sodium hydrogenocarbonate	9.2–10.8	10.33	$a_1 = 2.557 \times 10^3$ $a_2 = -4.285$ $a_3 = 1.919 \times 10^{-2}$	10.012

so-called Good's buffers; relevant chemical information is provided in Table 2.4. The constitutive chemicals of these buffers are neutral molecules, yet bearing both positive and negative charges (for being zwitterions); they are known for cell membrane impermeability, further to chemical inertness.

Good's buffers belong to one of six distinct chemical families: (1) acetamido, based on center C with carbonyl, amine, and another substituent, and encompassing ACES (or 2-[(2-amino-2-oxoethyl)amino]ethanesulfonic acid) and ADA (or 2,2′-[(2-amino-2-oxoethyl)imino]diacetic acid); (2) bis(2-hydroxyethyl)amine, based on open carbon backbone with intermediate N, and encompassing BES (or 2-[bis(2-hydroxyethyl)amino]ethanesulfonic acid), Bicine (or N,N-bis(2-hydroxyethyl)glycine), Bis-Tris (or 2-[bis(2-hydroxyethyl) amino]-2-(hydroxymethyl)-1,3-propanediol), DIPSO (or 3-[bis(hydroxyethyl)amino]-2-hydroxy-1-propanesulfonic acid), and TEA (or 2,2′,2″-nitrilotriethanol); (3) cyclohexylamino, based on hexagonal ring with amine substituent, and encompassing CABS (or 4-(cyclohexylamino)-1-butanesulfonic acid), CAPS (or 3-(cyclohexylamino)-1-propanesulfonic acid), CAPSO (or 3-(cyclohexylamino)-2-hydroxy-1-propanesulfonic acid), and CHES (or 2-(cyclohexylamino)ethanesulfonic acid); (4) morpholinic, based on heterocyclic hexagonal ring containing O and N, and encompassing MES (or 2-(4-morpholinyl)ethane sulfonic acid), MOBS (or 4-(4-morpholinyl)-1-butanesulfonic acid), MOPS (or 3-(4-morpholinyl)-1-propanesulfonic acid), and MOPSO (or 2-hydroxy-3-(4-morpholinyl)-1-propanesulfonic acid); (5) piperazinic, based on heterocyclic hexagonal ring containing N twice, and encompassing EPPS (or 3-[4-(2-hydroxyethyl)-1-piperazinyl]-1-propanesulfonic acid), HEPBS (or 4-[4-[2-hydroxyethyl)-1-piperazinyl]-1-butanesulfonic acid), HEPES (or 2-[4-(2-hydroxyethyl)-1-piperazinyl]ethanesulfonic acid), HEPPSO (or 2-hydroxy-3-[4-(2-hydroxyethyl)-1-piperazinyl]-1-propanesulfonic acid), PIPES (or 2,2′-(1,4-piperazinediyl)diethanesulfonic acid), and POPSO (or 3,3′-(1,4-piperazinediyl) bis(2-hydroxy-1-propanesulfonic acid); (6) propanol, based on open carbon backbone with alcohol and another two substituents, and encompassing AMP (or 2-amino-2-methyl-1-propanol), AMPD (or 2-amino-2-methyl-1,3-propanediol), and AMPSO (or 2-hydroxy-3-[(1-hydroxy-2-methyl-2-propanyl)amino]-1-propanesulfonic acid); and (7) TRIS, based on center C with three alcohol and one amine substituents, and encompassing BTP (or 2,2′-(1,3)-propanediyldiimino)bis[2-(hydroxymethyl)-1,3-propanediol]), TABS (or 4-{[1,3-dihydroxy-2-(hydroxymethyl)-2-propanyl]amino}-1-butanesulfonic acid), TAPS (or 3-{[1,3-dihydroxy-2-(hydroxymethyl)-2-propanyl]amino}-1-propanesulfonic acid), TAPSO (or 3-{[1,3-dihydroxy-2-(hydroxymethyl)-2-propanyl]amino}-2-hydroxy-1-propanesulfonic acid), TES (or 2-{[1,3-dihydroxy-2-(hydroxymethyl)-2-propanyl]amino}ethanesulfonic acid), and Tricine (or N-[1,3-dihydroxy-2-(hydroxymethyl)-2-propanyl]glycine). MES is not metabolized by bacteria and eukaryotic cells nor is by plant cells, but can be toxic at concentrations above 10^{-2} M; ACES, MOPS, and MOPSO have been successfully tested with yeasts, and MES, MOPS, and PIPES with mammalian cells. The reason why animal cells are more sensitive to presence of buffer arises from their lacking a cell wall; in particular, HEPES affects membrane potentials in neuronal cells, and TAPS inhibits connection channel activity in liver cells.

Table 2.4 Chemical structure of Good's buffers, with indication of center motif and substituents (single, R, or double, R_1 and R_2)

Chemical family	Center motif	Buffer	Substituents
Acetamido	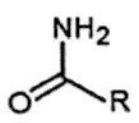	ACES	R: 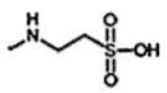
		ADA	R: 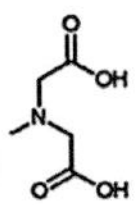
Bis(2-hydroxethyl) amine	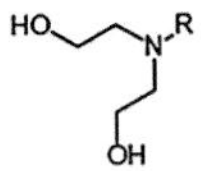	BES	R: 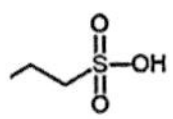
		Bicine	R: 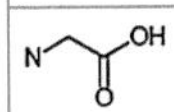
		Bis-Tris	R: 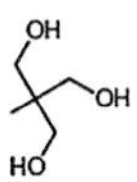
		DIPSO	R: 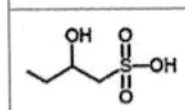
Cyclo-hexylamino	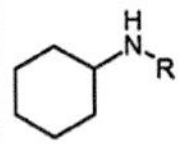	CABS	R: 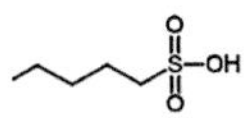
		CAPS	R: 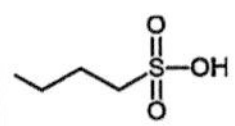
		CAPSO	R: 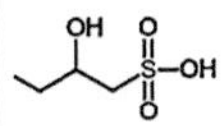
		CHES	R: 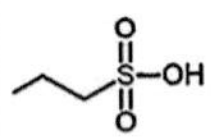
Morpholinic	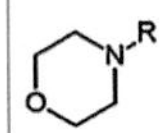	MES	R: 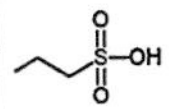
		MOBS	R: 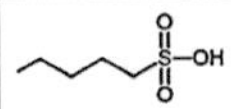
		MOPS	R: 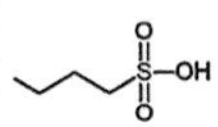
		MOPSO	R: 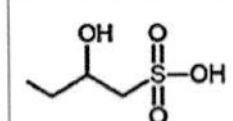
Piperazinic	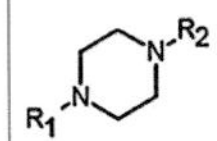	EPPS	R_1:  R_2: 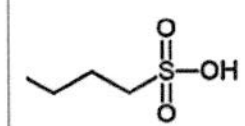
		HEPBS	R_1:  R_2: 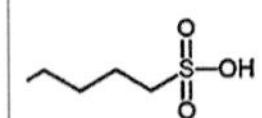
		HEPES	R_1:  R_2: 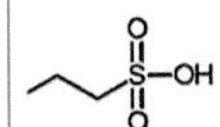
		HEPPSO	R_1:  R_2: 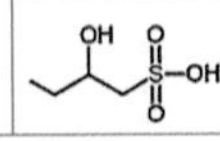
Propanol	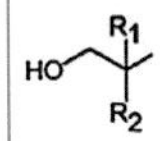	AMP	R_1: NH_2 R_2: CH_3
		AMPD	R_1: NH_2 R_2: 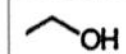
		AMPSO	R_1: CH_3 R_2: 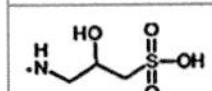
TRIS	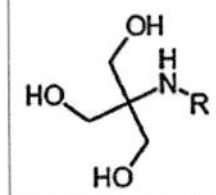	BTP	R: 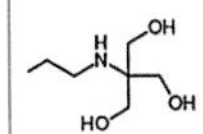
		TABS	R: 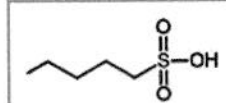
		TAPS	R: 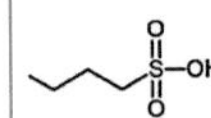
		TAPSO	R: 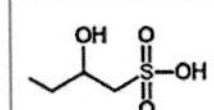

		TEA	R 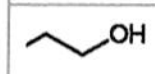					PIPES	R_1 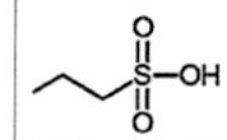 R_2 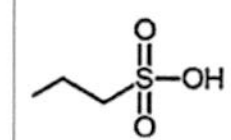			TES	R 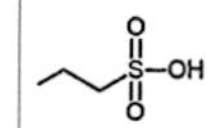
								POPSO	R_1 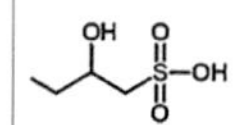 R_2 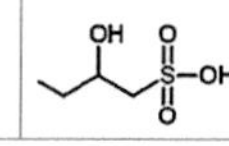			Tricine	R 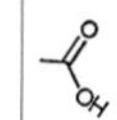

The useful pH ranges of the aforementioned buffers are depicted in Fig. 2.8, together with their characteristic protolysis constants, at one or two temperatures of practical interest. In most cases, $pK_a^*|_{37\,^\circ\mathrm{C}} < pK_a^*|_{25\,^\circ\mathrm{C}}$, because release of a hydrogen ion tends to be an exothermic process; only AMPSO and CAPS buffers contradict this rule, since supplementary molecular reconfiguration accompanying protolysis takes up more enthalpy than the latter.

In view of the relevance of keeping pH steady in bioreactors and the simplicity of doing so via chemical tools accounted for by buffering agents, the mechanism behind pH buffering is of the utmost importance, and selection of the germane chemicals can often be done separately from choice of medium nutrients, namely, from the above list of Good's buffers. Upon release of H^+, a weak acid HA, characterized by protolysis constant K_a, turns indeed to its conjugate base A^-, according to

$$HA + H_2O \underset{K_a}{\leftrightarrow} H_3O^+ + A^- \tag{2.211}$$

as chemical equation; the protolysis constant, K_a, is in turn defined by

$$K_a \equiv \frac{a_{H_3O^+} a_{A^-}}{a_{HA} a_{H_2O}} \tag{2.212}$$

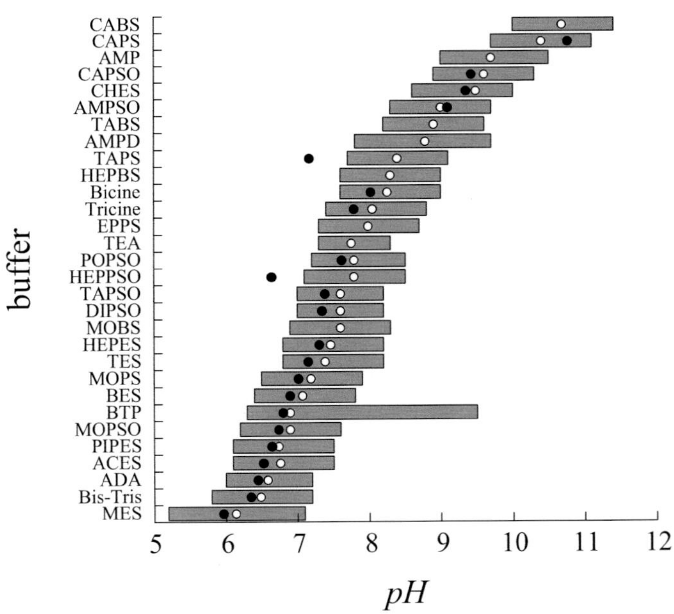

Fig. 2.8 Range of useful pH of Good's buffers (▬), with indication of pK_a^* at 25 °C (○) and 37 °C (●)

as per Guldberg and Waage's law of mass action, with a_i denoting activity of i-th species. Recalling that

$$a_i \equiv \gamma_i C_i, \tag{2.213}$$

where γ_i denotes activity coefficient and C_i denotes molar concentration of i-th species, one ends up with

$$K_a = \frac{\gamma_{H_3O^+} C_{H_3O^+} \gamma_{A^-} C_{A^-}}{\gamma_{HA} C_{HA} \gamma_{H_2O} C_{H_2O}} \tag{2.214}$$

stemming from Eq. (2.212); algebraic rearrangement unfolds

$$K_a = \frac{\gamma_{H_3O^+} \gamma_{A^-}}{\gamma_{HA} \gamma_{H_2O}} \frac{1}{C_{H_2O}} \frac{C_{H_3O^+} C_{A^-}}{C_{HA}}. \tag{2.215}$$

For most cases of current interest encompassing relatively dilute solutions, however, the first factor in the right-hand side of Eq. (2.215) is essentially constant and close to unity, whereas C_{H_2O} is almost constant (ca. 55.5 mol L^{-1}) over wide ranges of operation because water serves as solvent medium. This prompts collapse of such terms with K_a to yield

$$K_a^* = \frac{C_{H^+} C_{A^-}}{C_{HA}}, \tag{2.216}$$

as long as

$$K_a^* \equiv \frac{\gamma_{HA} \gamma_{H_2O}}{\gamma_{H_3O^+} \gamma_{A^-}} C_{H_2O} K_a \tag{2.217}$$

stands for (dimensional) equilibrium constant. Recall the extremely high charge density of H^+, for being a proton devoid of electronic orbital around it; this is incompatible with its existence as independent species, thus justifying its immediate abstraction by a water molecule contributed by its surroundings (where it is present in great excess), at the expense of dative covalent attachment of the form 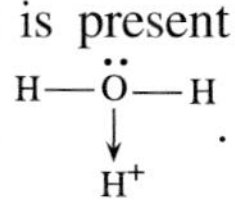. Nevertheless, C_{H^+} will hereafter be utilized as shorthand notation for $C_{H_3O^+}$, thus following traditional practice. Furthermore, the chemical process of protolysis is normally so fast that it can be considered to have reached equilibrium by any time period of interest, so a kinetic approach proves redundant. Water is also a (very) weak acid, meaning that concentrations of H^+ and OH^- resulting from its autoprotolysis abide to

$$K_w^* = C_{H^+} C_{OH^-}, \tag{2.218}$$

associated to the equilibrium represented by chemical equation

$$H_2O \underset{K_w}{\leftrightarrow} H^+ + OH^-; \tag{2.219}$$

this (spontaneous) process justifies the marginal conductivity of pure water. In the case of a strong acid, K_a^* is extremely large, which accommodates $C_{HA} \approx 0$ in Eq. (2.216); this phenomenon is usually referred to as full protolysis.

In view of the very low values for C_{H^+} usually found, it is handier to work with pH, formally defined as

$$pH \equiv -\log_{10} a_{H^+}, \tag{2.220}$$

where a_{H^+} denotes activity of H^+ in solution; or, in alternative,

$$pH = -\log_{10} \gamma_{H^+} C_{H^+}, \tag{2.221}$$

where γ_{H^+} denotes activity coefficient of H^+. The pH of pure water depends on temperature as a consequence of C_{H^+} being controlled by K_w, which relates, in turn, to K_w^* via

$$K_w^* \equiv \frac{\gamma_{H_2O}}{\gamma_{H^+} \gamma_{OH^-}} C_{H_2O} K_w; \tag{2.222}$$

being an equilibrium constant, K_w obeys

$$K_w = a_{H^+} a_{OH^-} \tag{2.223}$$

in agreement with Eq. (2.219), and consequently

$$K_w \equiv e^{-\frac{\Delta g_{ion}^{\theta}\{T\}}{\mathcal{R}T}} \tag{2.224}$$

with Δg_{ion} denoting (temperature-dependent) molar Gibbs' energy accompanying ionization. Since H^+ and OH^- in pure water come only from dissociation of water molecules, at 1:1 stoichiometry as per Eq. (2.219), and assuming the corresponding activity coefficients to be identical, one obtains

$$a_{H^+}{}^2 = K_w \tag{2.225}$$

from Eq. (2.223); this result is based on the common postulate of a unit activity of water as per Randall's rule, since water is in great excess for serving as solvent. Equation (2.225) can be rephrased as

$$a_{H^+} = \sqrt{K_w} \tag{2.226}$$

after having isolated a_{H^+}; or else

$$-\log_{10} a_{H^+} = -\log_{10}\sqrt{K_w} = -\frac{1}{2}\log_{10} K_w, \tag{2.227}$$

where insertion of Eqs. (2.220) and (2.224) leads to

$$pH = -\frac{1}{2}\log_{10} \mathrm{e}^{-\frac{\Delta g^{\theta}_{ion}\{T\}}{\mathcal{R}T}}. \tag{2.228}$$

After recalling the rule of change of base of a logarithm, Eq. (2.228) will instead appear as

$$pH = -\frac{1}{2}\log_{\mathrm{e}} \mathrm{e}^{-\frac{\Delta g^{\theta}_{ion}\{T\}}{\mathcal{R}T}} \log_{10}\mathrm{e}, \tag{2.229}$$

which simplifies to

$$pH = -\frac{1}{2}\left(-\frac{\Delta g^{\theta}_{ion}\{T\}}{\mathcal{R}T}\right)\log_{10}\mathrm{e} = \frac{\log_{10}\mathrm{e}}{2}\,\frac{\Delta g^{\theta}_{ion}\{T\}}{\mathcal{R}T} \tag{2.230}$$

or, equivalently,

$$pH|_w = 0.2171\,\frac{\Delta g^{\theta}_{ion}\{T\}}{\mathcal{R}T}; \tag{2.231}$$

remember that $\Delta g^{\theta}_{ion} = 79.9$ kJ mol^{-1}, thus implying that $pH|_w = 7.00$ at 25 °C in agreement with Eq. (2.231), but also $pH|_w = 7.64$ at 0 °C and $pH|_w = 5.59$ at 100 °C. Upon dissolution in water, an acid decreases its pH to below 7 (at 25 °C) because C_{H^+} increases on account of release of H^+ after Eq. (2.211), at the expense of a decrease in C_{OH^-} compatible with Eq. (2.218); conversely, a classical base releases OH^- when dissolved, so C_{H^+} must increase in order to keep the product $C_{H^+}C_{OH^-}$ constant as well, thus justifying why pH increases in this case.

The definition conveyed by Eq. (2.220) is useful because an ion-selective electrode is normally utilized to measure pH as an electric potential, E_{H^+}, relative to a standard potential associated to a reference electrode, E^θ; the former responds to activity of H^+ (relative to H, thus involving transfer of a single electron) according to Nernst's equation, namely,

$$E_{H^+} = E^\theta + \frac{\mathcal{R}T}{\mathcal{F}} \ln a_{H^+}, \tag{2.232}$$

where $\mathcal{F}$ denotes Faraday's constant ($\mathcal{F} = 9.6485 \times 10^4$ C mol^{-1}). Equation (2.232) may be reformulated as

$$E_{H^+} = E^\theta - \frac{\mathcal{R}T}{\mathcal{F}}(-\ln a_{H^+}) = E^\theta - \frac{\mathcal{R}T}{\mathcal{F}}(-\log_{10} a_{H^+}) \ln 10, \tag{2.233}$$

again after exchanging the base of the original logarithm, or else

$$E_{H^+} = E^\theta - 2.302 \frac{\mathcal{R}T}{\mathcal{F}} pH, \tag{2.234}$$

following insertion of Eq. (2.220).

Visual and photometric methods can also be used to measure pH, which rely on color changes of specific organic pigments; the former usually resort to pH test strips, while photometric determination involves shining a light beam through the sample and measuring its absorbance. Their application is, nevertheless, quite limited in bioreactor operation; in fact, photometric measurements will be unreliable if the medium is cloudy or possesses an inherent background color, while some components of the medium may even destroy the indicator chemicals in the test strips through redox reactions. A pH (glass) electrode is indeed a much better approach because of its high sensitivity and ease of utilization; it consists of a measuring electrode, with potential proportional to medium pH as per Eq. (2.234), to be combined with a reference electrode, with potential independent of medium pH. When these two electrodes are connected to each other and put in contact with an aqueous medium, a difference of potential, $E_{H^+} - E^\theta$, will arise between measuring and reference electrodes. Because of its intrinsic complexity, the system must be calibrated in advance to compensate for spurious or cumulative deviations over time, namely, asymmetry potentials that develop at the glass/liquid interface due to imperfections of the glass material developed during manufacture, mechanical abrasion due to usage, and chemical etching on the outer surface, as well as slow deterioration due to diffusion in medium and out of reference electrolyte. A temperature sensor is often considered, for more accurate measurement of pH, through elimination of thermal effects. Due to the linearity between pH and $E_{H^+} - E^0$ apparent in Eq. (2.234), two calibration points suffice, and universal standards are available for pH 4.01 and 10.01 (see Fig. 2.7f); the (theoretical) slope can then be obtained following algebraic rearrangement of Eq. (2.234) as

$$\frac{E_{H^+} - E^0}{pH} = -2.302\frac{\mathscr{R}T}{\mathcal{F}}, \tag{2.235}$$

which leads to −58.12 or − 60.10 mV/pH unit, at 20 °C and 30 °C, respectively.

In most cases, the aforementioned three probes are combined into a single device—termed combined pH electrode; this is schematized in Fig. 2.9. The heart of the electrode is a thin bulb of pH-sensitive glass, characterized by a molecular matrix of silicon oxide with

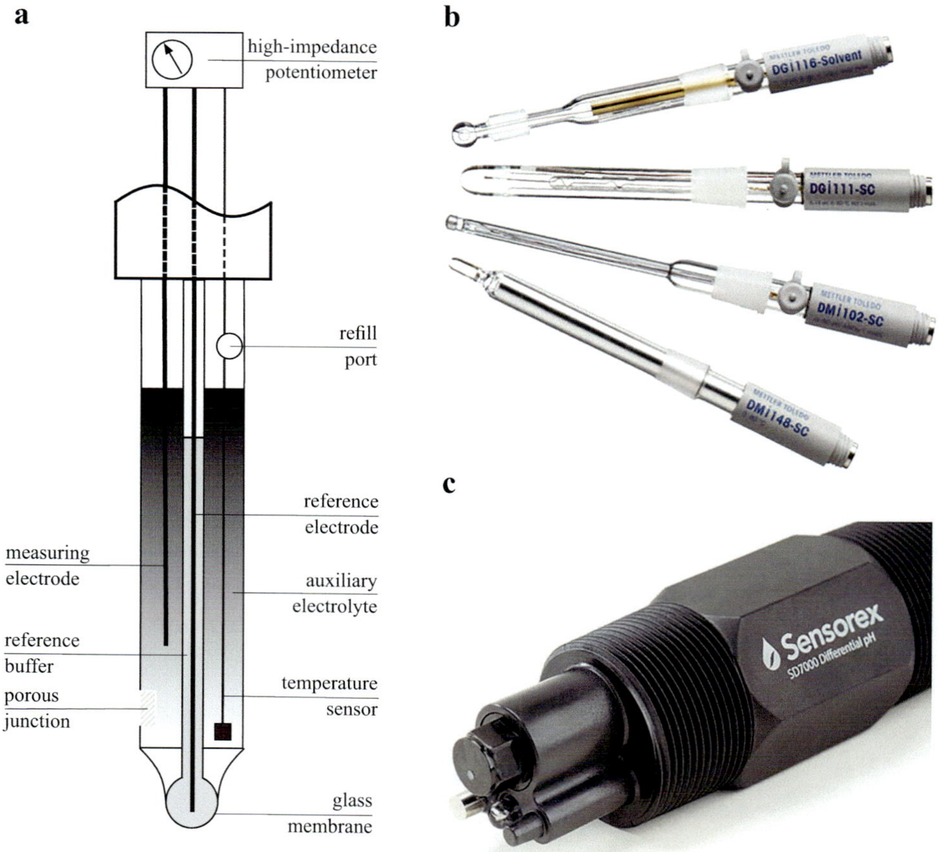

Fig. 2.9 Combined pH glass electrode as (**a**) schematic representation and examples (**b**) bearing distinct tip configurations (courtesy of Mettler Toledo, Columbus OH, USA) or (**c**) encased in sturdy framework (courtesy of Ryan Herco Flow Solutions, Burbank CA, USA)

deliberate addition of selected metal oxides; for instance, Corning™ glass contains 72% SiO_2, 22% Na_2O, and 6%(w/w) CaO. Each silicon atom is bound to three oxygen atoms along a plane, and each one is further bound to another oxygen atom above (or below) that plane. Therefore, the glass consists of an infinite, three-dimensional network of SiO groups, with each Si bound to four O's and each O shared by two Si's; within the interstices of this structure, there are sufficient cations to balance the negative charge of said silicate groups.

The atomic structure of this glass is designed to allow some mobility of singly-charged cations, as is notably the case of Na^+; such a mobility within the lattice is responsible for its electrical conductivity. Oxygen atoms within the lattice that are not bound to two silicon atoms possess a negative charge, to which Na^+ can ion-pair; hence, this cation will undergo slow diffusion through the lattice, moving from one charge pair site to another. While the glass membrane resistance is quite high (ca. 100 MΩ), such a displacement of cations within the glass creates a measurable potential between its two sides; this is known as Cremer's effect. When a glass of this type contacts an aqueous solution, the interfacial surface becomes hydrated as water enters a short distance (ca. 10 nm) into the crystal lattice and causes it to swell, while the bulk of the glass material remains dry; the glass membrane shape is designed to maximize moistening rate/extent. Some of the metal ions within the hydrated gel on the surface will diffuse out of the glass into solution; while some H^+ ions from solution will counter-diffuse into the hydrated gel, and thus charge-pair with oxygen anions near the glass surface. Hence, an ion exchange equilibrium will be established between fixed negative sites on the glass surface and H^+ ions of the contacting solution, according to

$$\mathrm{H^{+\,-}O-Si-(s)} \leftrightarrow \mathrm{H^+(aq)} + \mathrm{{}^-O-Si-(s)}; \tag{2.236}$$

here, $H^{+-}O - Si-$ and $^-O - Si-$ represent oxygen sites at the glass membrane that are ion-paired or unpaired with H^+, respectively. Note that H^+ cannot cross through the glass membrane, so the change in Gibbs' energy associated to this phenomenon is due solely to Na^+; therefore, C_{H^+} prevailing on the outside of the membrane is relayed as C_{Na^+} to its inside. To function as a pH sensor, a layer of pH-sensitive glass is to be placed between two solutions containing H^+—hereafter referred to by subscripts $_{in}$ and $_{out}$; as only the glass closest to either solution becomes hydrated, two individual equilibria will indeed be established, with potentials given by

$$E_{H^+,in} = E^\theta - 2.302\frac{\mathcal{R}T}{\mathcal{F}}pH_{in} \tag{2.237}$$

and

$$E_{H^+,out} = E^\theta - 2.302\frac{\mathcal{R}T}{\mathcal{F}}pH_{out}. \tag{2.238}$$

Ordered subtraction of Eq. (2.238) from Eq. (2.237) unfolds

$$E_{H^+,out} - E_{H^+,in} = E^\theta - 2.302\frac{\mathcal{R}T}{\mathcal{F}}pH_{out} - \left(E^\theta - 2.302\frac{\mathcal{R}T}{\mathcal{F}}pH_{in}\right), \tag{2.239}$$

where cancellation of E^θ with its negative and factoring out of $2.302\mathcal{R}T/\mathcal{F}$ leave

$$E_{H^+,out} - E_{H^+,in} = 2.302\frac{\mathcal{R}T}{\mathcal{F}}(pH_{in} - pH_{out}); \tag{2.240}$$

isolation of pH_{out} finally yields

$$pH_{out} = pH_{in} - \frac{E_{H^+,out} - E_{H^+,in}}{2.302\frac{\mathcal{R}T}{\mathcal{F}}}, \tag{2.241}$$

so pH of the outer solution can be directly obtained from (the known) pH of the inner solution and (the measured) difference of potential between the two sides of the glass membrane.

The glass membrane is blown onto the end of a length of glass tubing and sealed to the electrode; it is filled with a reference buffer solution with fixed concentration of H^+, typically at pH 7 brought about by a 10^{-7} M solution of HCl(aq). Spherical and conical membrane tips are the most common, yet cylindrical and dome-, flat-, or spear-shaped tips can fulfill unique process/medium specifications in terms of mechanical robustness, textural smoothness, or cleaning susceptibility. A silver wire is immersed in said reference buffer, so some AgCl will precipitate as a thin film on its surface (in agreement with $K^*_{AgCl} \equiv C_{Ag^+}(\text{aq})C_{Cl^-}(\text{aq}) = 1.8 \times 10^{-10}\text{mol}^2\ \text{L}^{-2}$); it accordingly serves as reference electrode. A second Ag/AgCl electrode is immersed in an electrolyte solution, usually KCl, contained in a compartment surrounding the reference solution compartment. The outer compartment contacts the solution under scrutiny on the outside of the probe, via a porous plug (or junction) that serves as salt bridge; many materials and mechanical designs exist for such a junction, namely, open hole, porous pin, ring/sleeve diaphragm, or fiber/metal bundle. Diffusion potentials at the junction affect measurement accuracy; to keep those disruption potentials to a low level, the junction should guarantee as large and consistent as possible outflow of electrolyte. Conversely, the junction must only be slightly permeable, otherwise the electrolyte will escape quickly; a compromise is thus to be reached by resorting to specific junction materials and mesh sizes. Ceramic junctions take advantage of porosity of unglazed ceramic, with an outflow rate ≤ 0.2 mL d^{-1} but a high electrical resistance (>1 kΩ), whereas platinum junctions consist of fine, twisted platinum filaments between which the electrolyte flows along precisely defined microchannels, with a maximum outflow rate of 1 mL d^{-1} and an intermediate electrical resistance (ca. 0.5 kΩ); as alternative, ground-joint junctions resort to the thin gap of unlubricated ground glass as outflow opening for the electrolyte, with outflow rates of ca. 3 mL d^{-1} and lowest electrical resistance (<0.1 kΩ).

Choice of Ag/AgCl for either electrode does not raise toxicity problems during operation or eventual disposal; it also possesses a wide range of operating temperature, so such electrodes can undergo sterilization without any problem. The hydrostatic pressure of the electrolyte should, nevertheless, be balanced as much as possible with that of the outer medium, so as to avoid excessive outflow/inflow; a refill hole is, anyway, provided to make up for electrolyte lost if used in liquid form. Although no outflow will occur if the electrolyte is embedded within a gel matrix, electrolyte refill reduces its contamination, thus permitting operation on the long run, unlike its gel version. The standard concentration of KCl(aq) is 3 mol L^{-1}, but it can include extra crystals to guarantee saturation over a wide temperature interval, while damping dilution by outer medium inflow over time; if, in addition, AgCl is added in solid form, then dissolution of that salt is slowed down. On the other hand, KCl has been widely elected for electrolyte because it possesses good electrical conductivity and is chemically neutral, so the amount leaking through the junction during measurement will not react with the medium or be toxic to the cells therein; since its ions are equally mobile, a diffusion potential will not build up for this reason either.

It should be emphasized that the junction is the most problematic part of the probe; in fact, should it clog due to particles suspended in the medium going into the porous area or because crystals of electrolyte forming inside the junction, stabilization of potential would take much longer. To reduce clogging, two or three junctions can be used in parallel, although at the expense of a higher overall outflow; such junctions may instead be placed in series as long as two separate electrolyte compartments are provided, thus minimizing contamination of the inner KCl (and enhancing electrode stability), while permitting use of a distinct outer electrolyte (useful if the medium is sensitive to chloride ions).

The combined electrode operates like a galvanic cell, schematically represented as

$$\mathrm{Ag(s)|AgCl(s)|K^{+}(aq),Cl^{-}(aq)}\lfloor \mathit{junction} \rfloor \mathrm{H^{+}_{\mathit{out}}(aq)}\|\mathbf{glass}\|\mathrm{H^{+}_{\mathit{in}}(aq),Cl^{-}(aq)|AgCl(s)|Ag(s)}. \tag{2.242}$$

Here, $\mathrm{Ag(s)|AgCl(s)|K^{+}(aq),Cl^{-}(aq)}\lfloor \mathit{junction} \rfloor \mathrm{H^{+}_{\mathit{out}}(aq)}\|$ represents the measuring electrode, where the junction assures contact between inner electrolyte K^{+}(aq), Cl^{-}(aq), and outer (unknown) H^{+}_{out}(aq); on the other hand, $\|\mathrm{H^{+}_{\mathit{in}}(aq),Cl^{-}(aq)|AgCl(s)|Ag(s)}$ represents the reference electrode, with those two cells separated by the glass membrane $\|\mathbf{glass}\|$. The diffusive barriers prevent (in $\|\mathbf{glass}\|$) or slow down (in $\lfloor \mathit{junction} \rfloor$) mixing of the solutions at stake; since similar electrodes are employed, any potentials generated at interfaces cancel out each other, so the voltage measured becomes essentially dependent on the difference between concentrations of $H^+_{out}(aq)$ and $H^+_{in}(aq)$ prevailing on the two sides of the glass membrane. All *pH* glass electrodes are characterized by an extremely high electrical resistance (50–500 MΩ), so such type of electrode can be utilized only with a high-impedance voltmeter. Unfortunately, at very high *pH* (>10), most glass electrodes become responsive also to $Na^+(aq)$, so the measured *pH* will be lower than its actual value; conversely, at too low *pH* (< 1), glass membranes become susceptible to saturation by H^+, so they will produce *pH* readings above the correct one.

Problem 2.6

When a weak acid, HA, is dissolved in water, the concentration of ions of *N* different types, resulting from dissociation thereof, may be relatively high; in this case, the major contribution to non-unit activity coefficients will come from the electric field generated by the (solvated) charges. Under these circumstances, γ_i may be approximately described by

$$\ln \gamma_i = -\alpha {z_i}^2 \sqrt{\vartheta}; \quad i = 1, 2, \ldots, N, \tag{2.243}$$

classically known as Debye and Huckel's limiting theory, where α denotes a constant and z_i denotes individual charge of *i*-th ion species. Ionic strength, ϑ, is in turn defined by

$$\vartheta \equiv \frac{1}{2} \sum_{j=1}^{N} C_j {z_j}^2, \tag{2.244}$$

and thus represents an average square charge, using half mole concentration, $C_j/2$, of *j*-th charged species as weight factor.

(a) Obtain a general expression for K_a, in the case of a weak acid.
(b) Knowing that $C_{H_2O} = 55.5$ mol L^{-1} and $\alpha = 1.172$ L$^{1/2}$ mol$^{-1/2}$, estimate the correction factor necessary to convert K_a^* to K_a, on the hypothesis that $C_{H^+} = C_{A^-} = 10^{-5}$mol L^{-1} and $C_{OH^-} = 10^{-9}$mol L^{-1}.

Solution

(a) There are three charged species in solution, H^+ from acid and water, OH^- from water, and A^- from acid; hence, Eq. (2.215) will read

(continued)

Problem 2.6 (continued)

$$K_a = \frac{\gamma_{H^+}\gamma_{A^-}}{C_{H_2O}} \frac{C_{H^+}C_{A^-}}{C_{HA}}, \tag{2.245}$$

since H_2O and HA are not ionic species. The prevailing ionic strength reads

$$\vartheta = \frac{1}{2}\left(C_{H^+}{z_{H^+}}^2 + C_{OH^-}{z_{OH^-}}^2 + C_{A^-}{z_{A^-}}^2\right) \tag{2.246}$$

stemming from Eq. (2.244), where realization that $z_{H^+} = 1$ and $z_{OH^-} = z_{A^-} = -1$ supports transformation to

$$\vartheta = \frac{1}{2}\left(C_{H^+}1^2 + C_{OH^-}(-1)^2 + C_{A^-}(-1)^2\right) = \frac{1}{2}(C_{H^+} + C_{OH^-} + C_{A^-}); \tag{2.247}$$

insertion of Eq. (2.247) in Eq. (2.243) gives then rise to

$$\ln\gamma_i = -\alpha {z_i}^2\sqrt{\frac{1}{2}(C_{H^+} + C_{OH^-} + C_{A^-})} = -\frac{\alpha}{\sqrt{2}}{z_i}^2\sqrt{C_{H^+} + C_{OH^-} + C_{A^-}}, \tag{2.248}$$

which specifically supports

$$\ln\gamma_{H^+} = -\frac{\alpha}{\sqrt{2}}1^2\sqrt{C_{H^+} + C_{OH^-} + C_{A^-}} = -\frac{\alpha}{\sqrt{2}}\sqrt{C_{H^+} + C_{OH^-} + C_{A^-}} \tag{2.249}$$

in the case of H^+,

$$\ln\gamma_{OH^-} = -\frac{\alpha}{\sqrt{2}}(-1)^2\sqrt{C_{H^+} + C_{OH^-} + C_{A^-}} = -\frac{\alpha}{\sqrt{2}}\sqrt{C_{H^+} + C_{OH^-} + C_{A^-}} \tag{2.250}$$

encompassing OH^-, and

$$\ln\gamma_{A^-} = -\frac{\alpha}{\sqrt{2}}(-1)^2\sqrt{C_{H^+} + C_{OH^-} + C_{A^-}} = -\frac{\alpha}{\sqrt{2}}\sqrt{C_{H^+} + C_{OH^-} + C_{A^-}} \tag{2.251}$$

(continued)

Problem 2.6 (continued)

pertaining to A^-. Once in possession of Eqs. (2.249)–(2.251), one may retrieve Eq. (2.245) as

$$K_a = \frac{e^{-\frac{\alpha}{\sqrt{2}}\sqrt{C_{H^+}+C_{OH^-}+C_{A^-}}}\, e^{-\frac{\alpha}{\sqrt{2}}\sqrt{C_{H^+}+C_{OH^-}+C_{A^-}}}}{C_{H_2O}} \frac{C_{H^+}C_{A^-}}{C_{HA}}, \tag{2.252}$$

where the exponential functions will be pooled together to get

$$K_a = \frac{e^{-2\frac{\alpha}{\sqrt{2}}\sqrt{C_{H^+}+C_{OH^-}+C_{A^-}}}}{C_{H_2O}} \frac{C_{H^+}C_{A^-}}{C_{HA}} = \frac{e^{-\sqrt{2}\,\alpha\sqrt{C_{H^+}+C_{OH^-}+C_{A^-}}}}{C_{H_2O}} \frac{C_{H^+}C_{A^-}}{C_{HA}}; \tag{2.253}$$

therefore, K_a depends not only on the product of C_{H^+} by C_{A^-}, but also on the exponential of the square root of their sum with C_{OH^-}. Note that $\ln \gamma_i < 0$ in all cases labeled as Eqs. (2.249)–(2.251), thus implying $\gamma_i < 1$, and consequently $a_i < C_i$ as per Eq. (2.213); this happens because electrostatic attraction between opposite charges and solvation of ions constrain their mobility and availability in solution, as if their actual concentration were lower.

(b) The correction factor under scrutiny is given by

$$\frac{K_a}{K_a^*} = \frac{\gamma_{H^+}\gamma_{A^-}}{C_{H_2O}}, \tag{2.254}$$

in view of Eq. (2.217) and $\gamma_{HA} = \gamma_{H_2O} = 1$ by hypothesis; insertion of Eqs. (2.216) and (2.253) then unfolds

$$\frac{K_a}{K_a^*} = \frac{e^{-\sqrt{2}\alpha\sqrt{C_{H^+}+C_{OH^-}+C_{A^-}}}}{C_{H_2O}} \tag{2.255}$$

from Eq. (2.254). Combination with the values provided for α and C_{H_2O} transforms Eq. (2.255) to

$$\frac{K_a}{K_a^*} = \frac{e^{-\sqrt{2}\cdot 1.172\sqrt{10^{-5}+10^{-9}+10^{-5}}}}{55.5} = \frac{e^{-1.658\cdot 0.004472}}{55.5} = \frac{0.9926}{55.5}$$
$$= 0.01788 \text{ L mol}^{-1}. \tag{2.256}$$

(continued)

Problem 2.6 (continued)

Note that the contribution of the ratio of activity coefficients, $\gamma_{H^+}\gamma_{A^-}/\gamma_{H_2O}=\gamma_{H^+}\gamma_{A^-}/1=\gamma_{H^+}\gamma_{A^-}$, to this value is a mere 0.9926 (see Eqs. (2.249), (2.251), and (2.256)), thus justifying why it is so often considered as approximately unity.

2.3.1 Strong Acid

When a strong acid, hereafter referred to by subscript $_{sa}$, at initial concentration $C_{HA,0}$ is dissolved in water,

$$HA \rightarrow H^+ + OH^- \tag{2.257}$$

emerges as descriptive chemical equation; the concentrations of the three chemical species at stake accordingly obey

$$C_{H^+,sa} = C_{HA,0} \tag{2.258}$$

and likewise

$$C_{A^-} = C_{A^-,sa} = C_{HA,0}, \tag{2.259}$$

together with

$$C_{HA} = C_{HA,sa} = 0 \tag{2.260}$$

since the original HA is assumed to have been quantitatively converted to A^-. Under these circumstances, the only germane equilibrium arises from Eq. (2.219), with associated dimensional equilibrium constant reading

$$K_w^* = C_{H^+}C_{OH^-}, \tag{2.261}$$

identical to Eq. (2.218); more specifically, Eq. (2.261) should read

$$K_w^* = \left(C_{H^+,w} + C_{H^+,sa}\right) C_{OH^-,w} \tag{2.262}$$

with subscript $_w$ referring to water, and upon combination with

$$C_{H^+} = C_{H^+,w} + C_{H^+,sa} \tag{2.263}$$

describing a mass balance to hydrogen ions in solution. Since OH^- is contributed solely by water, the stoichiometry underlying Eq. (2.219) implies

$$C_{OH^-} = C_{OH^-,w} = C_{H^+,w}; \tag{2.264}$$

combination with Eq. (2.264) converts Eq. (2.262) to

$$K_w^* = \left(C_{H^+,w} + C_{H^+,sa}\right) C_{H^+,w}, \tag{2.265}$$

whereas further insertion of Eq. (2.258) unfolds

$$K_w^* = C_{H^+,w}\left(C_{HA,0} + C_{H^+,w}\right). \tag{2.266}$$

It should be stressed that all H^+ ions released by the strong acid as it dissolves remain as such in solution, but will modulate release of similar ions from water via autoprotolysis, knowing that one OH^- moiety will be removed by each H^+ taken up back to H_2O form; however, $C_{H^+,w}$ will necessarily remain positive, for compatibility of Eq. (2.266) with $K_w^* > 0$. In other words, there will always be H^+ in solution originated previously from water, while any H^+ coming from the strong acid would never be taken up since that would be enforced only upon depletion of all H^+ coming from water.

Elimination of parenthesis transforms Eq. (2.266) to

$$K_w^* = {C_{H^+,w}}^2 + C_{HA,0} C_{H^+,w}, \tag{2.267}$$

or else

$${C_{H^+,w}}^2 + C_{HA,0} C_{H^+,w} - K_w^* = 0 \tag{2.268}$$

as canonical form of the underlying quadratic equation; the classical solving formula will then be invoked to get

$$C_{H^+,w} = \frac{-C_{HA,0} \pm \sqrt{C_{HA,0}^2 - 4(-K_w^*)}}{2} = \frac{-C_{HA,0} \pm \sqrt{C_{HA,0}^2 + 4K_w^*}}{2}, \tag{2.269}$$

along with elimination of consecutive negative signs. Since $\sqrt{C_{HA,0}^2 + 4K_w^*} > \sqrt{C_{HA,0}^2} = |C_{HA,0}| = C_{HA,0}$ due to $K_w^* > 0$, one realizes that only the plus sign preceding the square root in Eq. (2.269) will produce a physically meaningful solution (as the other would entail $C_{H^+,w} < 0$); therefore,

$$C_{H^+,w} = \frac{\sqrt{C_{HA,0}^2 + 4K_w^*} - C_{HA,0}}{2} \tag{2.270}$$

will hereafter be employed in lieu of Eq. (2.269). Insertion of Eqs. (2.258) and (2.270) justifies reformulation of Eq. (2.263) to

$$C_{H^+} = \frac{\sqrt{C_{HA,0}^2 + 4K_w^*} - C_{HA,0}}{2} + C_{HA,0}, \tag{2.271}$$

which may instead read

$$C_{H^+} = \sqrt{\left(\frac{C_{HA,0}}{2}\right)^2 + K_w^*} - \frac{C_{HA,0}}{2} + C_{HA,0} \tag{2.272}$$

after the fraction has been split and 2 taken under the square root; when terms alike are pooled together, Eq. (2.272) reduces to

$$C_{H^+} = \sqrt{\left(\frac{C_{HA,0}}{2}\right)^2 + K_w^*} + \frac{C_{HA,0}}{2}. \tag{2.273}$$

Since C_{H^+} to be dealt with in current practice is normally small, $\gamma_{H^+} \approx 1$ in Eq. (2.221), so Eq. (2.220) should to advantage be rephrased as

$$pH = -\log_{10} C_{H^+}, \tag{2.274}$$

obtained after setting $\gamma_{H^+} = 1$ in Eq. (2.221); this is equivalent to writing

$$C_{H^+} = 10^{-pH}, \tag{2.275}$$

upon isolation of C_{H^+}. Insertion of Eq. (2.275) converts Eq. (2.273) to

$$10^{-pH} = \sqrt{\left(\frac{C_{HA,0}}{2}\right)^2 + K_w^*} + \frac{C_{HA,0}}{2}, \tag{2.276}$$

where application of decimal logarithms to both sides gives rise to

$$-pH = \log_{10}\left\{\sqrt{\left(\frac{C_{HA,0}}{2}\right)^2 + K_w^*} + \frac{C_{HA,0}}{2}\right\}, \tag{2.277}$$

which readily degenerates to

$$pH|_{w+sa} = -\log_{10}\left\{\sqrt{\left(\frac{C_{HA,0}}{2}\right)^2 + K_w^*} + \frac{C_{HA,0}}{2}\right\} \tag{2.278}$$

as negatives are taken of both sides. The variation of pH with $C_{HA,0}$, for a solution of a strong acid in water, is plotted in Fig. 2.10. Note the plateau prevailing at low initial concentration of weak acid, with $pH = 7$ serving as vertical intercept due to the prominent water contribution; and the sharp transition to an inclined, decreasing straight line when the overall concentration of acid molecules at startup is sufficiently large.

If the initial concentration of (strong) acid is large compared to K_w^*, then Eq. (2.278) reduces to

$$\begin{aligned} pH\Bigg|_{\substack{w+sa \\ C_{HA,0} >> \sqrt{K_w^*}}} &\approx -\log_{10}\left\{\sqrt{\left(\frac{C_{HA,0}}{2}\right)^2} + \frac{C_{HA,0}}{2}\right\} = -\log_{10}\left\{\left|\frac{C_{HA,0}}{2}\right| + \frac{C_{HA,0}}{2}\right\} \\ &= -\log_{10}\left\{\frac{C_{HA,0}}{2} + \frac{C_{HA,0}}{2}\right\} \end{aligned} \tag{2.279}$$

since $(C_{HA,0}/2)^2 + K_w^* \approx (C_{HA,0}/2)^2$; this degenerates to merely

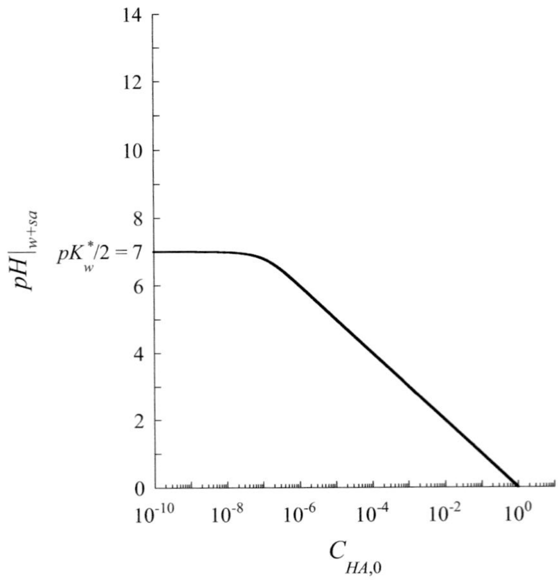

Fig. 2.10 Variation of pH of solution of strong acid, $pH|_{w+sa}$, in water, characterized by autoprotolysis constant $pK_w^* = 14$, versus initial molar concentration of acid, $C_{HA,0}$

$$pH\Big|_{\substack{w+sa \\ C_{HA,0} >> \sqrt{K_w^*}}} = -\log_{10} C_{HA,0} \tag{2.280}$$

after lumping common terms. Equation (2.280) describes the straight line in Fig. 2.10 prevailing at large $C_{HA,0}$, which exhibits -1 for slope as per the logarithmic scale chosen; hence, variation in pH proves proportional to (logarithmic) initial concentration of strong acid, since H^+ produced by full dissociation of said acid will override H^+ released by dissociation of water.

The opposite asymptotic pattern entertained by Eq. (2.272) reads

$$pH\Big|_{\substack{w+sa \\ C_{HA,0} << \sqrt{K_w^*}}} \approx -\log_{10}\sqrt{K_w^*} = -\frac{1}{2}\log_{10}K_w^* = \frac{1}{2}\left(-\log_{10}K_w^*\right), \tag{2.281}$$

because $C_{HA,0} << \sqrt{K_w^*}$ readily implies $(C_{HA,0}/2)^2 + K_w^* \approx K_w^*$, and likewise $\sqrt{K_w^*} + C_{HA,0}/2 \approx \sqrt{K_w^*}$, where the operational rule of calculation of the logarithm of a power was recalled; after setting

$$p\Xi \equiv -\log_{10}\Xi, \tag{2.282}$$

as condensed notation for a generic quantity Ξ using Eq. (2.274) for template, Eq. (2.281) will be rephrased to

$$\left. pH \right|_{\substack{w+sa \\ C_{HA,0} << \sqrt{K_w^*}}} = \frac{1}{2} pK_w^*. \tag{2.283}$$

Equation (2.283) describes the horizontal straight line in Fig. 2.10, with vertical intercept 7 on account of $K_w^* = 10^{-14}$ (at room temperature), and thus $pK_w^*/2 = 14/2 = 7$; in this case, solution pH is fixed by protolysis of water itself, since the strong acid dissolved therein upon dissociation is too dilute to play a role.

Finally, it is instructive to ascertain the variation in (equilibrium) concentration of all relevant species in solution with increasing initial concentration of strong acid, as done in Fig. 2.11; Eqs. (2.258), (2.259), (2.264), and (2.270) were utilized for $C_{H^+,sa}$, $C_{A^-,sa}$, $C_{OH^-,w}$, and $C_{H^+,w}$, respectively. Full dissociation implies indeed a variation in both $C_{H^+,sa}$ and $C_{A^-,sa}$ coincident with that of $C_{HA,0}$, thus justifying omission of $C_{HA,sa}$ in Fig. 2.11 for being trivially nil; conversely, $C_{H^+,w}$ and $C_{OH^-,w}$ remain essentially constant

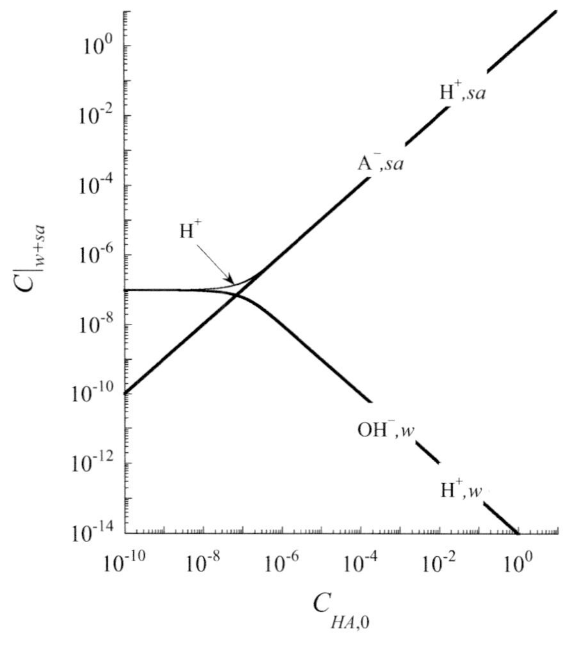

Fig. 2.11 Variation of molar concentration in aqueous solution of strong acid, $C|_{w+sa}$, of hydrogen ion originated from water, H^+,w (—), or acid, H^+,sa (—), and sum thereof, H^+ (___), as well as of anionic base originated from water, OH^-,w (—), or acid, A^-,sa (—), versus initial molar concentration of acid, $C_{HA,0}$

(and identical to each other) at the values characteristic of pure water as long as the amount of acid added is irrelevant, whereas they drop as necessary to abide to Eq. (2.262) when $C_{HA,0}$ is so high that $C_{H^+,sa}$ dominates the inventory of H⁺ in solution. Consequently, the thin curve accounting for $C_{H^+,w} + C_{H^+,sa}$, and driven by $C_{H^+,w}$ at low $C_{HA,0}$ and $C_{H^+,sa}$ at high $C_{HA,0}$, essentially mimics the curve in Fig. 2.10, in view of the similar logarithmic scales elected for ordinate, although turned upside down as per the minus sign in Eq. (2.274).

Problem 2.7

Derive the asymptotic patterns for C_{H^+}, at low and high $C_{HA,0}$, directly from Eqs. (2.263) and (2.270).

Solution

In order to use Eq. (2.270) as starting point, it is more convenient to first redo it as

$$C_{H^+,w} = \frac{\left(\sqrt{C_{HA,0}^2 + 4K_w^*} - C_{HA,0}\right)\left(\sqrt{C_{HA,0}^2 + 4K_w^*} + C_{HA,0}\right)}{2\left(\sqrt{C_{HA,0}^2 + 4K_w^*} + C_{HA,0}\right)}, \tag{2.284}$$

following multiplication of both numerator and denominator by $\sqrt{C_{HA,0}^2 + 4K_w^*} + C_{HA,0}$; since the product of two conjugate binomials coincides with the difference of the squares of its terms, one obtains

$$C_{H^+,w} = \frac{\left(\sqrt{C_{HA,0}^2 + 4K_w^*}\right)^2 - C_{HA,0}^2}{2\left(\sqrt{C_{HA,0}^2 + 4K_w^*} + C_{HA,0}\right)}, \tag{2.285}$$

which reduces to

$$C_{H^+,w} = \frac{C_{HA,0}^2 + 4K_w^* - C_{HA,0}^2}{2\left(\sqrt{C_{HA,0}^2 + 4K_w^*} + C_{HA,0}\right)} \tag{2.286}$$

after composing square power with square root in numerator. Symmetrical terms cancel out, so Eq. (2.286) simplifies to

$$C_{H^+,w} = \frac{4K_w^*}{2\left(\sqrt{C_{HA,0}^2 + 4K_w^*} + C_{HA,0}\right)}, \tag{2.287}$$

where 2 will in turn drop off both numerator and denominator as

(continued)

Problem 2.7 (continued)

$$C_{H^+,w} = \frac{2K_w^*}{\sqrt{C_{HA,0}^2 + 4K_w^*} + C_{HA,0}}. \tag{2.288}$$

If the initial concentration of strong acid is small, then Eq. (2.288) will be driven by

$$\lim_{C_{HA,0} \to 0} C_{H^+,w} = \frac{2K_w^*}{\sqrt{4K_w^*}}, \tag{2.289}$$

which degenerates to

$$\lim_{C_{HA,0} \to 0} C_{H^+,w} = \frac{2K_w^*}{2\sqrt{K_w^*}} = \sqrt{K_w^*} \tag{2.290}$$

upon taking the square root of 4, and then cancelling out $\sqrt{K_w^*}$ between numerator and denominator. On the other hand, one realizes that

$$\lim_{C_{HA,0} \to 0} C_{H^+} = \lim_{C_{HA,0} \to 0} \left(C_{H^+,w} + C_{H^+,sa}\right) = \lim_{C_{HA,0} \to 0} C_{H^+,w} + \lim_{C_{HA,0} \to 0} C_{H^+,sa} \tag{2.291}$$

stems from Eq. (2.263), coupled with linearity of the limit operator; hence, Eqs. (2.258) and (2.290) will be inserted to get

$$\lim_{C_{HA,0} \to 0} C_{H^+} = \sqrt{K_w^*} + \lim_{C_{HA,0} \to 0} C_{HA,0} = \sqrt{K_w^*} + 0, \tag{2.292}$$

where negative logarithms taken of both sides unfold

$$-\log_{10}\left\{\lim_{C_{HA,0} \to 0} C_{H^+}\right\} = \lim_{C_{HA,0} \to 0} (-\log_{10} C_{H^+}) = -\log_{10}\sqrt{K_w^*} = \frac{1}{2}(-\log_{10} K_w), \tag{2.293}$$

compatible with Eq. (2.283) after taking Eqs. (2.274) and (2.282) on board. By the same token, Eq. (2.288) gives rise to

(continued)

Problem 2.7 (continued)

$$\lim_{C_{HA,0}\to\infty} C_{H^+,w} = \frac{2K_w^*}{\sqrt{C_{HA,0}^{\,2}} + C_{HA,0}} \tag{2.294}$$

when $C_{HA,0}$ grows unbounded, where square root being the inverse of square power justifies simplification to

$$\lim_{C_{HA,0}\to\infty} C_{H^+,w} = \frac{2K_w^*}{C_{HA,0} + C_{HA,0}}; \tag{2.295}$$

after lumping terms alike, and then dropping 2 from both numerator and denominator, Eq. (2.295) reduces to just

$$\lim_{C_{HA,0}\to\infty} C_{H^+,w} = \frac{2K_w^*}{2C_{HA,0}} = \frac{K_w^*}{C_{HA,0}}. \tag{2.296}$$

Equation (2.263) can now be revisited as

$$\lim_{C_{HA,0}\to\infty} C_{H^+} = \lim_{C_{HA,0}\to\infty} \left(C_{H^+,w} + C_{H^+,sa}\right) = \lim_{C_{HA,0}\to\infty} C_{H^+,w} + \lim_{C_{HA,0}\to\infty} C_{H^+,sa}, \tag{2.297}$$

as per the classical theorems on limits, where insertion of Eqs. (2.258) and (2.296) produces

$$\lim_{C_{HA,0}\to\infty} C_{H^+} = \lim_{C_{HA,0}\to\infty} \frac{K_w^*}{C_{HA,0}} + \lim_{C_{HA,0}\to\infty} C_{HA,0}. \tag{2.298}$$

Since the last term in Eq. (2.298) overrides its reciprocal when $C_{HA,0}$ is very large, one is left with

$$\lim_{C_{HA,0}\to\infty} C_{H^+} = \frac{K_w^*}{\infty} + \lim_{C_{HA,0}\to\infty} C_{HA,0} = 0 + \lim_{C_{HA,0}\to\infty} C_{HA,0}, \tag{2.299}$$

or else

$$-\log_{10}\left\{\lim_{C_{HA,0}\to\infty} C_{H^+}\right\} = \lim_{C_{HA,0}\to\infty} \left(-\log_{10} C_{H^+}\right) = -\log_{10} C_{HA,0} \tag{2.300}$$

once logarithms and negatives are sequentially taken of both sides; Eq. (2.300) leads directly to Eq. (2.280) upon combination with Eq. (2.274), as expected.

2.3.2 Weak Acid

Consider now the case of an aqueous solution of a weak acid, hereafter referred to by subscript $_{wa}$, undergoing dissociation according to

$$\mathrm{HA} \underset{K_a^*}{\leftrightarrow} \mathrm{H}^+ + \mathrm{A}^-, \tag{2.301}$$

which reflects Eq. (2.211) but resorts to a simpler nomenclature. Although the underlying stoichiometry supports

$$C_{A^-} = C_{A^-,wa} = C_{H^+,wa}, \tag{2.302}$$

as also happens with a strong acid, one realizes that

$$C_{HA} = C_{HA,wa} = C_{HA,0} - C_{H^+,wa}, \tag{2.303}$$

with $C_{HA,wa} > 0$ due to incomplete dissociation, while the global concentration of hydrogen ions satisfies

$$C_{H^+} = C_{H^+,w} + C_{H^+,wa} \tag{2.304}$$

as alias to Eq. (2.263). Recall again Eq. (2.216), namely,

$$K_a^* = \frac{C_{H^+} C_{A^-}}{C_{HA}}; \tag{2.305}$$

insertion of Eqs. (2.302)–(2.304) leads to

$$K_a^* = \frac{\left(C_{H^+,w} + C_{H^+,wa}\right) C_{H^+,wa}}{C_{HA,0} - C_{H^+,wa}}, \tag{2.306}$$

where two unknowns emerge, that is, $C_{H^+,w}$ and $C_{H^+,wa}$. On the other hand,

$$K_w^* = C_{H^+} C_{OH^-} \quad (2.307)$$

describes protolysis of water in a way similar to Eq. (2.261), where insertion of Eqs. (2.264) and (2.304) gives rise to

$$K_w^* = \left(C_{H^+,w} + C_{H^+,wa}\right) C_{OH^-,w}; \quad (2.308)$$

since

$$C_{OH^-,w} = C_{H^+,w} \quad (2.309)$$

also in parallel to Eq. (2.264), one is led to

$$K_w^* = \left(C_{H^+,w} + C_{H^+,wa}\right) C_{H^+,w} \quad (2.310)$$

stemming from Eqs. (2.308) and (2.309) as descriptor of protolysis of water, which entails the very same unknowns, that is, $C_{H^+,w}$ and $C_{H^+,wa}$, already found in Eq. (2.306). Isolation of $C_{H^+,wa}$ from Eq. (2.310) is in order, namely,

$$C_{H^+,wa} = \frac{K_w^*}{C_{H^+,w}} - C_{H^+,w} \quad (2.311)$$

after dividing both sides by $C_{H^+,w}$, followed by subtraction of $C_{H^+,w}$ therefrom; insertion of Eq. (2.311) converts Eq. (2.304) to

$$C_{H^+} = C_{H^+,w} + \frac{K_w^*}{C_{H^+,w}} - C_{H^+,w}, \quad (2.312)$$

which breaks down to

$$C_{H^+} = \frac{K_w^*}{C_{H^+,w}} \quad (2.313)$$

upon cancellation of symmetrical terms. Equation (2.311) may instead be inserted in Eq. (2.306) to get

$$K_a^* = \frac{\left(C_{H^+,w} + \frac{K_w^*}{C_{H^+,w}} - C_{H^+,w}\right)\left(\frac{K_w^*}{C_{H^+,w}} - C_{H^+,w}\right)}{C_{HA,0} - \left(\frac{K_w^*}{C_{H^+,w}} - C_{H^+,w}\right)}; \tag{2.314}$$

cancellation of $C_{H^+,w}$ with its negative in numerator leaves

$$K_a^* = \frac{\frac{K_w^*}{C_{H^+,w}}\left(\frac{K_w^*}{C_{H^+,w}} - C_{H^+,w}\right)}{C_{HA,0} - \frac{K_w^*}{C_{H^+,w}} + C_{H^+,w}}, \tag{2.315}$$

along with removal of parenthesis in denominator. Reduction in number of fraction levels, complemented by application of the distributive property convert Eq. (2.315) to

$$K_a^* C_{HA,0} - \frac{K_a^* K_w^*}{C_{H^+,w}} + K_a^* C_{H^+,w} = \frac{K_w^{*2}}{C_{H^+,w}{}^2} - K_w^*, \tag{2.316}$$

while an extra multiplication of both sides by $C_{H^+,w}{}^2$ unfolds

$$K_a^* C_{HA,0} C_{H^+,w}{}^2 - K_a^* K_w^* C_{H^+,w} + K_a^* C_{H^+,w}{}^3 = K_w^{*2} - K_w^* C_{H^+,w}{}^2; \tag{2.317}$$

algebraic regrouping finally leads to

$$K_a^* C_{H^+,w}{}^3 + \left(K_a^* C_{HA,0} + K_w^*\right) C_{H^+,w}{}^2 - K_a^* K_w^* C_{H^+,w} - K_w^{*2} = 0 \tag{2.318}$$

as plotted in Fig. 2.12, produced with the aid of Newton and Raphson's method as tool to numerically find its solution. The decrease in actual pH when the initial concentration of acid increases becomes shallower for higher pK_a^*, because fewer and fewer of the original acid molecules will be able to release protons due to the finite magnitude of K_a^*. Equation (2.318) degenerates to

$$\left(K_a^* C_{H^+,w}{}^3 + K_a^* C_{HA,0} C_{H^+,w}{}^2 - K_a^* K_w^* C_{H^+,w} = 0\right)\Big|_{K_a^* \to \infty} \tag{2.319}$$

should K_a^* grow unbounded, thus making K_w^{*2} negligible; this prompts division of both sides by $K_a^* C_{H^+,w}$ to get

Fig. 2.12 Variation of pH of solution of weak acid, $pH|_{w+wa}$, of selected protolysis constant, pK^*_a, in water, characterized by autoprotolysis constant $pK^*_w = 14$, versus initial molar concentration of acid, $C_{HA,0}$

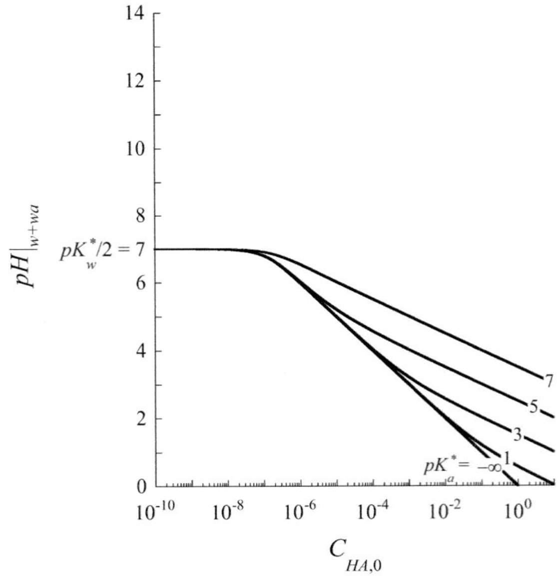

$$\left(C_{H^+,w}{}^2 + C_{HA,0}C_{H^+,w} - K^*_w = 0\right)\Big|_{K^*_a \to \infty}, \tag{2.320}$$

which coincides with Eq. (2.268) characteristic of a strong acid (described by a very large protolysis constant), so the curve labeled as $pK^*_a = -\infty$ coincides with that in Fig. 2.10.

If $K^*_a \to \sqrt{K^*_w}$, then Eq. (2.318) degenerates to

$$\left(\sqrt{K^*_w}C_{H^+,w}{}^3 + \left(\sqrt{K^*_w}C_{HA,0} + K^*_w\right)C_{H^+,w}{}^2 - \sqrt{K^*_w}K^*_w C_{H^+,w} - K^{*2}_w = 0\right)\Big|_{K^*_a \to \sqrt{K^*_w}}, \tag{2.321}$$

where division of both sides by $\sqrt{K^*_w}$ leaves

$$\left(C_{H^+,w}{}^3 + \left(C_{HA,0} + \sqrt{K^*_w}\right)C_{H^+,w}{}^2 - K^*_w C_{H^+,w} - K^{*\frac{3}{2}}_w = 0\right)\Big|_{K^*_a \to \sqrt{K^*_w}}; \tag{2.322}$$

in the vicinity of $C_{H^+,w} = \sqrt{K^*_w}$, the left-hand side of Eq. (2.322) would be driven by

$$\left.\left(C_{H^+,w}{}^3+\left(C_{HA,0}+\sqrt{K_w^*}\right)C_{H^+,w}{}^2-K_w^*C_{H^+,w}-K_w^{*\frac{3}{2}}\right)\right|_{\substack{K_a^*\to\sqrt{K_w^*}\\ C_{H^+,w}\to\sqrt{K_w^*}}}$$
$$=K_w^{*\frac{3}{2}}+K_w^*C_{HA,0}+K_w^{*\frac{3}{2}}-K_w^{*\frac{3}{2}}-K_w^{*\frac{3}{2}} \tag{2.323}$$

upon elimination of parenthesis, so cancelling of symmetrical terms ensues as

$$\left.\left(C_{H^+,w}{}^3+\left(C_{HA,0}+\sqrt{K_w^*}\right)C_{H^+,w}{}^2-K_w^*\ C_{H^+,w}-K_w^{*\frac{3}{2}}\right)\right|_{\substack{K_a^*\to\sqrt{K_w^*}\\ C_{H^+,w}\to\sqrt{K_w^*}}}$$
$$=K_w^*C_{HA,0}>0. \tag{2.324}$$

Inspection of the right-hand side of Eq. (2.324), vis à vis with that of Eq. (2.322), indicates that $\sqrt{K_w^*}$ is never reached by $C_{H^+,w}$, consistent with the curves in Fig. 2.12 lying above the high-$C_{HA,0}$ oblique asymptote, even when pK_a^* equals $7=pK_w^*/2$; therefore, a finite fraction of acid will always be present in dissociated form.

When $C_{HA,0}$ is small compared to $C_{H^+,w}$, then $C_{H^+,wa}<C_{HA,0}$ is even smaller in view of Eq. (2.303), so Eq. (2.306) can be well approximated by

$$\left.\left(K_a^*\approx\frac{C_{H^+,wa}C_{H^+,w}}{C_{HA,0}-C_{H^+,wa}}\right)\right|_{C_{H^+,wa}<C_{HA,0}<<C_{H^+,w}}, \tag{2.325}$$

as per $C_{H^+,w}+C_{H^+,wa}\approx C_{H^+,w}$, and Eq. (2.310) will likewise approach

$$\left.\left(K_w^*\approx C_{H^+,w}{}^2\right)\right|_{C_{H^+,wa}<C_{HA,0}<<C_{H^+,w}} \tag{2.326}$$

in view of $C_{H^+,w}+C_{H^+,wa}\approx C_{H^+,w}$. Elimination of denominators transforms Eq. (2.325) to

$$K_a^*C_{HA,0}-K_a^*C_{H^+,wa}=C_{H^+,wa}C_{H^+,w}, \tag{2.327}$$

whereas square roots taken of both sides of Eq. (2.326) convert it to

$$C_{H^+,w}=\sqrt{K_w^*}. \tag{2.328}$$

Isolation of $C_{H^+,wa}$ from Eq. (2.327) unfolds

$$C_{H^+,wa} = \frac{K_a^* C_{HA,0}}{K_a^* + C_{H^+,w}}, \tag{2.329}$$

while insertion of Eq. (2.328) justifies

$$C_{H^+,wa} = \frac{K_a^* C_{HA,0}}{K_a^* + \sqrt{K_w^*}}; \tag{2.330}$$

in view of Eqs. (2.328) and (2.330), it becomes possible to redo Eq. (2.304) as

$$C_{H^+} = \sqrt{K_w^*} + \frac{K_a^*}{K_a^* + \sqrt{K_w^*}} C_{HA,0}. \tag{2.331}$$

Realization that $C_{HA,0} << C_{H^+,w} = \sqrt{K_w^*}$, with the aid of Eq. (2.328), justifies reduction of Eq. (2.331) to

$$\left(C_{H^+} \approx \sqrt{K_w^*} \right) \Big|_{C_{H^+,wa} < C_{HA,0} << C_{H^+,w}}, \tag{2.332}$$

because $K_a^* / \left(K_a^* + \sqrt{K_w^*} \right)$ is of the order of unity (when $\sqrt{K_w^*}$ is small) or even below (when $\sqrt{K_w^*}$ is large); in logarithmic form, Eq. (2.332) looks like

$$\log_{10} C_{H^+} = \frac{1}{2} \log_{10} K_w^*, \tag{2.333}$$

and thus

$$pH \Big|_{\substack{w+wa \\ C_{HA,0} << \sqrt{K_w^*}}} = \frac{1}{2} pK_w^* \tag{2.334}$$

at the expense of Eqs. (2.274) and (2.282) after having taken negatives of both sides, where Eq. (2.328) justifies the change of condition in subscript. The K_a^*-independent horizontal asymptote in Fig. 2.12 is described by Eq. (2.334), which accordingly drives all curves at low $C_{HA,0}$. This trend mimics that shown in Fig. 2.10 at low $C_{HA,0}$, which is an obvious consequence of both processes being controlled by water autoprotolysis rather than acid protolysis.

To shed further light into this point, it is convenient to inspect how concentration of each chemical species in solution responds to initial concentration of (weak) acid—as done in Fig. 2.13; hence, $C_{A^-,wa}$, $C_{HA,wa}$, $C_{OH^-,w}$, $C_{H^+,wa}$, and $C_{H^+,w}$ were retrieved via

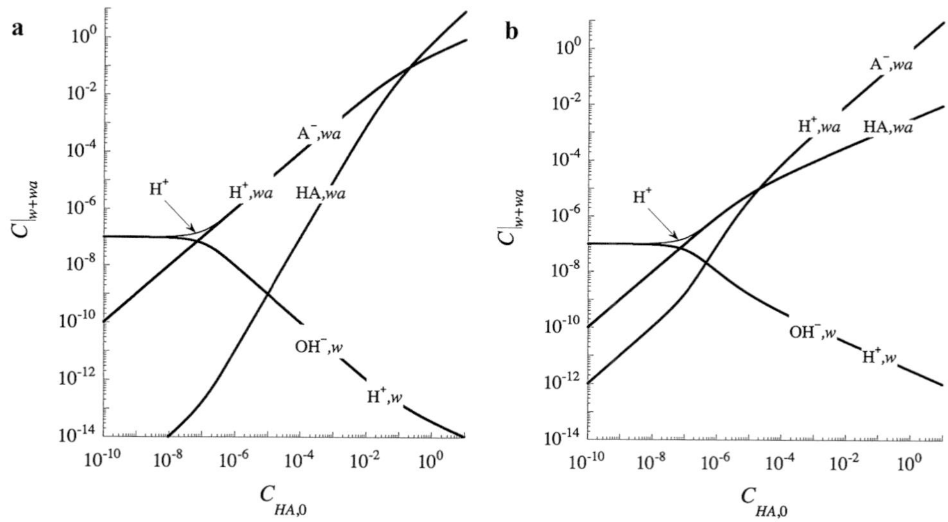

Fig. 2.13 Variation of molar concentrations in aqueous solution of weak acid, $C|_{w+wa}$, characterized by protolysis constant (**a**) $pK_a^*=1$ and (**b**) $pK_a^*=5$, of undissociated acid, HA,wa (—), hydrogen ion originated from water, H^+,w (—) or acid, H^+,wa (—), and sum thereof, H^+ (___), as well as anionic base originated from water, OH^-,w, or acid, A^-,sa, versus initial molar concentration of acid, $C_{HA,0}$

Eqs. (2.302), (2.303), (2.309), (2.311), and (2.318), respectively. As expected, both $C_{HA,\,wa}$ and $C_{A^-,wa}$ increase with $C_{HA,0}$ as per their common source, which also justifies their sum always coinciding with $C_{HA,0}$, with $C_{HA,\,wa}$ approaching $C_{HA,0}$ at large values of the latter; the profile is steeper when K_a^* is larger (or pK_a^* is smaller, for that matter) because protolysis is accordingly promoted. Both being formed from H_2O, at 1:1 stoichiometry, justifies coincidence of the curves representing $C_{H^+,w}$ and $C_{OH^-,w}$, and the same holds with regard to $C_{H^+,wa}$ and $C_{A^-,wa}$ as products of dissociation of HA. Once again, the sum of $C_{H^+,w}$ with $C_{H^+,wa}$ produces a curve similar to that in Fig. 2.12 labeled with the same pK_a^*, but flipped vertically around the horizontal line of vertical intercept 10^{-7}, as per equivalence between the logarithmic scale chosen for C_{H^+} and the definition of pH conveyed by Eq. (2.274).

In attempts to ascertain the behavior of pH with large $C_{HA,0}$, one should to advantage start by solving Eq. (2.310) with regard to $C_{H^+,w}$, instead of $C_{H^+,wa}$ as done in Eq. (2.311). Toward this deed, Eq. (2.310) is to be revisited as

$$C_{H^+,w}{}^2 + C_{H^+,wa}C_{H^+,w} - K_w^* = 0, \tag{2.335}$$

that is, a quadratic equation in $C_{H^+,w}$, which is similar in functional form to Eq. (2.268) provided that $C_{HA,0}$ is exchanged with $C_{H^+,wa}$; hence, one may directly jump to

$$C_{H^+,w} = \frac{\sqrt{C_{H^+,wa}{}^2 + 4K_w^*} - C_{H^+,wa}}{2} \tag{2.336}$$

as analogue to Eq. (2.270). Once in possession of Eq. (2.336), it becomes possible to rewrite Eq. (2.306) as

$$K_a^* = \frac{\left(\frac{\sqrt{C_{H^+,wa}{}^2 + 4K_w^*} - C_{H^+,wa}}{2} + C_{H^+,wa}\right) C_{H^+,wa}}{C_{HA,0} - C_{H^+,wa}}, \tag{2.337}$$

where the linear terms on $C_{H^+,wa}$ are to be lumped in numerator, and 2 in denominator moved into the square root to get

$$K_a^* = \frac{C_{H^+,wa}\left(\sqrt{\left(\frac{C_{H^+,wa}}{2}\right)^2 + K_w^*} + \frac{C_{H^+,wa}}{2}\right)}{C_{HA,0} - C_{H^+,wa}}; \tag{2.338}$$

since $C_{HA,0}$ is hereby considered as large, Eq. (2.338) essentially reduces to

$$\left.\left(K_a^* \approx \frac{C_{H^+,wa}\left(\sqrt{\left(\frac{C_{H^+,wa}}{2}\right)^2 + K_w^*} + \frac{C_{H^+,wa}}{2}\right)}{C_{HA,0}}\right)\right|_{C_{HA,0} >> C_{H^+,wa}}, \tag{2.339}$$

as per $C_{HA,0} - C_{H^+,wa} \approx C_{HA,0}$ under such circumstances. After getting rid of denominators, Eq. (2.339) becomes

$$K_a^* C_{HA,0} = C_{H^+,wa}\left(\sqrt{\left(\frac{C_{H^+,wa}}{2}\right)^2 + K_w^*} + \frac{C_{H^+,wa}}{2}\right), \tag{2.340}$$

where division of both sides by $C_{H^+,wa}$ unfolds

$$K_a^* \frac{C_{HA,0}}{C_{H^+,wa}} = \sqrt{\left(\frac{C_{H^+,wa}}{2}\right)^2 + K_w^*} + \frac{C_{H^+,wa}}{2} \tag{2.341}$$

or, after subtracting $C_{H^+,wa}/2$ from both sides,

$$K_a^* \frac{C_{HA,0}}{C_{H^+,wa}} - \frac{C_{H^+,wa}}{2} = \sqrt{\left(\frac{C_{H^+,wa}}{2}\right)^2 + K_w^*}. \tag{2.342}$$

Once squares are taken of both sides, Eq. (2.342) turns to

$$\left(K_a^* \frac{C_{HA,0}}{C_{H^+,wa}} - \frac{C_{H^+,wa}}{2}\right)^2 = \left(\frac{C_{H^+,wa}}{2}\right)^2 + K_w^*, \tag{2.343}$$

where expansion of the square of the binomial gives rise to

$$\left(K_a^* \frac{C_{HA,0}}{C_{H^+,wa}}\right)^2 - 2K_a^* \frac{C_{HA,0}}{C_{H^+,wa}} \frac{C_{H^+,wa}}{2} + \left(\frac{C_{H^+,wa}}{2}\right)^2 = \left(\frac{C_{H^+,wa}}{2}\right)^2 + K_w^*; \tag{2.344}$$

upon cancelling $2C_{H^+,wa}$ between numerator and denominator of the second term in the left-hand side, complemented by dropping of common squares between sides, Eq. (2.344) simplifies to

$$\left(K_a^* \frac{C_{HA,0}}{C_{H^+,wa}}\right)^2 - K_a^* C_{HA,0} = K_w^*. \tag{2.345}$$

Elimination of parenthesis and addition of $K_a^* C_{HA,0}$ to both sides convert Eq. (2.345) to

$$K_a^{*2} \frac{C_{HA,0}{}^2}{C_{H^+,wa}{}^2} = K_w^* + K_a^* C_{HA,0}; \tag{2.346}$$

isolation of $C_{H^+,wa}{}^2$ ensues as

$$C_{H^+,wa}{}^2 = \frac{K_a^{*2} C_{HA,0}{}^2}{K_w^* + K_a^* C_{HA,0}}, \tag{2.347}$$

while square roots taken of both sides leave

$$C_{H^+,wa} = \frac{K_a^* C_{HA,0}}{\sqrt{K_w^* + K_a^* C_{HA,0}}}. \tag{2.348}$$

Since only large values of $C_{HA,0}$ are of interest here, Eq. (2.348) reduces to

$$\left(C_{H^+,wa} \approx \frac{K_a^* C_{HA,0}}{\sqrt{K_a^* C_{HA,0}}} \right) \Bigg|_{C_{HA,0} >> \sqrt{K_w^*}} \tag{2.349}$$

on account of $K_w^* + K_a^* C_{HA,0} \approx K_a^* C_{HA,0}$, which breaks down to

$$C_{H^+,wa} = \sqrt{K_a^* C_{HA,0}}; \tag{2.350}$$

if Eq. (2.350) is rewritten as

$$C_{HA,0} = \frac{{C_{H^+,wa}}^2}{K_a^*} \tag{2.351}$$

following isolation of $C_{HA,0}$, then one realizes that $C_{HA,0} >> C_{H^+,wa}$ since $K_a^* << 1$ for weak acids, thus confirming equivalence between the condition conveyed as subscript in Eq. (2.339) and $C_{HA,0} >> \sqrt{K_w^*}$ used for subscript in Eq. (2.349). Insertion of Eq. (2.350) then converts Eq. (2.336) to

$$C_{H^+,w} = \frac{\sqrt{\left(\sqrt{K_a^* C_{HA,0}}\right)^2 + 4K_w^*} - \sqrt{K_a^* C_{HA,0}}}{2} = \frac{\sqrt{K_a^* C_{HA,0} + 4K_w^*} - \sqrt{K_a^* C_{HA,0}}}{2}, \tag{2.352}$$

where square root and square power meanwhile cancelled out for being the inverse of each other. Following multiplication and division of the right-hand side by the conjugate of its numerator, Eq. (2.352) leads to

$$C_{H^+,w} = \frac{\left(\sqrt{K_a^* C_{HA,0} + 4K_w^*}\right)^2 - \left(\sqrt{K_a^* C_{HA,0}}\right)^2}{2\left(\sqrt{K_a^* C_{HA,0} + 4K_w^*} + \sqrt{K_a^* C_{HA,0}}\right)} = \frac{K_a^* C_{HA,0} + 4K_w^* - K_a^* C_{HA,0}}{2\left(\sqrt{K_a^* C_{HA,0} + 4K_w^*} + \sqrt{K_a^* C_{HA,0}}\right)}, \tag{2.353}$$

again taking advantage of composition of square power and square root; symmetrical terms cancel out in numerator, and factor 2 drops off both numerator and denominator afterward, so Eq. (2.353) will appear as merely

$$C_{H^+,w} = \frac{4K_w^*}{2\left(\sqrt{K_a^* C_{HA,0} + 4K_w^*} + \sqrt{K_a^* C_{HA,0}}\right)} = \frac{2K_w^*}{\sqrt{K_a^* C_{HA,0} + 4K_w^*} + \sqrt{K_a^* C_{HA,0}}}. \tag{2.354}$$

Remember that $C_{HA,0}$ is large by hypothesis, so Eq. (2.354) reduces, in turn, to

$$\left(C_{H^+,w} \approx \frac{2K_w^*}{\sqrt{K_a^* C_{HA,0}} + \sqrt{K_a^* C_{HA,0}}} \right) \Bigg|_{C_{HA,0} >> \sqrt{K_w^*}} \tag{2.355}$$

or, equivalently,

$$C_{H^+,w} = \frac{2K_w^*}{2\sqrt{K_a^* C_{HA,0}}} = \frac{K_w^*}{\sqrt{K_a^* C_{HA,0}}} \tag{2.356}$$

after lumping terms in denominator, and then dividing both numerator and denominator by 2. Upon combination with Eq. (2.350) representing $C_{H^+,wa}$ and Eq. (2.356) representing $C_{H^+,w}$, Eq. (2.304) will look like

$$C_{H^+} = \frac{K_w^*}{\sqrt{K_a^* C_{HA,0}}} + \sqrt{K_a^* C_{HA,0}}, \tag{2.357}$$

which degenerates to

$$\left(C_{H^+} \approx \sqrt{K_a^* C_{HA,0}} \right) \Bigg|_{C_{HA,0} >> \sqrt{K_w^*}}, \tag{2.358}$$

due to the too low magnitude of $K_w^*/\sqrt{K_a^* C_{HA,0}}$ compared to that of $\sqrt{K_a^* C_{HA,0}}$. After taking decimal logarithms of both sides, Eq. (2.358) gives rise to

$$\log_{10} C_{H^+} = \log_{10} \sqrt{K_a^* C_{HA,0}} = \frac{1}{2} \log_{10} K_a^* C_{HA,0} = \frac{1}{2} \log_{10} K_a^* + \frac{1}{2} \log_{10} C_{HA,0} \tag{2.359}$$

with the aid of the operational rules pertaining to logarithms, which becomes

$$-\log_{10} C_{H^+} = \frac{1}{2} \left(-\log_{10} K_a^* \right) - \frac{1}{2} \log_{10} C_{HA,0} \tag{2.360}$$

once negatives are taken of both sides; Eq. (2.360) can thus be rephrased as

$$pH \Bigg|_{\substack{w+wa \\ C_{HA,0} >> \sqrt{K_w^*}}} = \frac{1}{2} pK_a^* - \frac{1}{2} \log_{10} C_{HA,0} \tag{2.361}$$

at the expense of Eqs. (2.274) and (2.282) for definition of pH and pK_a^*, respectively. Equation (2.361) justifies $-1/2$ for slope of every high-$C_{HA,0}$ asymptote in Fig. 2.12, unlike

-1 in Fig. 2.10 encompassing an aqueous solution of a strong acid. The representative curve is translated upward when pK_a^* increases, via its effect upon the vertical intercept, that is, $pK_a^*/2$; which in turn serves as ordinate when abscissa is nil, that is, $\log_{10} C_{HA,0} = 0$ or $C_{HA,0} = 10^0 = 1$. A smooth transition from the strong acid-type of behavior, prevailing in the vicinity of $C_{HA,0} = 10^{-7}$, to the weak-acid characteristic pattern occurs at a higher and higher $C_{HA,0}$ as $pK_a^* \to -\infty$ —as expected, since the very large K_a^* describing a strong acid is equivalent to a very negative value for pK_a^*.

Problem 2.8

Devise an alternative strategy to analytically solve Eqs. (2.306) and (2.310) for $C_{H^+,w}$ and $C_{H^+,wa}$ and discuss the result vis-à-vis with Fig. 2.12.

Solution

Remember that Eqs. (2.306) and (2.310) yield $K_a^* \equiv K_a^*\{C_{H^+,w}, C_{H^+,wa}\}$ and $K_w^* \equiv K_w^*\{C_{H^+,w}, C_{H^+,wa}\}$, respectively, as implicit functions of $C_{H^+,wa}$ or $C_{H^+,w}$ (or vice versa), with the latter meanwhile manipulated to produce an explicit relationship of the type $C_{H^+,wa} \equiv C_{H^+,wa}\{C_{H^+,w}\}$, labeled as Eq. (2.311). One will instead reformulate Eq. (2.310) as

$$C_{H^+,w}{}^2 + C_{H^+,wa}C_{H^+,w} - K_w^* = 0 \tag{2.362}$$

as previously done with Eq. (2.335), which supports

$$C_{H^+,w} = \frac{-C_{H^+,wa} \pm \sqrt{C_{H^+,wa}{}^2 + 4K_w^*}}{2} \tag{2.363}$$

after applying the classical solving formula for a quadratic equation; in view of $C_{H^+,wa} \geq 0$ being a requirement for physical feasibility, Eq. (2.363) should be replaced by just

$$C_{H^+,w} = \frac{\sqrt{C_{H^+,wa}{}^2 + 4K_w^*} - C_{H^+,wa}}{2} \tag{2.364}$$

since $\sqrt{C_{H^+,wa}{}^2 + 4K_w^*} > \sqrt{C_{H^+,wa}{}^2} = C_{H^+,wa}$ and in agreement with Eq. (2.336), or, equivalently,

(continued)

Problem 2.8 (continued)

$$C_{H^+,w} = \frac{\sqrt{C_{H^+,wa}{}^2 + 4K_w^*} - \sqrt{C_{H^+,wa}{}^2}}{2}. \tag{2.365}$$

Application of the concept of (first-order) differential to Eq. (2.365) supports

$$C_{H^+,w} \approx \frac{\left.\frac{d}{dC_{H^+,wa}{}^2}\left(\sqrt{C_{H^+,wa}{}^2 + 4K_w^*}\right)\right|_{C_{H^+,wa}{}^2}\left(\left(C_{H^+,wa}{}^2 + 4K_w^*\right) - C_{H^+,wa}{}^2\right)}{2}, \tag{2.366}$$

in view of $(f\{x+\delta\} - f\{x\})/((x+\delta) - x) \approx (df/dx)|_x$ being applicable to a generic function f of a generic variable x, along with a small increment δ of x; Eq. (2.366) gives rise to

$$C_{H^+,w} = \frac{\left.\frac{1}{2\sqrt{C_{H^+,wa}{}^2 + 4K_w^*}}\right|_{C_{H^+,wa}{}^2}\left(C_{H^+,wa}{}^2 + 4K_w^* - C_{H^+,wa}{}^2\right)}{2} \tag{2.367}$$

as per the rule of differentiation of a square root. Simplification of the content of the parenthesis, complemented by replacement of $C_{H^+,wa}{}^2 + 4K_w$ by $C_{H^+,wa}{}^2$ as indicated, justifies simplification of Eq. (2.367) to

$$C_{H^+,w} = \frac{\frac{1}{2\sqrt{C_{H^+,wa}{}^2}}4K_w^*}{2}, \tag{2.368}$$

where composition with the square root cancels the square power as

$$C_{H^+,w} = \frac{\frac{4K_w^*}{2C_{H^+,wa}}}{2} = \frac{4K_w^*}{4C_{H^+,wa}} = \frac{K_w^*}{C_{H^+,wa}}, \tag{2.369}$$

along with dropping of 4 in numerator with 2 (twice) in denominator. Insertion of Eq. (2.369) in Eq. (2.306) is now in order, according to

$$K_a^* = \frac{\left(\frac{K_w^*}{C_{H^+,wa}} + C_{H^+,wa}\right)C_{H^+,wa}}{C_{HA,0} - C_{H^+,wa}}, \tag{2.370}$$

which prompts elimination of denominators as

(continued)

Problem 2.8 (continued)

$$K_a^*\left(C_{HA,0} - C_{H^+,wa}\right) = C_{H^+,wa}\left(\frac{K_w^*}{C_{H^+,wa}} + C_{H^+,wa}\right) \tag{2.371}$$

and where parenthesis can be removed to leave

$$K_a^* C_{HA,0} - K_a^* C_{H^+,wa} = K_w^* + C_{H^+,wa}{}^2; \tag{2.372}$$

in canonical form, Eq. (2.372) looks like

$$C_{H^+,wa}{}^2 + K_a^* C_{H^+,wa} + K_w^* - K_a^* C_{HA,0} = 0. \tag{2.373}$$

The solving formula for this quadratic equation on $C_{H^+,wa}$ may again be invoked, namely,

$$C_{H^+,wa} = \frac{-K_a^* \pm \sqrt{K_a^{*2} - 4\left(K_w^* - K_a^* C_{HA,0}\right)}}{2} = \frac{-K_a^* \pm \sqrt{K_a^{*2} + 4\left(K_a^* C_{HA,0} - K_w^*\right)}}{2}; \tag{2.374}$$

for the cases of practical interest here, $K_a^* C_{HA,0} > K_w^*$, which implies $K_a^* C_{HA,0} - K_w^* > 0$ and, consequently, $\sqrt{K_a^{*2} + 4\left(K_a^* C_{HA,0} - K_w^*\right)} > \sqrt{K_a^{*2}} = K_a^*$. Therefore, the only physically realizable form of Eq. (2.374) looks like

$$C_{H^+,wa} = \frac{\sqrt{K_a^{*2} + 4\left(K_a^* C_{HA,0} - K_w^*\right)} - K_a^*}{2} = \sqrt{\left(\frac{K_a^*}{2}\right)^2 + K_a^* C_{HA,0} - K_w^*} - \frac{K_a^*}{2}, \tag{2.375}$$

where the fraction dash was meanwhile removed. The total concentration of H^+ will accordingly read

$$C_{H^+} = \frac{K_w^*}{C_{H^+,wa}} + C_{H^+,wa}, \tag{2.376}$$

based on Eq. (2.304) upon combination with Eq. (2.369); one then attains

(continued)

Problem 2.8 (continued)

$$C_{H^+} = \frac{K_w^*}{\sqrt{\left(\frac{K_a^*}{2}\right)^2 + K_a^* C_{HA,0} - K_w^*} - \frac{K_a^*}{2}} + \sqrt{\left(\frac{K_a^*}{2}\right)^2 + K_a^* C_{HA,0} - K_w^*} - \frac{K_a^*}{2}, \tag{2.377}$$

following insertion of Eq. (2.375), where lumping terms in the right-hand side generates

$$C_{H^+} = \frac{K_w^* + \left(\sqrt{\left(\frac{K_a^*}{2}\right)^2 + K_a^* C_{HA,0} - K_w^*} - \frac{K_a^*}{2}\right)^2}{\sqrt{\left(\frac{K_a^*}{2}\right)^2 + K_a^* C_{HA,0} - K_w^*} - \frac{K_a^*}{2}}. \tag{2.378}$$

Recalling Newton's binomial theorem, Eq. (2.378) may be redone as

$$C_{H^+} = \frac{K_w^* + \left(\sqrt{\left(\frac{K_a^*}{2}\right)^2 + K_a^* C_{HA,0} - K_w^*}\right)^2 - 2\sqrt{\left(\frac{K_a^*}{2}\right)^2 + K_a^* C_{HA,0} - K_w^*}\,\frac{K_a^*}{2} + \left(\frac{K_a^*}{2}\right)^2}{\sqrt{\left(\frac{K_a^*}{2}\right)^2 + K_a^* C_{HA,0} - K_w^*} - \frac{K_a^*}{2}}, \tag{2.379}$$

which readily simplifies to

$$C_{H^+} = \frac{K_w^* + \left(\frac{K_a^*}{2}\right)^2 + K_a^* C_{HA,0} - K_w^* - K_a^*\sqrt{\left(\frac{K_a^*}{2}\right)^2 + K_a^* C_{HA,0} - K_w^*} + \left(\frac{K_a^*}{2}\right)^2}{\sqrt{\left(\frac{K_a^*}{2}\right)^2 + K_a^* C_{HA,0} - K_w^*} - \frac{K_a^*}{2}}; \tag{2.380}$$

terms alike will then be lumped to give

(continued)

Problem 2.8 (continued)

$$C_{H^+} = \frac{2\frac{K_a^{*2}}{4} + K_a^* C_{HA,0} - K_a^* \sqrt{\left(\frac{K_a^*}{2}\right)^2 + K_a^* C_{HA,0} - K_w^*}}{\sqrt{\left(\frac{K_a^*}{2}\right)^2 + K_a^* C_{HA,0} - K_w^*} - \frac{K_a^*}{2}}, \tag{2.381}$$

which breaks down to

$$C_{H^+} = K_a^* \frac{\frac{K_a^*}{2} + C_{HA,0} - \sqrt{\left(\frac{K_a^*}{2}\right)^2 + K_a^* C_{HA,0} - K_w^*}}{\sqrt{\left(\frac{K_a^*}{2}\right)^2 + K_a^* C_{HA,0} - K_w^*} - \frac{K_a^*}{2}} \tag{2.382}$$

as K_a^* is factored out. When decimal logarithms are taken of both sides, Eq. (2.382) becomes

$$\log_{10} C_{H^+} = \log_{10} K_a^* + \log_{10} \frac{\frac{K_a^*}{2} + C_{HA,0} - \sqrt{\left(\frac{K_a^*}{2}\right)^2 + K_a^* C_{HA,0} - K_w^*}}{\sqrt{\left(\frac{K_a^*}{2}\right)^2 + K_a^* C_{HA,0} - K_w^*} - \frac{K_a^*}{2}}, \tag{2.383}$$

which can be condensed to

$$pH = pK_a^* + \log_{10} \frac{\sqrt{\left(\frac{K_a^*}{2}\right)^2 + K_a^* C_{HA,0} - K_w^*} - \frac{K_a^*}{2}}{\frac{K_a^*}{2} + C_{HA,0} - \sqrt{\left(\frac{K_a^*}{2}\right)^2 + K_a^* C_{HA,0} - K_w^*}} \tag{2.384}$$

with the aid of Eqs. (2.274) and (2.282), after having taken negatives of both sides and then replacing the negative of a logarithm by the logarithm of the reciprocal of its argument; a graphical interpretation of Eq. (2.384) is provided Fig. 2.14. The analytical solution conveyed by Eq. (2.384) is indeed quite accurate when the initial concentration of (weak) acid lies above 10^{-6} mol L^{-1}, as concluded from comparison of Fig. 2.14 with Fig. 2.12; below that threshold, the differential approximation $\left(\sqrt{C_{H^+,wa}{}^2 + 4K_w^*} - C_{H^+,wa}\right)/2 \approx K_w^*/C_{H^+,wa}$ fails, see Eq. (2.369) vis-à-vis with Eq. (2.365).

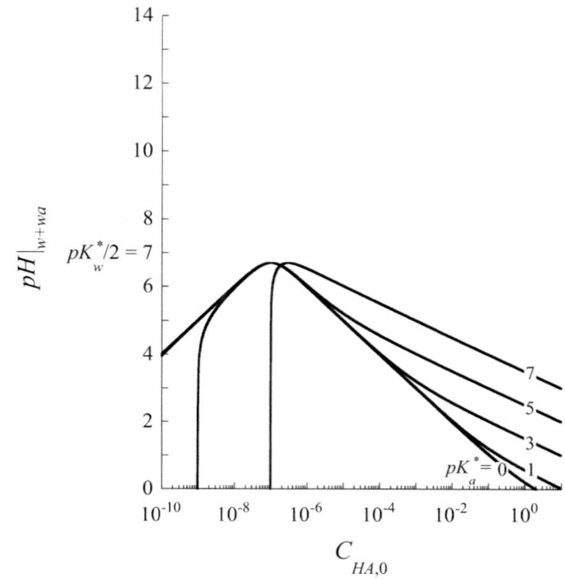

Fig. 2.14 Variation of pH of solution of weak acid, $pH|_{w+wa}$, of selected protolysis constant, pK_a^*, in water, characterized by autoprotolysis constant pK_w^*, versus initial molar concentration of acid, $C_{HA,0}$, using the differential linear approximation

2.3.3 Weak Acid and Conjugate Base

Consider finally an aqueous solution containing a weak acid, HA, with K_a^* denoting its protolysis constant, at initial concentration $C_{HA,0}$, as well as a completely soluble salt, MA, of its conjugate base A^-, at initial concentration $C_{M,0}$. The process of dissolution typically satisfies stoichiometric equation

$$MA \rightarrow M^+ + A^-, \tag{2.385}$$

for which explicitation of an equilibrium constant proves redundant because the fully dissociated salt, as solvated M^+ (or cationic) species and A^- (or anionic) species, materializes its stable form in solution. The concentrations of other germane species also in solution abide to

$$C_{HA} + C_{A^-} = C_{HA,0} + C_{MA,0} \tag{2.386}$$

as per a mass balance to A-containing moieties, which may appear as either HA (of total concentration C_{HA}) or A (of total concentration C_{A^-}, resulting from partial dissociation of HA and full dissociation of MA); coupled with

$$C_{M^+} = C_{M^+,st} = C_{MA,0}, \tag{2.387}$$

with subscript $_{st}$ referring to salt, because cations M^+, at concentration C_{M^+}, are (normally) inert from a chemical point of view, and ionization of MA is quantitative (by hypothesis). Since electroneutrality in solution must be satisfied,

$$C_{A^-} + C_{OH^-} = C_{H^+} + C_{M^+} \tag{2.388}$$

holds true; this gives rise to

$$C_{A^-} + C_{OH^-} = C_{H^+} + C_{MA,0}, \tag{2.389}$$

on account of Eq. (2.387). On the other hand, two chemical equilibria control the concentrations of species in solution, described by

$$K_a^* = \frac{C_{H^+} C_{A^-}}{C_{HA}} \tag{2.390}$$

that reflects Eq. (2.216) pertaining to protolysis of HA, and

$$K_w^* = C_{H^+} C_{OH^-} \tag{2.391}$$

that echoes Eq. (2.218) encompassing autoprotolysis of water; the system formed by four independent equations, that is, Eqs. (2.386) and (2.389)–(2.391), accordingly suffices to calculate its four unknowns, that is, C_{HA}, C_{A^-}, C_{H^+}, and C_{OH^-}.

One should to advantage start by solving Eq. (2.391) for C_{OH^-}, namely,

$$C_{OH^-} = \frac{K_w^*}{C_{H^+}} \tag{2.392}$$

that conveys $C_{OH^-} \equiv C_{OH^-}\{C_{H^+}\}$, and likewise Eq. (2.389) with regard to C_{A^-} as

$$C_{A^-} = C_{MA,0} + C_{H^+} - C_{OH^-}; \tag{2.393}$$

insertion of Eq. (2.392) then generates

$$C_{A^-} = C_{MA,0} + C_{H^+} - \frac{K_w^*}{C_{H^+}} \tag{2.394}$$

from Eq. (2.393), with $C_{A^-} \equiv C_{A^-}\{C_{H^+}\}$. Solution of Eq. (2.386) for C_{HA} is now in order, according to

$$C_{HA} = C_{HA,0} + C_{MA,0} - C_{A^-}, \tag{2.395}$$

where combination with Eq. (2.394) yields

$$C_{HA} = C_{HA,0} + C_{MA,0} - C_{MA,0} - C_{H^+} + \frac{K_w^*}{C_{H^+}}; \tag{2.396}$$

after cancelling out $C_{MA,0}$ with its negative, Eq. (2.396) leads to

$$C_{HA} = C_{HA,0} - C_{H^+} + \frac{K_w^*}{C_{H^+}}, \tag{2.397}$$

again in the form $C_{HA} \equiv C_{HA}\{C_{H^+}\}$. Equation (2.390) eventually gives rise to

$$K_a^* = \frac{C_{H^+}\left(C_{MA,0} + C_{H^+} - \frac{K_w^*}{C_{H^+}}\right)}{C_{HA,0} - C_{H^+} + \frac{K_w^*}{C_{H^+}}} \tag{2.398}$$

at the expense of Eqs. (2.394) and (2.397), which conveys an implicit relationship on C_{H^+}; after getting rid of denominators and parenthesis, Eq. (2.398) becomes

$$K_a^* C_{HA,0} - K_a^* C_{H^+} + \frac{K_a^* K_w^*}{C_{H^+}} = C_{MA,0} C_{H^+} + C_{H^+}{}^2 - K_w^*, \tag{2.399}$$

where multiplication of both sides by C_{H^+} produces

$$K_a^* C_{HA,0} C_{H^+} - K_a^* C_{H^+}{}^2 + K_a^* K_w^* = C_{MA,0} C_{H^+}{}^2 + C_{H^+}{}^3 - K_w^* C_{H^+} \tag{2.400}$$

or, after lumping terms alike,

$$C_{H^+}{}^3 + \left(K_a^* + C_{MA,0}\right)C_{H^+}{}^2 - \left(K_w^* + K_a^* C_{HA,0}\right)C_{H^+} - K_a^* K_w^* = 0. \tag{2.401}$$

Once C_{H^+} has been obtained, via numerical solution of Eq. (2.401), it should be redone in logarithmic form with the aid of Eq. (2.274); the results are depicted in Fig. 2.15. As expected, the curve labeled as $C_{MA,0} = 0$ coincides with the curve in Fig. 2.12 associated with the same value of pK_a^*, since Eq. (2.318) is a particular case of Eq. (2.401). In fact, Eq. (2.401) degenerates to

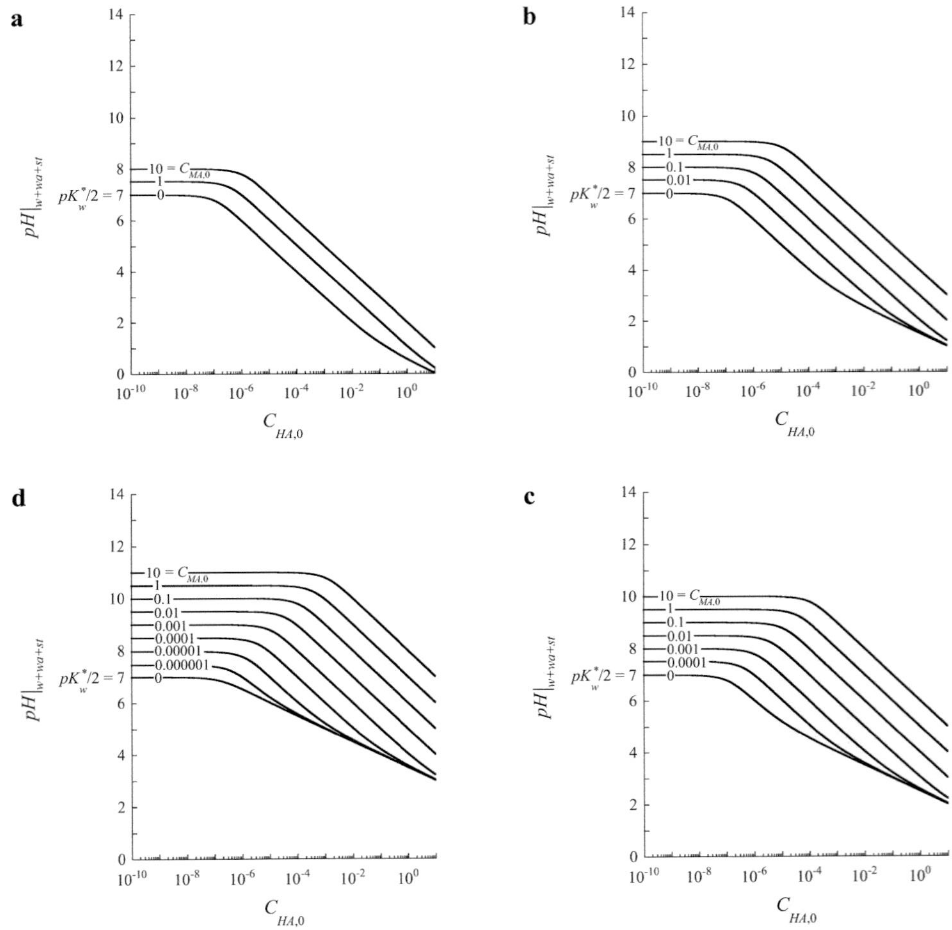

Fig. 2.15 Variation of pH of solution of weak acid and its conjugate base in soluble salt form, $pH|_{w+wa+st}$, characterized by protolysis constant (**a**) $pK_a^* = 1$, (**b**) $pK_a^* = 3$, (**c**) $pK_a^* = 5$, and (**d**) $pK_a^* = 7$, in water, characterized by autoprotolysis constant $pK_w^* = 14$, versus initial molar concentration of acid, $C_{HA,0}$, and selected initial concentrations of salt, $C_{MA,0}$

$$\left(C_{H^+}{}^3 + K_a^* C_{H^+}{}^2 - \left(K_w^* + K_a^* C_{HA,0} \right) C_{H^+} - K_a^* K_w^* = 0 \right) \Big|_{C_{MA,0}=0}, \tag{2.402}$$

where C_{H^+} ought to be replaced by a suitable function of $C_{H^+,w}$ for comparability. After realizing that

$$C_{H^+,w} = C_{OH^-,w} = C_{OH^-}, \tag{2.403}$$

due to the stoichiometry underlying Eq. (2.219) and to water being the sole source of OH^- ions, one can revisit Eq. (2.392) as

$$C_{H^+,w} = \frac{K_w^*}{C_{H^+}}; \tag{2.404}$$

this is equivalent to writing

$$C_{H^+} = \frac{K_w^*}{C_{H^+,w}}, \tag{2.405}$$

upon exchange of C_{H^+} and $C_{H^+,w}$. Insertion of Eq. (2.405) converts Eq. (2.402) to

$$\left(\frac{K_w^*}{C_{H^+,w}} \right)^3 + K_a^* \left(\frac{K_w^*}{C_{H^+,w}} \right)^2 - \left(K_w^* + K_a^* C_{HA,0} \right) \frac{K_w^*}{C_{H^+,w}} - K_a^* K_w^* = 0, \tag{2.406}$$

which prompts multiplication of both sides by $C_{H^+,w}^3$ as

$$K_w^{*3} + K_a^* K_w^{*2} C_{H^+,w} - K_w^* \left(K_w^* + K_a^* C_{HA,0} \right) C_{H^+,w}{}^2 - K_a^* K_w^* C_{H^+,w}{}^3 = 0; \tag{2.407}$$

algebraic rearrangement unfolds

$$K_a^* K_w^* C_{H^+,w}{}^3 + K_w^* \left(K_w^* + K_a^* C_{HA,0} \right) C_{H^+,w}{}^2 - K_a^* K_w^{*2} C_{H^+,w} - K_w^{*3} = 0, \tag{2.408}$$

where division of both sides by K_w^* leads finally to

$$\left(K_a^* C_{H^+,w}{}^3 + \left(K_w^* + K_a^* C_{HA,0} \right) C_{H^+,w}{}^2 - K_a^* K_w^* C_{H^+,w} - K_w^{*2} = 0 \right) \Big|_{C_{MA,0}=0} \tag{2.409}$$

that coincides with Eq. (2.318)—as expected.

The pH at very low $C_{HA,0}$ varies substantially with $C_{MA,0}$ because the A^- released via full dissociation of salt will act as a base, and thus abstract H^+ from solvent; this is why pH increases consistently with $C_{MA,0}$. It is remarkable, however, that the initial pH remains stable within a wider and wider range of values for $C_{HA,0}$ when $C_{MA,0}$ increases. Under these circumstances, Eq. (2.400) will be retrieved as

$$\left(K_a^* + C_{MA,0}\right) C_{H^+}{}^2 - K_a^* K_w^* = 0, \tag{2.410}$$

where only the square and independent terms were retained. This approximation is reasonable when C_{H^+} is very low compared to $\sqrt[3]{K_a^* K_w^*}$, thus implying $C_{H^+}{}^3 - K_a^* K_w^* \approx -K_a^* K_w^*$, together with $C_{HA,0}$ being very low as well, which implies $K_w^* + K_a^* C_{HA,0} \approx K_w^*$, further to $C_{H^+,w} << K_a^*$ supporting $-K_w^* C_{H^+} + K_w^* K_a^* \approx K_w^* K_a^*$. Isolation of $C_{H^+}{}^2$ from Eq. (2.410) prompts

$$C_{H^+}{}^2 = \frac{K_a^* K_w^*}{K_a^* + C_{MA,0}}, \tag{2.411}$$

which simplifies to

$$C_{H^+}{}^2 = \frac{K_a^* K_w^*}{C_{MA,0}} \tag{2.412}$$

for $K_a^* << C_{MA,0}$ as case of practical interest; after taking logarithms of both sides, Eq. (2.412) yields

$$2\log_{10} C_{H^+} = \log_{10} K_a^* + \log_{10} K_w^* - \log_{10} C_{MA,0} \tag{2.413}$$

as per the operational rules of logarithms, where negatives taken of both sides leave, in turn,

$$2(-\log_{10} C_{H^+}) = -\log_{10} K_a^* - \log_{10} K_w^* + \log_{10} C_{MA,0}. \tag{2.414}$$

Recalling Eqs. (2.274) and (2.282), one may redo Eq. (2.414) as

$$2pH = pK_a^* + pK_w^* + \log_{10} C_{MA,0}, \tag{2.415}$$

or else

$$pH\Big|_{\substack{w+wa+st\\ C_{HA,0}<<\sqrt{K_w^*}}} = \frac{1}{2}\left(pK_a^* + pK_w^* + \log_{10}C_{MA,0}\right) \tag{2.416}$$

upon division of both sides by 2; Eq. (2.416) describes the vertical intercept of the horizontal asymptotes that drive the curves within the low-$C_{HA,0}$ range in Fig. 2.15.

The limiting case of small $C_{MA,0}$ cannot be handled directly from Eq. (2.416) and should instead resort to the simplified version of Eq. (2.400) when $C_{MA,0} \rightarrow 0$, that is,

$$K_a^* C_{H^+}{}^2 - K_a^* K_w^* = 0 \tag{2.417}$$

in lieu of Eq. (2.410); after dividing both sides by K_a^*, one gets

$$C_{H^+}{}^2 = K_w^*, \tag{2.418}$$

or else

$$2(-\log_{10}C_{H^+}) = -\log_{10}K_w^* \tag{2.419}$$

in (negative) logarithmic form, which is equivalent to writing

$$pH\Big|_{\substack{w+wa+st\\ C_{HA,0}<<\sqrt{K_w^*}\\ C_{MA,0}\rightarrow 0}} = \frac{1}{2}pK_w^* \tag{2.420}$$

in view of Eqs. (2.274) and (2.282) once more. Besides describing the vertical intercept of the lowest curves in Fig. 2.15, Eq. (2.420) appears as alias to Eq. (2.334) because the solution is now restricted to only a solubilized weak acid.

Conversely, if $C_{HA,0}$ is sufficiently high, then pH decreases with an increase in $C_{HA,0}$, following essentially similar patterns, despite the specific value of $C_{MA,0}$ (see Fig. 2.15). A mathematical expression for this asymptotic behavior follows again from Eq. (2.400), but now reduced to

$$\left(K_a^* + C_{MA,0}\right)C_{H^+}{}^2 - \left(K_w^* + K_a^* C_{HA,0}\right)C_{H^+} = 0 \tag{2.421}$$

after realizing that ${C_{H^+}}^3$ and $K_a^* K_w^*$ are sufficiently small compared to the outstanding terms, they can be neglected; within the range under scrutiny, $K_a^* \ll C_{MA,0}$ and $K_w^* \ll K_a^* C_{HA,0}$, which justify further simplification of Eq. (2.421) to

$$C_{MA,0} {C_{H^+}}^2 - K_a^* C_{HA,0} C_{H^+} = 0. \tag{2.422}$$

Upon division of both sides by C_{H^+}, Eq. (2.422) becomes

$$C_{MA,0} C_{H^+} - K_a^* C_{HA,0} = 0, \tag{2.423}$$

which can be solved for C_{H^+} as

$$C_{H^+} = K_a^* \frac{C_{HA,0}}{C_{MA,0}}; \tag{2.424}$$

upon application of logarithms to both sides, Eq. (2.424) unfolds

$$\log_{10} C_{H^+} = \log_{10} K_a^* + \log_{10} \frac{C_{HA,0}}{C_{MA,0}}, \tag{2.425}$$

yet negatives will still be taken of both sides to get

$$-\log_{10} C_{H^+} = -\log_{10} K_a^* - \log_{10} \frac{C_{HA,0}}{C_{MA,0}}. \tag{2.426}$$

After bringing Eqs. (2.274) and (2.282) on board, and taking the reciprocal of the argument of the last logarithm at the expense of swapping its sign to positive, one obtains

$$pH \Bigg|_{\substack{w+wa+st \\ C_{HA,0} >> \sqrt{K_w^*}}} = pK_a^* + \log_{10} \frac{C_{MA,0}}{C_{HA,0}} \tag{2.427}$$

from Eq. (2.426), classically known as Henderson and Hasselbach's equation; after rewriting it as

$$pH = pK_a^* + \log_{10} C_{MA,0} - \log_{10} C_{HA,0}, \tag{2.428}$$

a direct descriptor is produced for straight lines bearing $pK_a^* + \log_{10} C_{MA,0}$ as vertical intercept (or ordinate at $C_{HA,0} = 10^0 = 1$) and -1 for slope in a bilogarithmic plot, as observed in the region of large $C_{HA,0,0}$ in Fig. 2.15.

Once again, the asymptotic case of $C_{MA,\,0} \rightarrow 0$ cannot be derived from Eq. (2.427), but one should instead depart from

$$C_{H^+}{}^3 - \left(K_w^* + K_a^* C_{HA,0}\right) C_{H^+} = 0 \tag{2.429}$$

in lieu of Eq. (2.417), holding exactly the terms in the original Eq. (2.401) that were discarded in Eq. (2.417); after realizing that $K_w^* + K_a^* C_{HA,0} \approx K_a^* C_{HA,0}$ in the region of interest, Eq. (2.429) reduces to

$$C_{H^+}{}^3 = K_a^* C_{HA,0} C_{H^+}, \tag{2.430}$$

which prompts division of both sides by C_{H^+} as

$$C_{H^+}{}^2 = K_a^* C_{HA,0}. \tag{2.431}$$

Application of logarithms to both sides, followed by multiplication thereof by -1 transform Eq. (2.431) to

$$2(-\log_{10} C_{H^+}) = -\log_{10} K_a^* - \log_{10} C_{HA,0}, \tag{2.432}$$

where a more condensed notation is possible, namely,

$$2pH = pK_a^* - \log_{10} C_{HA,0} \tag{2.433}$$

as per Eqs. (2.274) and (2.282); a final division of both sides by 2 leaves

$$pH\Bigg|_{\substack{w+wa+st \\ C_{HA,0} >> \sqrt{K_w^*} \\ C_{MA,0} \rightarrow 0}} = \frac{1}{2}\left(pK_a^* - \log_{10} C_{HA,0}\right), \tag{2.434}$$

which retrieves Eq. (2.361) as expected.

The justification for the graphical patterns observed in Fig. 2.15 lies on the distinct concentrations of the various chemical species at stake in response to actual values of $C_{HA,0}$ or $C_{MA,0}$; after recalling C_{OH^-}, C_{A^-}, and C_{HA} from Eqs. (2.392), (2.394), and (2.397), respectively, together with C_{H^+} obtained upon solution of Eq. (2.401), it is possible to generate the curves plotted in Fig. 2.16. The curve representing C_{H^+} logically coincides with the corresponding curve in Fig. 2.15, after horizontal flipping around the horizontal

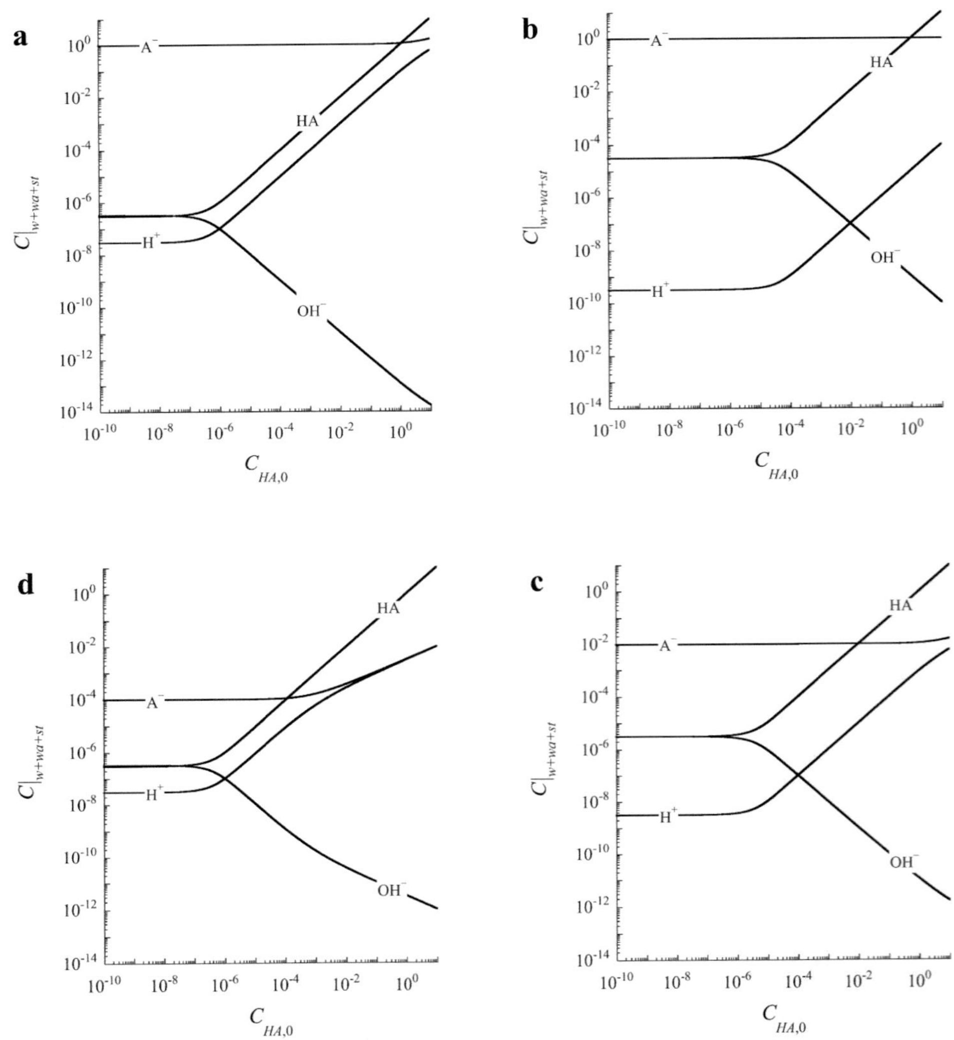

Fig. 2.16 Variation of molar concentrations in aqueous solution of weak acid and conjugate base in the form of soluble salt, $C|_{w+wa+st}$, characterized by protolysis constant (**a**) $pK_a^* = 1$ and (**b–d**) $pK_a^* = 5$, of undissociated acid, HA, conjugate base, A^-, and hydrogen ion, H^+, as well as hydroxide ion, OH^-, versus initial molar concentration of acid, $C_{HA,0}$, for selected values of initial molar concentration of base (**a, b**) $C_{MA,0} = 1$, (**c**) $C_{MA,0} = 0.01$, and (**d**) $C_{MA,0} = 0.0001$

line with 10^{-7} for vertical intercept as implied by Eq. (2.274); the concentration of A^- is quite stable at $C_{MA,0}$ for low initial concentrations of HA, but it increases with $C_{HA,0}$ especially when $C_{MA,0}$ is low and $C_{HA,0}$ is high. The curve corresponding to C_{OH^-} is constrained by its C_{H^+} counterpart, so that their product remains constant at $K_w^* = 10^{-14}$ M^2 as per Eq. (2.218); and C_{HA} also increases with $C_{HA,0}$ at higher concentrations of the latter, but is fixed by pK_a^*, pK_w^*, and $C_{MA,0}$ in the lower range of $C_{HA,0}$—see Eq. (2.416) pertaining to C_{H^+}, yet constraining C_{HA} through Eq. (2.390).

Problem 2.9

Consider again the stoichiometry underlying chemical equations labeled as Eqs. (2.219) and (2.301), carried out at room temperature in the presence of fully dissociated (conjugate base) salt, see Eq. (2.385).

(a) Obtain expressions for the extent of either reaction, as univariate functions of $C_{H+} \equiv C_{H+}\{C_{HA,0}, C_{MA,0}, K_a^*, K_w^*\}$; plot them versus $C_{HA,0}$, for selected values of $C_{MA,0}$ and $pK_a^* = 5$, and discuss the results obtained.

(b) Derive expressions for the concentration of all species in solution according to source, as a function of the reaction extents referred to in *a*) pertaining to the two (reversible) protolysis reactions—besides K_w^*, $C_{HA,0}$, and $C_{MA,0}$.

Solution

(a) The definition of reaction extent, χ_j, has it that

$$\chi_j \equiv \frac{\Delta C_{i,j}}{\nu_{i,j}}, \tag{2.435}$$

with $\Delta C_{i,j}$ denoting change in molar concentration of i-th compound brought about by j-th reaction, and $\nu_{i,j}$ denotes algebraic stoichiometric coefficient of i-th compound in j-th reaction. Should χ_1 and χ_2 denote extent of reactions described by Eqs. (2.219) and (2.301), respectively, then

$$\Delta C_{H^+} = \chi_1 + \chi_2 \tag{2.436}$$

will hold regarding hydrogen ion concentration; by the same token,

$$\Delta C_{OH^-} = \chi_1, \tag{2.437}$$

$$\Delta C_{HA} = -\chi_2, \tag{2.438}$$

and

(continued)

Problem 2.9 (continued)

$$\Delta C_{A^-} = \chi_2 \tag{2.439}$$

will be satisfied by hydroxide ion, undissociated (weak) acid, and conjugated base/dissociated acid cation, respectively, on the assumption that the salt completely and irreversibly dissociates from the very beginning. However, not all four equations are independent of each other, and thus able to carry significant information on χ_1 and χ_2. In fact, one obtains

$$\Delta C_{H^+} - \Delta C_{A^-} = \chi_1 \tag{2.440}$$

upon ordered subtraction of Eq. (2.439) from Eq. (2.436), which may be rewritten as

$$\Delta C_{H^+} - \Delta C_{A^-} = \Delta C_{OH^-} \tag{2.441}$$

upon combination with Eq. (2.437); and

$$\Delta C_{HA} + \Delta C_{A^-} = 0 \tag{2.442}$$

similarly emerges upon ordered addition of Eqs. (2.438) and (2.439). On the other hand, application of the finite variation-operator Δ to Eq. (2.389) unfolds

$$\Delta C_{A^-} + \Delta C_{OH^-} = \Delta(C_{A^-} + C_{OH^-}) = \Delta(C_{H^+} + C_{MA,0}) = \Delta C_{H^+}, \tag{2.443}$$

which coincides with Eq. (2.441) because $C_{MA,0}$ is a constant; and similar application of Δ to Eq. (2.386) yields

$$\Delta C_{HA} + \Delta C_{A^-} = \Delta(C_{HA} + C_{A^-}) = \Delta(C_{HA,0} + C_{MA,0}) = 0, \tag{2.444}$$

which coincides with Eq. (2.442) since both $C_{HA,0}$ and $C_{MA,0}$ are constant. After realizing that χ_1 and χ_2 do not appear explicitly in Eqs. (2.441) and (2.442), one concludes that the original Eqs. (2.435) and (2.437) are not useful for the purpose here; and the same applies to Eqs. (2.386) and (2.389), for being equivalent thereto. One is therefore left with only Eqs. (2.437) and (2.439), which immediately convey χ_1 and χ_2, respectively; and which indirectly correspond to Eqs. (2.390) and (2.391), and thus convey the

(continued)

Problem 2.9 (continued)

underlying (and relevant) thermodynamic constraints. In view of Eq. (2.264) valid at large,

$$\Delta C_{OH^-} \equiv C_{OH^-} - C_{OH^-,0} = C_{OH^-} - C_{H^+,0} \tag{2.445}$$

applies, knowing that

$$C_{H^+,0} C_{OH^-,0} = C_{H^+,0} C_{H^+,0} = C_{H^+,0}{}^2 = K_w^* \tag{2.446}$$

holds at starting conditions, referred to by subscript $_0$ and entailing pure water, in full agreement with Eqs. (2.219) and (2.261); hence,

$$C_{H^+,0} = \sqrt{K_w^*} \tag{2.447}$$

from Eq. (2.446), which will be inserted in Eq. (2.445) to get

$$\Delta C_{OH^-} = C_{OH^-} - \sqrt{K_w^*} \tag{2.448}$$

or, after revisiting Eq. (2.437),

$$\chi_1 = C_{OH^-} - \sqrt{K_w^*}. \tag{2.449}$$

Combination with Eq. (2.392), as alias to Eq. (2.391), finally converts Eq. (2.449) to

$$\chi_1\{C_{H^+}\} = \frac{K_w^*}{C_{H^+}} - \sqrt{K_w^*}, \tag{2.450}$$

thus unfolding $\chi_1\{C_{H^+}; K_w^*\}$, with $C_{H^+}\{C_{HA,0}, C_{MA,0}, K_a^*, K_w^*\}$ as per Eq. (2.401). On the other hand,

$$C_{HA} = C_{HA,0} + C_{MA,0} - C_{A^-}, \tag{2.451}$$

stemming from Eq. (2.386), can be used to obtain

(continued)

Problem 2.9 (continued)

$$K_a^* = \frac{C_{H^+}C_{A^-}}{C_{HA,0} + C_{MA,0} - C_{A^-}} \tag{2.452}$$

from Eq. (2.390); elimination of denominators unfolds

$$K_a^*(C_{HA,0} + C_{MA,0}) - K_a^* C_{A^-} = C_{H^+}C_{A^-}, \tag{2.453}$$

where lumping of common terms leads to

$$K_a^*(C_{HA,0} + C_{MA,0}) = \left(K_a^* + C_{H^+}\right)C_{A^-}, \tag{2.454}$$

and consequently

$$C_{A^-} = \frac{K_a^*(C_{HA,0} + C_{MA,0})}{K_a^* + C_{H^+}} \tag{2.455}$$

upon isolation of C_{A^-}. Recalling the definition of ΔC_{A^-}, that is,

$$\Delta C_{A^-} \equiv C_{A^-} - C_{MA,0}, \tag{2.456}$$

one readily obtains

$$\Delta C_{A^-} = \frac{K_a^*(C_{HA,0} + C_{MA,0})}{K_a^* + C_{H^+}} - C_{MA,0} \tag{2.457}$$

with the aid of Eq. (2.454); this is equivalent to

$$\chi_2\{C_{H^+}\} = \frac{K_a^*(C_{HA,0} + C_{MA,0})}{K_a^* + C_{H^+}} - C_{MA,0}, \tag{2.458}$$

in view of Eq. (2.439), which gives rise to $\chi_2\left\{C_{H^+}; C_{HA,0}, C_{MA,0}, K_a^*, K_w^*\right\}$, again with $C_{H^+}\left\{C_{HA,0}, C_{MA,0}, K_a^*, K_w^*\right\}$. The variation of χ_1 and χ_2 is plotted in Fig. 2.17, on the left and the right, respectively, as a (continuous) function of $C_{HA,0}$ and a (discrete) function of $C_{MA,0}$, for $K_w^* = 10^{-14}$ M^2 and $K_a^* = 10^{-pK_a^*} = 10^{-5}$ M, following combination with the solution of Eq. (2.401) for C_{H^+}. When the initial concentration of weak acid increases, the extent of autoprotolysis of water and the extent of protolysis of acid remain essentially constant, being positive and negative, respectively, until $pH \approx 6$–7 is reached; a sudden decrease in χ_1 is then observed, concomitant

(continued)

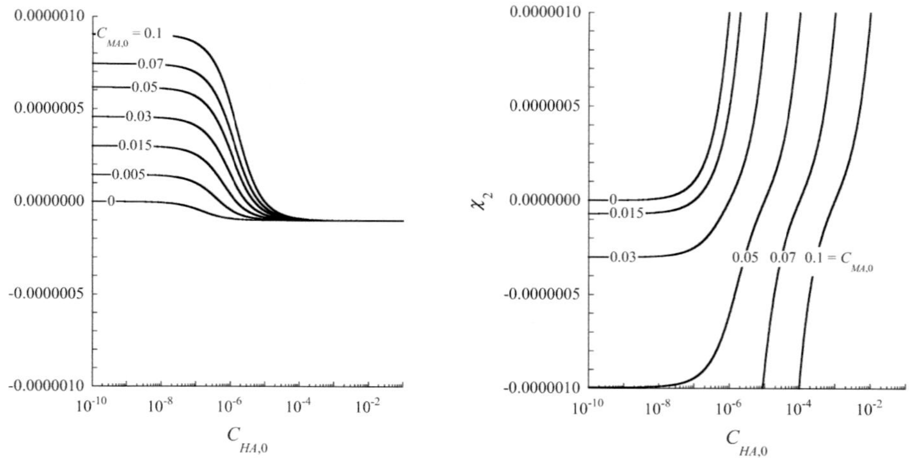

Fig. 2.17 Variation of molar extent of dissociation of water, χ_1 (left), characterized by protolysis constant $pK_w^* = 14$, and of dissociation of weak acid, χ_2 (right), characterized by protolysis constant $pK_a^* = 5$, in water, versus initial molar concentration of acid, $C_{HA,0}$, and selected initial molar concentrations of salt, $C_{MA,0}$

Problem 2.9 (continued)

with a sudden increase in χ_2 as a consequence of H^+ released by acid becoming dominant, although χ_1 rapidly stabilizes while χ_2 goes on growing. This pattern is consistent with Fig. 2.16b–d, since the change in C_{H^+} given by Eq. (2.401) has been taken into account to generate Eqs. (2.450) and (2.458). Presence of conjugate base magnifies the value of $\chi_1 > 0$ and $\chi_2 < 0$ within the low $C_{HA,0}$ range, but is not discriminatory in terms of the former within the high-$C_{HA,0}$ range; whereas a higher $C_{MA,0}$ delays evolution of χ_2 toward positive values, as $C_{HA,0}$ is raised.

(b) There are four relevant species in solution, namely, H^+, OH^-, HA, and A^-, and five relevant concentrations thereof, that is, $C_{H^+,w}$ and $C_{H^+,wa}$, $C_{OH^-,w}$, $C_{HA,wa}$, and $C_{A^-,wa}$, with known $C_{HA,0}$ and $C_{MA,0}$. The stoichiometry of Eq. (2.219) enforces

$$\Delta C_{H^+,w} = \Delta C_{OH^-,w} \tag{2.459}$$

consistent with Eq. (2.264) referring to water; a similar reasoning based on Eq. (2.301) has it that

(continued)

Problem 2.9 (continued)

$$\Delta C_{H^+,wa} = \Delta C_{A^-,wa} = -\Delta C_{HA,wa}, \tag{2.460}$$

consistent with Eqs. (2.302) and (2.303). Water serves as sole source of hydroxide ions, so one may revisit Eq. (2.449) as

$$C_{OH^-,w}\{\chi_1; K_w^*\} = \sqrt{K_w^*} + \chi_1. \tag{2.461}$$

Combination of Eqs. (2.459) and (2.264) unfolds

$$\Delta C_{H^+,w} = \Delta C_{OH^-}, \tag{2.462}$$

while Eq. (2.436) supports conversion to

$$\Delta C_{H^+,w} = \chi_1; \tag{2.463}$$

since the initial concentration of H^+ comes only from water, one may write

$$C_{H^+,w} - C_{H^+,w,0} = C_{H^+,w} - \sqrt{K_w^*} \tag{2.464}$$

as per Eq. (2.447), or else

$$\begin{aligned} C_{H^+,w}\{\chi_1; K_w^*\} &= \sqrt{K_w^*} + C_{H^+,w} - C_{H^+,w,0} \\ &= \sqrt{K_w^*} + \Delta C_{H^+,w} = \sqrt{K_w^*} + \chi_1 \end{aligned} \tag{2.465}$$

also with the aid of Eq. (2.463). Based on realization that

$$\Delta C_{HA} = \Delta C_{HA,wa} \equiv C_{HA,wa} - C_{HA,0}, \tag{2.466}$$

one may retrieve Eq. (2.438) to get

$$-\chi_2 = C_{HA,wa} - \Delta C_{HA,0}; \tag{2.467}$$

Equation (2.467) can be solved for $C_{HA,wa}$ as

$$C_{HA,wa}\{\chi_2; C_{HA,0}\} = C_{HA,0} - \chi_2. \tag{2.468}$$

A balance to A^- supports, in turn,

(continued)

Problem 2.9 (continued)

$$\Delta C_{A^-} = \Delta C_{A^-,wa}, \tag{2.469}$$

where combination with Eq. (2.439) unfolds

$$\chi_2 = \Delta C_{A^-,wa}; \tag{2.470}$$

after realizing that

$$\Delta C_{A^-,wa} \equiv C_{A^-,wa} - C_{MA,0} \tag{2.471}$$

due to the initial quantitative (and irreversible) dissociation of salt, Eq. (2.470) becomes

$$\chi_2 = C_{A^-,wa} - C_{MA,0}, \tag{2.472}$$

and thus

$$C_{A^-,wa}\{\chi_2; C_{MA,0}\} = C_{MA,0} + \chi_2 \tag{2.473}$$

upon isolation of $C_{A^-,wa}$. A balance to hydrogen ions justifies

$$\Delta C_{H^+} = \Delta C_{H^+,w} + \Delta C_{H^+,wa}, \tag{2.474}$$

where insertion of Eqs. (2.436) and (2.463) generates

$$\chi_1 + \chi_2 = \chi_1 + \Delta C_{H^+,wa}; \tag{2.475}$$

after cancelling identical terms between sides, Eq. (2.475) leaves

$$\sqrt{K_w^*} + \Delta C_{H^+,wa} = \chi_2. \tag{2.476}$$

Equation (2.476) may be rephrased as

$$C_{H^+,wa} - C_{H^+,wa,0} = C_{H^+,wa} - 0 = C_{H^+,wa}\{\chi_2\} = \chi_2 \tag{2.477}$$

that conveys the concentration of the outstanding relevant species H^+,*wa*.

2.3.4 Buffer Stability

The effect of initial concentration of acid upon solution pH was ascertained previously—in the case of a strong acid, for which (equilibrium) pH declines rapidly with its overall concentration (see Fig. 2.10); and in the case of a weak acid, for which (equilibrium) pH declines more smoothly, for the same overall concentration (see Fig. 2.12). When its conjugate base is added, pH becomes essentially stable over a wider range of concentrations of overall acid (see Fig. 2.15), but such a pH plateau is raised and widened when the concentration of conjugate base added increases; therefore, it is of the utmost interest to investigate what happens for a given overall concentration of A-containing moieties, say C_{bf}, when only their relative proportion is changed, using

$$C_{bf} \equiv C_{HA,0} + C_{MA,0} \tag{2.478}$$

for definition, and where $C_{HA,0}$ and $C_{MA,0}$ denote initial concentration of acid HA and salt MA, respectively. Under these circumstances, Eq. (2.401) should be revisited as

$$C_{H^+}{}^3 + \left(K_a^* + C_{bf}\left(1 - \frac{C_{HA,0}}{C_{bf}}\right)\right)C_{H^+}{}^2 - \left(K_w^* + K_a^* C_{bf}\frac{C_{HA,0}}{C_{bf}}\right)C_{H^+} - K_a^* K_w^* = 0, \tag{2.479}$$

where $C_{MA,0}$ was replaced by $C_{bf} - C_{HA,0}$ as per Eq. (2.478) and C_{bf} factored out afterward; after taking negatives of the (numerical) solution to Eq. (2.479), and recalling the definition of pH conveyed by Eq. (2.274), one obtains the plot labeled as Fig. 2.18, in the form of a continuous function of $C_{HA,0}/C_{bf}$, for preselected values of K_a^* and C_{bf}. It is remarkable that an inflection point appears in every curve, at exactly the same location in terms of abscissa, that is, 0.5, and with ordinate pH coinciding with the corresponding pK_a^*; in addition, pH is particularly stable in the neighborhood of said point, that is, when $C_{HA,\,0}$ and $C_{MA,\,0}$ are close to each other, for a given sum thereof (see C_{bf} in Eq. (2.478)).

In attempts to mathematically confirm the above feature, one should first recall that

$$\frac{dpH}{d\left(\frac{C_{HA,0}}{C_{bf}}\right)} \equiv \frac{d(-\log_{10}C_{H^+})}{d\left(\frac{C_{HA,0}}{C_{bf}}\right)} = -\frac{d(\log_e C_{H^+}\log_{10}e)}{d\left(\frac{C_{HA,0}}{C_{bf}}\right)} \equiv -\frac{d(\ln C_{H^+}\log_{10}e)}{d\left(\frac{C_{HA,0}}{C_{bf}}\right)} = -\frac{d\ln C_{H^+}}{d\left(\frac{C_{HA,0}}{C_{bf}}\right)}\log_{10}e \tag{2.480}$$

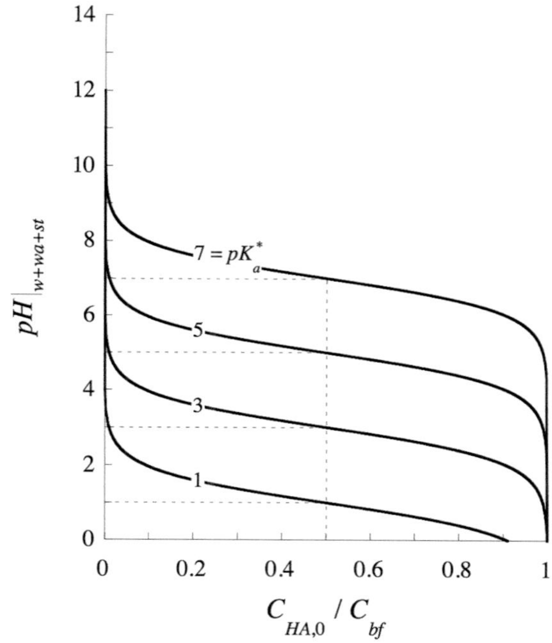

Fig. 2.18 Variation of pH of solution of weak acid and its conjugate base in soluble salt form, $pH|_{w+wa+st}$, characterized by selected values of protolysis constant, pK_a^*, in water, characterized by autoprotolysis constant $pK_w^* = 14$, versus initial concentration of acid, $C_{HA,0}$, normalized by total buffer concentration $C_{bf} = 1$ mol L^{-1}

written with the aid of Eq. (2.274), together with a convenient change of logarithm base and realization that a differential operator is linear; the rules of differentiation of a (natural) logarithm and of a composite function lead then to

$$\frac{dpH}{d\left(\frac{C_{HA,0}}{C_{bf}}\right)} = -\frac{1}{C_{H^+}} \frac{dC_{H^+}}{d\left(\frac{C_{HA,0}}{C_{bf}}\right)} \log_{10}e. \tag{2.481}$$

The second-order derivative of pH with regard to $C_{HA,0}/C_{bf}$ will accordingly look like

$$\begin{aligned} \frac{d^2pH}{d\left(\frac{C_{HA,0}}{C_{bf}}\right)^2} &\equiv \frac{d}{d\left(\frac{C_{HA,0}}{C_{bf}}\right)} \left(\frac{dpH}{d\left(\frac{C_{HA,0}}{C_{bf}}\right)}\right) = \frac{d}{d\left(\frac{C_{HA,0}}{C_{bf}}\right)} \left(-\frac{1}{C_{H^+}} \frac{dC_{H^+}}{d\left(\frac{C_{HA,0}}{C_{bf}}\right)} \log_{10}e\right) \\ &= -\frac{d}{d\left(\frac{C_{HA,0}}{C_{bf}}\right)} \left(\frac{1}{C_{H^+}} \frac{dC_{H^+}}{d\left(\frac{C_{HA,0}}{C_{bf}}\right)}\right) \log_{10}e, \end{aligned} \tag{2.482}$$

based on its definition and upon combination with Eq. (2.481), where constant factors were meanwhile taken off the differential operator; application of the rules of differentiation of a product and a reciprocal is in order, namely,

$$\frac{d^2pH}{d\left(\frac{C_{HA,0}}{C_{bf}}\right)^2} = -\left(-\frac{1}{{C_{H^+}}^2}\frac{dC_{H^+}}{d\left(\frac{C_{HA,0}}{C_{bf}}\right)}\frac{dC_{H^+}}{d\left(\frac{C_{HA,0}}{C_{bf}}\right)} + \frac{1}{C_{H^+}}\frac{d}{d\left(\frac{C_{HA,0}}{C_{bf}}\right)}\left(\frac{dC_{H^+}}{d\left(\frac{C_{HA,0}}{C_{bf}}\right)}\right)\right)\log_{10}e, \tag{2.483}$$

where combination of factors alike and simplification of notation leave

$$\frac{d^2pH}{d\left(\frac{C_{HA,0}}{C_{bf}}\right)^2} = \left(\frac{1}{{C_{H^+}}^2}\left(\frac{dC_{H^+}}{d\left(\frac{C_{HA,0}}{C_{bf}}\right)}\right)^2 - \frac{1}{C_{H^+}}\frac{d^2C_{H^+}}{d\left(\frac{C_{HA,0}}{C_{bf}}\right)^2}\right)\log_{10}e. \tag{2.484}$$

Since Eq. (2.479) does not convey $C_{H^+} \equiv C_{H^+}\{C_{HA,0}/C_{bf}\}$ explicitly, one should start by rewriting it as

$$\Phi\left\{C_{H^+}, \frac{C_{HA,0}}{C_{bf}}\right\} = 0, \tag{2.485}$$

provided that

$$\Phi\left\{C_{H^+}, \frac{C_{HA,0}}{C_{bf}}\right\} \equiv {C_{H^+}}^3 + \left(K_a^* + C_{bf}\left(1 - \frac{C_{HA,0}}{C_{bf}}\right)\right){C_{H^+}}^2 - \left(K_w^* + K_a^* C_{bf}\frac{C_{HA,0}}{C_{bf}}\right)C_{H^+} - K_a^* K_w^*; \tag{2.486}$$

the first-order derivative of C_{H^+} with regard to $C_{HA,0}/C_{bf}$ will then become accessible through

$$\frac{dC_{H^+}}{d\left(\frac{C_{HA,0}}{C_{bf}}\right)} = -\frac{\frac{\partial \Phi}{\partial\left(\frac{C_{HA,0}}{C_{bf}}\right)}}{\frac{\partial \Phi}{\partial C_{H^+}}}, \tag{2.487}$$

at the expense of implicit differentiation. Recalling Eq. (2.486), one obtains

$$\frac{dC_{H^+}}{d\left(\frac{C_{HA,0}}{C_{bf}}\right)} = -\frac{-C_{bf}C_{H^+}{}^2 - K_a^* C_{bf} C_{H^+}}{3C_{H^+}{}^2 + 2\left(K_a^* + C_{bf}\left(1 - \frac{C_{HA,0}}{C_{bf}}\right)\right)C_{H^+} - \left(K_w^* + K_a^* C_{bf}\frac{C_{HA,0}}{C_{bf}}\right)} \tag{2.488}$$

from Eq. (2.487); negative signs in numerator will then be collapsed, and $C_{bf}C_{H^+}$ factored out to get

$$\frac{dC_{H^+}}{d\left(\frac{C_{HA,0}}{C_{bf}}\right)} = \frac{C_{bf}C_{H^+}\left(K_a^* + C_{H^+}\right)}{3C_{H^+}{}^2 + 2\left(K_a^* + C_{bf}\left(1 - \frac{C_{HA,0}}{C_{bf}}\right)\right)C_{H^+} - \left(K_w^* + K_a^* C_{bf}\frac{C_{HA,0}}{C_{bf}}\right)}. \tag{2.489}$$

The second-order derivative of C_{H^+} with regard to $C_{HA,0}/C_{bf}$ will be ascertained via

$$\begin{aligned}\frac{d^2C_{H^+}}{d\left(\frac{C_{HA,0}}{C_{bf}}\right)^2} &\equiv \frac{d}{d\left(\frac{C_{HA,0}}{C_{bf}}\right)}\left(\frac{dC_{H^+}}{d\left(\frac{C_{HA,0}}{C_{bf}}\right)}\right) \\ &= \frac{d}{d\left(\frac{C_{HA,0}}{C_{bf}}\right)}\left(\frac{C_{bf}\left(C_{H^+}{}^2 + K_a^* C_{H^+}\right)}{3C_{H^+}{}^2 + 2\left(K_a^* + C_{bf}\left(1 - \frac{C_{HA,0}}{C_{bf}}\right)\right)C_{H^+} - \left(K_w^* + K_a^* C_{bf}\frac{C_{HA,0}}{C_{bf}}\right)}\right),\end{aligned} \tag{2.490}$$

upon application of its definition to Eq. (2.488) with C_{bf} factored out for convenience; the classical rules of differentiation of a quotient and a composite function have it that

$$\frac{d^2C_{H^+}}{d\left(\frac{C_{HA,0}}{C_{bf}}\right)^2} = C_{bf}\frac{\left(\begin{aligned}&\left(2C_{H^+} + K_a^*\right)\frac{dC_{H^+}}{d\left(\frac{C_{HA,0}}{C_{bf}}\right)}\left(3C_{H^+}{}^2 + 2\left(K_a^* + C_{bf}\left(1 - \frac{C_{HA,0}}{C_{bf}}\right)\right)C_{H^+} - \left(K_w^* + K_a^* C_{bf}\frac{C_{HA,0}}{C_{bf}}\right)\right) \\ &-\left(C_{H^+}{}^2 + K_a^* C_{H^+}\right)\left(6C_{H^+}\frac{dC_{H^+}}{d\left(\frac{C_{HA,0}}{C_{bf}}\right)} - 2C_{bf}C_{H^+} + 2\left(K_a^* + C_{bf}\left(1 - \frac{C_{HA,0}}{C_{bf}}\right)\right)\frac{dC_{H^+}}{d\left(\frac{C_{HA,0}}{C_{bf}}\right)} - K_a^* C_{bf}\right)\end{aligned}\right)}{\left(3C_{H^+}{}^2 + 2\left(K_a^* + C_{bf}\left(1 - \frac{C_{HA,0}}{C_{bf}}\right)\right)C_{H^+} - \left(K_w^* + K_a^* C_{bf}\frac{C_{HA,0}}{C_{bf}}\right)\right)^2}, \tag{2.491}$$

after having taken out C_{bf} for its constancy and realizing that $d\phi\{\xi,\zeta\}/d\xi$ should be calculated as $(\partial\phi/\partial\xi)_\zeta + (\partial\phi/\partial\zeta)_\xi(d\zeta/d\xi)$. By definition, every inflection point in Fig. 2.18 should satisfy

$$\frac{d^2pH}{d\left(\frac{C_{HA,0}}{C_{bf}}\right)^2}=0, \tag{2.492}$$

which calls for reformulation of Eq. (2.484) to

$$\left(\frac{1}{C_{H^+}{}^2}\left(\frac{dC_{H^+}}{d\left(\frac{C_{HA,0}}{C_{bf}}\right)}\right)^2-\frac{1}{C_{H^+}}\frac{d^2C_{H^+}}{d\left(\frac{C_{HA,0}}{C_{bf}}\right)^2}\right)\log_{10}\mathrm{e}=0; \tag{2.493}$$

after multiplying both sides by $C_{H^+}{}^2/\log_{10}\mathrm{e}$, Eq. (2.493) will look like

$$\left(\frac{dC_{H^+}}{d\left(\frac{C_{HA,0}}{C_{bf}}\right)}\right)^2=C_{H^+}\frac{d^2C_{H^+}}{d\left(\frac{C_{HA,0}}{C_{bf}}\right)^2}. \tag{2.494}$$

Insertion of Eqs. (2.489) and (2.491) converts Eq. (2.494) to

$$\left(\frac{C_{bf}C_{H^+}\left(K_a^*+C_{H^+}\right)}{3C_{H^+}{}^2+2\left(K_a^*+C_{bf}\left(1-\frac{C_{HA,0}}{C_{bf}}\right)\right)C_{H^+}-\left(K_w^*+K_a^*C_{bf}\frac{C_{HA,0}}{C_{bf}}\right)}\right)^2$$
$$=C_{H^+}C_{bf}\frac{\left(\begin{array}{l}\left(2C_{H^+}+K_a^*\right)\dfrac{dC_{H^+}}{d\left(\frac{C_{HA,0}}{C_{bf}}\right)}\left(3C_{H^+}{}^2+2\left(K_a^*+C_{bf}\left(1-\dfrac{C_{HA,0}}{C_{bf}}\right)\right)C_{H^+}-\left(K_w^*+K_a^*C_{bf}\dfrac{C_{HA,0}}{C_{bf}}\right)\right)\\ -\left(C_{H^+}{}^2+K_a^*C_{H^+}\right)\left(6C_{H^+}\dfrac{dC_{H^+}}{d\left(\frac{C_{HA,0}}{C_{bf}}\right)}-2C_{bf}C_{H^+}+2\left(K_a^*+C_{bf}\left(1-\dfrac{C_{HA,0}}{C_{bf}}\right)\right)\dfrac{dC_{H^+}}{d\left(\frac{C_{HA,0}}{C_{bf}}\right)}-K_a^*C_{bf}\right)\end{array}\right)}{\left(3C_{H^+}{}^2+2\left(K_a^*+C_{bf}\left(1-\frac{C_{HA,0}}{C_{bf}}\right)\right)C_{H^+}-\left(K_w^*+K_a^*C_{bf}\frac{C_{HA,0}}{C_{bf}}\right)\right)^2}, \tag{2.495}$$

where elimination of denominators between sides leaves

$$C_{bf}^2 C_{H^+}^2 \left(C_{H^+} + K_a^*\right)^2$$
$$= C_{bf} C_{H^+} \left(2C_{H^+} + K_a^*\right) \frac{dC_{H^+}}{d\left(\frac{C_{HA,0}}{C_{bf}}\right)} \left(3C_{H^+}^2 + 2\left(K_a^* + C_{bf}\left(1 - \frac{C_{HA,0}}{C_{bf}}\right)\right) C_{H^+} - \left(K_w^* + K_a^* C_{bf} \frac{C_{HA,0}}{C_{bf}}\right)\right)$$
$$- C_{bf} C_{H^+} \left(C_{H^+}^2 + K_a^* C_{H^+}\right) \left(6C_{H^+} \frac{dC_{H^+}}{d\left(\frac{C_{HA,0}}{C_{bf}}\right)} - 2C_{bf} C_{H^+} + 2\left(K_a^* + C_{bf}\left(1 - \frac{C_{HA,0}}{C_{bf}}\right)\right) \frac{dC_{H^+}}{d\left(\frac{C_{HA,0}}{C_{bf}}\right)} - K_a^* C_{bf}\right); \tag{2.496}$$

after dropping $C_{bf} C_{H^+}$ from both sides, Eq. (2.496) reduces to

$$C_{bf} C_{H^+} \left(C_{H^+} + K_a^*\right)^2$$
$$= \left(2C_{H^+} + K_a^*\right) \frac{dC_{H^+}}{d\left(\frac{C_{HA,0}}{C_{bf}}\right)} \left(3C_{H^+}^2 + 2\left(K_a^* + C_{bf}\left(1 - \frac{C_{HA,0}}{C_{bf}}\right)\right) C_{H^+} - \left(K_w^* + K_a^* C_{bf} \frac{C_{HA,0}}{C_{bf}}\right)\right)$$
$$- C_{H^+} \left(C_{H^+} + K_a^*\right) \left(6C_{H^+} \frac{dC_{H^+}}{d\left(\frac{C_{HA,0}}{C_{bf}}\right)} + 2\left(K_a^* + C_{bf}\left(1 - \frac{C_{HA,0}}{C_{bf}}\right)\right) \frac{dC_{H^+}}{d\left(\frac{C_{HA,0}}{C_{bf}}\right)} - C_{bf}\left(2C_{H^+} + K_a^*\right)\right), \tag{2.497}$$

along with factoring out C_{H^+} or C_{bf} (as appropriate). Selective elimination of parentheses in Eq. (2.497), complemented by placement of derivative-free terms in the right-hand side along with factoring out of $dC_{H^+}/d\left(C_{HA,0}/C_{bf}\right)$ give rise to

$$\frac{dC_{H^+}}{d\left(\frac{C_{HA,0}}{C_{bf}}\right)} \left(2C_{H^+} + K_a^*\right) \left(3C_{H^+}^2 + 2\left(K_a^* + C_{bf}\left(1 - \frac{C_{HA,0}}{C_{bf}}\right)\right) C_{H^+} - \left(K_w^* + K_a^* C_{bf} \frac{C_{HA,0}}{C_{bf}}\right)\right)$$
$$- \frac{dC_{H^+}}{d\left(\frac{C_{HA,0}}{C_{bf}}\right)} C_{H^+} \left(C_{H^+} + K_a^*\right) \left(6C_{H^+} + 2\left(K_a^* + C_{bf}\left(1 - \frac{C_{HA,0}}{C_{bf}}\right)\right)\right)$$
$$= C_{bf} C_{H^+} \left(C_{H^+} + K_a^*\right)^2 - C_{bf} C_{H^+} \left(C_{H^+} + K_a^*\right)\left(2C_{H^+} + K_a^*\right); \tag{2.498}$$

a further lumping of differential terms in the left-hand side, coupled with factoring out of $C_{bf} C_{H^+} \left(C_{H^+} + K_a^*\right)$ in the right-hand side lead then to

$$\frac{dC_{H^+}}{d\left(\frac{C_{HA,0}}{C_{bf}}\right)}\left(\begin{array}{l}(2C_{H^+}+K_a^*)\left(3C_{H^+}{}^2+2\left(K_a^*+C_{bf}\left(1-\frac{C_{HA,0}}{C_{bf}}\right)\right)C_{H^+}-\left(K_w^*+K_a^*C_{bf}\frac{C_{HA,0}}{C_{bf}}\right)\right)\\ -C_{H^+}(C_{H^+}+K_a^*)\left(6C_{H^+}+2\left(K_a^*+C_{bf}\left(1-\frac{C_{HA,0}}{C_{bf}}\right)\right)\right)\end{array}\right)$$
$$=C_{bf}C_{H^+}(C_{H^+}+K_a^*)(C_{H^+}+K_a^*-2C_{H^+}-K_a^*). \tag{2.499}$$

Insertion of Eq. (2.489), followed by condensation of similar terms transform Eq. (2.499) to

$$\frac{C_{bf}C_{H^+}(C_{H^+}+K_a^*)}{3C_{H^+}{}^2+2\left(K_a^*+C_{bf}\left(1-\frac{C_{HA,0}}{C_{bf}}\right)\right)C_{H^+}-\left(K_w^*+K_a^*C_{bf}\frac{C_{HA,0}}{C_{bf}}\right)}$$
$$\times\left(\begin{array}{l}(2C_{H^+}+K_a^*)\left(\begin{array}{l}3C_{H^+}{}^2\\ +2\left(K_a^*+C_{bf}\left(1-\frac{C_{HA,0}}{C_{bf}}\right)\right)C_{H^+}\\ -\left(K_w^*+K_a^*C_{bf}\frac{C_{HA,0}}{C_{bf}}\right)\end{array}\right)\\ -C_{H^+}(C_{H^+}+K_a^*)\left(6C_{H^+}+2\left(K_a^*+C_{bf}\left(1-\frac{C_{HA,0}}{C_{bf}}\right)\right)\right)\end{array}\right)$$
$$=C_{bf}C_{H^+}(C_{H^+}+K_a^*)(-C_{H^+}), \tag{2.500}$$

which prompts division of both sides by $C_{bf}C_{H^+}(C_{H^+}+K_a^*)$ to get

$$\frac{(2C_{H^+}+K_a^*)\left(\begin{array}{l}3C_{H^+}{}^2\\ +2\left(K_a^*+C_{bf}\left(1-\frac{C_{HA,0}}{C_{bf}}\right)\right)C_{H^+}\\ -\left(K_w^*+K_a^*C_{bf}\frac{C_{HA,0}}{C_{bf}}\right)\end{array}\right)-C_{H^+}(C_{H^+}+K_a^*)\left(6C_{H^+}+2\left(K_a^*+C_{bf}\left(1-\frac{C_{HA,0}}{C_{bf}}\right)\right)\right)}{3C_{H^+}{}^2+2\left(K_a^*+C_{bf}\left(1-\frac{C_{HA,0}}{C_{bf}}\right)\right)C_{H^+}-\left(K_w^*+K_a^*C_{bf}\frac{C_{HA,0}}{C_{bf}}\right)}=-C_{H^+}; \tag{2.501}$$

elimination of denominators finally produces

$$C_{H^+}\left(C_{H^+}+K_a^*\right)\left(6C_{H^+}+2\left(K_a^*+C_{bf}\left(1-\frac{C_{HA,0}}{C_{bf}}\right)\right)\right)-\left(2C_{H^+}+K_a^*\right)\left(\begin{array}{l}3C_{H^+}{}^2\\+2\left(K_a^*+C_{bf}\left(1-\frac{C_{HA,0}}{C_{bf}}\right)\right)C_{H^+}\\-\left(K_w^*+K_a^*C_{bf}\frac{C_{HA,0}}{C_{bf}}\right)\end{array}\right)$$
$$=C_{H^+}\left(3C_{H^+}{}^2+2\left(K_a^*+C_{bf}\left(1-\frac{C_{HA,0}}{C_{bf}}\right)\right)C_{H^+}-\left(K_w^*+K_a^*C_{bf}\frac{C_{HA,0}}{C_{bf}}\right)\right), \tag{2.502}$$

where negatives of both sides were meanwhile taken. Inspection of the order of magnitude of the various terms in Eq. (2.502) indicates that $3C_{H^+}{}^2 \ll 2\left(K_a^* + C_{bf}\left(1 - C_{HA,0}/C_{bf}\right)\right)C_{H^+}$ and $6C_{H^+} \ll 2\left(K_a^* + C_{bf}\left(1 - C_{HA,0}/C_{bf}\right)\right)$ within the operating ranges of interest in practice, that is, with $C_{bf} \gg C_{H^+}$, so simplification ensues as

$$2C_{H^+}\left(C_{H^+}+K_a^*\right)\left(K_a^*+C_{bf}\left(1-\frac{C_{HA,0}}{C_{bf}}\right)\right)-\left(2C_{H^+}+K_a^*\right)\left(\begin{array}{l}2\left(K_a^*+C_{bf}\left(1-\frac{C_{HA,0}}{C_{bf}}\right)\right)C_{H^+}\\-\left(K_w^*+K_a^*C_{bf}\frac{C_{HA,0}}{C_{bf}}\right)\end{array}\right)$$
$$\approx C_{H^+}\left(2\left(K_a^*+C_{bf}\left(1-\frac{C_{HA,0}}{C_{bf}}\right)\right)C_{H^+}-\left(K_w^*+K_a^*C_{bf}\frac{C_{HA,0}}{C_{bf}}\right)\right); \tag{2.503}$$

one also realizes that $K_a^* \ll C_{bf}\left(1 - C_{HA,0}/C_{bf}\right)$and $K_w^* \ll K_a^* C_{bf} C_{HA,0}/C_{bf}$ in view of $C_{bf} \gg K_a^* \gg K_w^*$, which allows further simplification of Eq. (2.503) to

$$2C_{bf}C_{H^+}\left(C_{H^+}+K_a^*\right)\left(1-\frac{C_{HA,0}}{C_{bf}}\right)-\left(2C_{H^+}+K_a^*\right)\left(2C_{bf}\left(1-\frac{C_{HA,0}}{C_{bf}}\right)C_{H^+}-K_a^*C_{bf}\frac{C_{HA,0}}{C_{bf}}\right)$$
$$\approx C_{H^+}\left(2C_{bf}\left(1-\frac{C_{HA,0}}{C_{bf}}\right)C_{H^+}-K_a^*C_{bf}\frac{C_{HA,0}}{C_{bf}}\right), \tag{2.504}$$

and further suggests division of both sides by C_{bf}, namely,

$$2C_{H^+}\left(C_{H^+}+K_a^*\right)\left(1-\frac{C_{HA,0}}{C_{bf}}\right)-\left(2C_{H^+}+K_a^*\right)\left(2\left(1-\frac{C_{HA,0}}{C_{bf}}\right)C_{H^+}-K_a^*\frac{C_{HA,0}}{C_{bf}}\right)$$
$$=C_{H^+}\left(2\left(1-\frac{C_{HA,0}}{C_{bf}}\right)C_{H^+}-K_a^*\frac{C_{HA,0}}{C_{bf}}\right). \tag{2.505}$$

After lumping terms alike between sides, Eq. (2.505) reduces to

$$2C_{H^+}\left(C_{H^+}+K_a^*\right)\left(1-\frac{C_{HA,0}}{C_{bf}}\right)=\left(2C_{H^+}+K_a^*+C_{H^+}\right)\left(2\left(1-\frac{C_{HA,0}}{C_{bf}}\right)C_{H^+}-K_a^*\frac{C_{HA,0}}{C_{bf}}\right), \tag{2.506}$$

where $2C_{H^+}$ will be pooled together with C_{H^+} as

$$2C_{H^+}\left(C_{H^+}+K_a^*\right)\left(1-\frac{C_{HA,0}}{C_{bf}}\right)=\left(3C_{H^+}+K_a^*\right)\left(2\left(1-\frac{C_{HA,0}}{C_{bf}}\right)C_{H^+}-K_a^*\frac{C_{HA,0}}{C_{bf}}\right); \tag{2.507}$$

selective elimination of parentheses produces

$$2C_{H^+}\left(C_{H^+}+K_a^*\right)\left(1-\frac{C_{HA,0}}{C_{bf}}\right)=2C_{H^+}\left(3C_{H^+}+K_a^*\right)\left(1-\frac{C_{HA,0}}{C_{bf}}\right)-K_a^*\left(3C_{H^+}+K_a^*\right)\frac{C_{HA,0}}{C_{bf}}. \tag{2.508}$$

Factoring out of $2C_{H^+}\left(1-C_{HA,0}/C_{bf}\right)$ in Eq. (2.508) ensues as

$$2C_{H^+}\left(C_{H^+}+K_a^*-3C_{H^+}-K_a^*\right)\left(1-\frac{C_{HA,0}}{C_{bf}}\right)+K_a^*\left(3C_{H^+}+K_a^*\right)\frac{C_{HA,0}}{C_{bf}}=0 \tag{2.509}$$

or, upon lumping terms alike,

$$2C_{H^+}\left(-2C_{H^+}\right)\left(1-\frac{C_{HA,0}}{C_{bf}}\right)+K_a^*\left(3C_{H^+}+K_a^*\right)\frac{C_{HA,0}}{C_{bf}}=0; \tag{2.510}$$

Equation (2.510) breaks down to

$$4C_{H^+}{}^2\left(1-\frac{C_{HA,0}}{C_{bf}}\right)-K_a^*\left(3C_{H^+}+K_a^*\right)\frac{C_{HA,0}}{C_{bf}}=0, \tag{2.511}$$

once constants are collapsed and negatives taken of both sides. Following a similar rationale, Eq. (2.479) will degenerate to

$$\left(K_a^* + C_{bf}\left(1 - \frac{C_{HA,0}}{C_{bf}}\right)\right)C_{H^+}{}^2 - \left(K_w^* + K_a^* C_{bf}\frac{C_{HA,0}}{C_{bf}}\right)C_{H^+} \approx 0, \tag{2.512}$$

valid within the ranges of C_{H^+} and $C_{HA,0}/C_{bf}$ of practical interest, because $C_{H^+}{}^3 \ll \left(K_a^* + C_{bf}\left(1 - C_{HA,0}/C_{bf}\right)\right)C_{H^+}{}^2$ and $K_a^* K_w^* \ll \left(K_w^* + K_a^* C_{bf} C_{HA,0}/C_{bf}\right)C_{H^+}$ on account of $C_{H^+} \ll C_{bf}$ and $K_w^* \ll C_{HA,0}$, respectively—consistent with the approximations implemented previously; division of both sides by C_{H^+} then leaves

$$\left(K_a^* + C_{bf}\left(1 - \frac{C_{HA,0}}{C_{bf}}\right)\right)C_{H^+} - \left(K_w^* + K_a^* C_{bf}\frac{C_{HA,0}}{C_{bf}}\right) = 0. \tag{2.513}$$

In addition, $K_a^* \ll C_{bf}\left(1 - C_{HA,0}/C_{bf}\right)$ because $K_a^* \ll C_{bf}$, and $K_w^* \ll K_a^* C_{bf} C_{HA,0}/C_{bf}$ because $K_w^* \ll K_a^* C_{HA,0}$ support extra simplification of Eq. (2.513) to

$$C_{bf}\left(1 - \frac{C_{HA,0}}{C_{bf}}\right)C_{H^+} - K_a^* C_{bf}\frac{C_{HA,0}}{C_{bf}} \approx 0; \tag{2.514}$$

after diving both sides by C_{bf}, Eq. (2.514) becomes

$$\left(1 - \frac{C_{HA,0}}{C_{bf}}\right)C_{H^+} - K_a^* \frac{C_{HA,0}}{C_{bf}} = 0, \tag{2.515}$$

which prompts isolation of C_{H^+} as

$$C_{H^+} = \frac{\frac{C_{HA,0}}{C_{bf}}}{1 - \frac{C_{HA,0}}{C_{bf}}} K_a^*. \tag{2.516}$$

Insertion of Eq. (2.516) converts Eq. (2.511) to

$$4\left(\frac{\frac{C_{HA,0}}{C_{bf}}}{1 - \frac{C_{HA,0}}{C_{bf}}} K_a^*\right)^2 \left(1 - \frac{C_{HA,0}}{C_{bf}}\right) - K_a^*\left(3\frac{\frac{C_{HA,0}}{C_{bf}}}{1 - \frac{C_{HA,0}}{C_{bf}}} K_a^* + K_a^*\right)\frac{C_{HA,0}}{C_{bf}} = 0, \tag{2.517}$$

where powers of products will be spelled out and K_a^* factored out as

$$4K_a^{*2}\frac{\left(\frac{C_{HA,0}}{C_{bf}}\right)^2}{\left(1-\frac{C_{HA,0}}{C_{bf}}\right)^2}\left(1-\frac{C_{HA,0}}{C_{bf}}\right)-K_a^{*2}\left(1+\frac{3\frac{C_{HA,0}}{C_{bf}}}{1-\frac{C_{HA,0}}{C_{bf}}}\right)\frac{C_{HA,0}}{C_{bf}}=0; \tag{2.518}$$

after cancelling $1 - C_{HA,0}/C_{bf}$ between numerator and denominator of the first term, dropping $K_a^{*2}\, C_{HA,0}/C_{bf}$ from both sides, and eliminating parentheses, Eq. (2.518) turns to

$$4\frac{\frac{C_{HA,0}}{C_{bf}}}{1-\frac{C_{HA,0}}{C_{bf}}}-\frac{1-\frac{C_{HA,0}}{C_{bf}}+3\frac{C_{HA,0}}{C_{bf}}}{1-\frac{C_{HA,0}}{C_{bf}}}=0. \tag{2.519}$$

Multiplication of both sides by $1 - C_{HA,0}/C_{bf}$ transforms Eq. (2.519) to

$$4\frac{C_{HA,0}}{C_{bf}}=1-\frac{C_{HA,0}}{C_{bf}}+3\frac{C_{HA,0}}{C_{bf}}, \tag{2.520}$$

where similar terms will be collapsed as

$$2\frac{C_{HA,0}}{C_{bf}}=1; \tag{2.521}$$

isolation of $C_{HA,0}/C_{bf}$ leaves merely

$$\left.\frac{C_{HA,0}}{C_{bf}}\right|_{\frac{d^2pH}{d\left(\frac{C_{HA,0}}{C_{bf}}\right)^2}=0}=\frac{1}{2}, \tag{2.522}$$

thus confirming the horizontal location of the center of the plateau of every curve in Fig. 2.18. Combination with Eq. (2.522) finally converts Eq. (2.516) to

$$C_{H^+}=\frac{\frac{1}{2}}{1-\frac{1}{2}}K_a^*=\frac{\frac{1}{2}}{\frac{1}{2}}K_a^*, \tag{2.523}$$

which degenerates to

$$C_{H^+}=K_a^*; \tag{2.524}$$

after applying logarithms to both sides and then taking their negatives, Eq. (2.524) becomes

$$-\log_{10}C_{H^+} = -\log_{10}K_a^*, \tag{2.525}$$

or else

$$pH\Big|_{\frac{d^2pH}{d\left(\frac{C_{HA,0}}{C_{bf}}\right)^2}=0} = pK_a^* \tag{2.526}$$

in view of Eqs. (2.274) and (2.282), and consistent with the ordinates of the inflection points grasped in Fig. 2.18. Therefore, the strongest effect of a buffer, with overall concentration C_{bf}, occurs when half thereof is accounted for by $C_{HA,0}$ and the other half by $C_{MA,0}$, so only this possibility, that is,

$$C_{HA,0} = C_{MA,0} = \frac{1}{2}C_{bf} \tag{2.527}$$

consistent with Eq. (2.478), will be explored hereafter for the sake of bookkeeping. Note that when Eq. (2.527) is inserted in Eq. (2.479), one obtains

$$C_{H^+}{}^3 + \left(K_a^* + C_{bf}\left(1 - \frac{1}{2}\right)\right)C_{H^+}{}^3 - \left(K_w^* + K_a^*C_{bf}\frac{1}{2}\right)C_{H^+} - K_a^*K_w^* = 0 \tag{2.528}$$

or, equivalently,

$$C_{H^+}{}^3 + \left(K_a^* + \frac{1}{2}C_{bf}\right)C_{H^+}{}^2 - \left(K_w^* + \frac{1}{2}K_a^*C_{bf}\right)C_{H^+} - K_a^*K_w^* = 0 \tag{2.529}$$

after straightforward algebraic rearrangement; Eq. (2.529) drives the curves of Fig. 2.18 in the vicinity of their inflection point.

Remember that the usefulness of a buffer hinges on how efficiently it can resist changes in pH, namely, when a (strong) acid or base is added thereto. The former causes release of H^+ into solution via direct protolysis, while the latter triggers abstraction of H^+ from solution via reverse protolysis; in both cases, such phenomena are meant to counterbalance the extra value of C_{HA} or C_{A^-}, respectively, in general agreement with Le Châtelier's principle. A more quantitative statement is, however, convenient, in attempts to estimate C_{bf} necessary to avoid pH changes beyond a given threshold. Hence, one will consider an aqueous system containing a weak acid at initial concentration $C_{HA,0} = C_{bf}/2$ as per

Eq. (2.527), and its conjugate base at initial concentration $C_{MA,0} = C_{bf}/2$ as well, and further added with a strong base DOH, which promptly (and fully) dissociates as

$$DOH \rightarrow D^+ + OH^-, \tag{2.530}$$

thus leading to concentrations C_{D^+} and $C_{OH^-,sb}$ of D^+ and OH^-, respectively, such that

$$C_{D^+} = C_{DOH,0} \tag{2.531}$$

for species D^+ being chemically inert (by hypothesis), further to $C_{OH^-,sb} = C_{DOH,0}$. The two chemical equilibria ruling concentrations of species in solution, described by Eqs. (2.219) and (2.301), still obey

$$K_w^* = C_{H^+} C_{OH^-} \tag{2.532}$$

as alias to Eq. (2.218) and

$$K_a^* = \frac{C_{H^+} C_{A^-}}{C_{HA}} \tag{2.533}$$

as alias to Eq. (2.305), with K_w^* and K_a^* denoting the underlying protolysis constants (as before); a balance to A-containing moieties reads

$$C_{HA} + C_{A^-} = C_{bf}, \tag{2.534}$$

while a balance to charges takes the form

$$C_{A^-} + C_{OH^-} = C_{H^+} + C_{M^+} + C_{D^+} \tag{2.535}$$

meant to assure electroneutrality in solution, where Eq. (2.531), together with

$$C_{M^+} = \frac{1}{2} C_{bf} \tag{2.536}$$

consistent with Eqs. (2.387) and (2.527), allow conversion of Eq. (2.535) to

$$C_{A^-} + C_{OH^-} = C_{H^+} + \frac{1}{2}C_{bf} + C_{DOH,0}. \tag{2.537}$$

The buffering power, β, is conveniently defined by

$$\beta \equiv \frac{dC_{DOH,0}}{dpH} \tag{2.538}$$

or, after applying the chain differentiation rule,

$$\beta = \frac{dC_{DOH,0}}{dC_{H^+}} \frac{dC_{H^+}}{dpH}; \tag{2.539}$$

Equation (2.538) describes how high the concentration of added base, $C_{D^+} = C_{DOH,0}$, must be to bring about a unit change in solution *pH*, so all relevant concentrations are to be eventually expressed in terms of $C_{DOH,0}$ and C_{H^+}. Recalling the rule of differentiation of an inverse function, Eq. (2.539) can be rewritten as

$$\beta = \frac{\frac{dC_{DOH,0}}{dC_{H^+}}}{\frac{dpH}{dC_{H^+}}}, \tag{2.540}$$

whereas Eq. (2.274), redone as

$$pH = -\ln C_{H^+} \log_{10} e = -\frac{\ln C_{H^+}}{\ln 10} \tag{2.541}$$

with the aid of the operational rules of logarithms, will undergo differentiation as

$$\frac{dpH}{dC_{H^+}} = -\frac{1}{C_{H^+} \ln 10}; \tag{2.542}$$

Equation (2.542) supports conversion of Eq. (2.540) to

$$\beta = -C_{H^+} \frac{dC_{DOH,0}}{dC_{H^+}} \ln 10, \tag{2.543}$$

along with reduction in number of fraction levels.

After solving Eq. (2.532) for C_{OH^-}, that is,

$$C_{OH^-} = \frac{K_w^*}{C_{H^+}}, \tag{2.544}$$

one obtains $C_{OH^-} \equiv C_{OH^-}\{C_{H^+}\}$; Eq. (2.537) will then be solved for C_{A^-}, according to

$$C_{A^-} = C_{H^+} + \frac{1}{2}C_{bf} + C_{DOH,0} - C_{OH^-}, \tag{2.545}$$

where insertion of Eq. (2.544) gives rise to

$$C_{A^-} = \frac{1}{2}C_{bf} + C_{H^+} - \frac{K_w^*}{C_{H^+}} + C_{DOH,0} \tag{2.546}$$

and yields $C_{A^-} \equiv C_{A^-}\{C_{H^+}, C_{DOH,0}\}$. Combination of Eqs. (2.534) and (2.546) unfolds

$$C_{HA} + \frac{1}{2}C_{bf} + C_{H^+} - \frac{K_w^*}{C_{H^+}} + C_{DOH,0} = C_{bf}, \tag{2.547}$$

where lumping of terms alike, complemented by isolation of C_{HA} leave

$$C_{HA} = \frac{1}{2}C_{bf} - C_{H^+} + \frac{K_w^*}{C_{H^+}} - C_{DOH,0}; \tag{2.548}$$

once again, a relationship of the type $C_{HA} \equiv C_{HA}\{C_{H^+}, C_{DOH,0}\}$ has been attained. Equation (2.533), as only independent equation left at this stage, will now be rephrased as

$$K_a^* = \frac{C_{H^+}\left(\frac{1}{2}C_{bf} + C_{H^+} - \frac{K_w^*}{C_{H^+}} + C_{DOH,0}\right)}{\frac{1}{2}C_{bf} - C_{H^+} + \frac{K_w^*}{C_{H^+}} - C_{DOH,0}} \tag{2.549}$$

at the expense of Eqs. (2.546) and (2.548); upon multiplication of both sides by $C_{bf}/2 - C_{H^+} + K_w^*/C_{H^+} - C_{DOH,0}$, Eq. (2.549) turns to

$$\frac{1}{2}K_a^*C_{bf} - K_a^*C_{H^+} + \frac{K_a^*K_w^*}{C_{H^+}} - K_a^*C_{DOH,0} = \frac{1}{2}C_{bf}C_{H^+} + C_{H^+}{}^2 - K_w^* + C_{DOH,0}C_{H^+}, \tag{2.550}$$

along with elimination of parenthesis. Multiplication of both sides by C_{H^+} transforms Eq. (2.550) to

$$\frac{1}{2}K_a^*C_{bf}C_{H^+} - K_a^*C_{H^+}{}^2 + K_a^*K_w^* - K_a^*C_{DOH,0}C_{H^+} \\ = \frac{1}{2}C_{bf}C_{H^+}{}^2 + C_{H^+}{}^3 - K_w^*C_{H^+} + C_{DOH,0}C_{H^+}{}^2, \tag{2.551}$$

where terms alike will be pooled together as

$$C_{H^+}{}^3 + \left(K_a^* + \frac{1}{2}C_{bf} + C_{DOH,0}\right)C_{H^+}{}^2 - \left(K_w^* + \frac{1}{2}K_a^*C_{bf} - K_a^*C_{DOH,0}\right)C_{H^+} - K_a^*K_w^* = 0; \tag{2.552}$$

note that Eq. (2.552) reduces to

$$\left(C_{H^+}{}^3 + \left(K_a^* + \frac{1}{2}C_{bf}\right)C_{H^+}{}^2 - \left(K_w^* + \frac{1}{2}K_a^*C_{bf}\right)C_{H^+} - K_a^*K_w^* = 0\right)\Bigg|_{C_{DOH,0}\to 0} \tag{2.553}$$

when no extra base DOH is present, which coincides (as expected) with Eq. (2.529). Equation (2.552) may instead appear as

$$C_{H^+}{}^3 + \left(K_a^* + \frac{1}{2}C_{bf}\right)C_{H^+}{}^2 + C_{H^+}{}^2C_{DOH,0} - \left(K_w^* + \frac{1}{2}K_a^*C_{bf}\right)C_{H^+} \\ + K_a^*C_{H^+}C_{DOH,0} - K_a^*K_w^* = 0 \tag{2.554}$$

after having splitted selected parentheses, where $C_{DOH,0}$ will be factored out to obtain

$$\left(C_{H^+}{}^2 + K_a^*C_{H^+}\right)C_{DOH,0} \\ = -C_{H^+}{}^3 - \left(K_a^* + \frac{1}{2}C_{bf}\right)C_{H^+}{}^2 + \left(K_w^* + \frac{1}{2}K_a^*C_{bf}\right)C_{H^+} + K_a^*K_w^*; \tag{2.555}$$

isolation of $C_{DOH,0}$ finally yields

$$C_{DOH,0} = \frac{K_a^* K_w^* + \left(K_w^* + \frac{1}{2} K_a^* C_{bf}\right) C_{H^+} - \left(K_a^* + \frac{1}{2} C_{bf}\right) {C_{H^+}}^2 - {C_{H^+}}^3}{K_a^* C_{H^+} + {C_{H^+}}^2}, \tag{2.556}$$

thus providing a relationship of the type $C_{H^+} \equiv C_{H^+}\{C_{DOH,0}\}$, yet in implicit form. Differentiation of both sides of Eq. (2.556) with regard to C_{H^+} leads to

$$\frac{dC_{DOH,0}}{dC_{H^+}} = \frac{\left(\begin{array}{l} \left(K_w^* + \frac{1}{2} K_a^* C_{bf} - 2\left(K_a^* + \frac{1}{2} C_{bf}\right) C_{H^+} - 3{C_{H^+}}^2\right)\left(K_a^* C_{H^+} + {C_{H^+}}^2\right) \\ -\left(K_a^* K_w^* + \left(K_w^* + \frac{1}{2} K_a^* C_{bf}\right) C_{H^+} - \left(K_a^* + \frac{1}{2} C_{bf}\right) {C_{H^+}}^2 - {C_{H^+}}^3\right)\left(K_a^* + 2C_{H^+}\right) \end{array}\right)}{\left(K_a^* C_{H^+} + {C_{H^+}}^2\right)^2}, \tag{2.557}$$

where selective removal of parentheses and combination with Eq. (2.543) lead to

$$-\frac{\beta}{C_{H^+} \ln 10} = \frac{\left(\begin{array}{l} K_a^*\left(K_w^* + \frac{1}{2} K_a^* C_{bf}\right) C_{H^+} + \left(K_w^* + \frac{1}{2} K_a^* C_{bf}\right) {C_{H^+}}^2 - 2K_a^*\left(K_a^* + \frac{1}{2} C_{bf}\right) {C_{H^+}}^2 \\ -2\left(K_a^* + \frac{1}{2} C_{bf}\right) {C_{H^+}}^3 - 3K_a^* {C_{H^+}}^3 - 3{C_{H^+}}^4 - K_a^{*2} K_w^* - 2K_a^* K_w^* C_{H^+} - K_a^*\left(K_w^* + \frac{1}{2} K_a^* C_{bf}\right) C_{H^+} \\ -2\left(K_w^* + \frac{1}{2} K_a^* C_{bf}\right) {C_{H^+}}^2 + K_a^*\left(K_a^* + \frac{1}{2} C_{bf}\right) {C_{H^+}}^2 + 2\left(K_a^* + \frac{1}{2} C_{bf}\right) {C_{H^+}}^3 + K_a^* {C_{H^+}}^3 + 2{C_{H^+}}^4 \end{array}\right)}{\left(K_a^* C_{H^+} + {C_{H^+}}^2\right)^2}; \tag{2.558}$$

a final lumping of terms alike leaves

$$-\frac{\beta}{C_{H^+} \ln 10} = \frac{\left(\begin{array}{l} (2-3){C_{H^+}}^4 + \left(-2K_a^* - C_{bf} - 3K_a^* + 2K_a^* + C_{bf} + K_a^*\right) {C_{H^+}}^3 \\ +\left(K_w^* + \frac{1}{2} K_a^* C_{bf} - 2K_a^{*2} - K_a^* C_{bf} - 2K_w^* - K_a^* C_{bf} + K_a^{*2} + \frac{1}{2} K_a^* C_{bf}\right) {C_{H^+}}^2 \\ +\left(K_a^* K_w^* + \frac{1}{2} K_a^{*2} C_{bf} - 2K_a^* K_w^* - K_a^* K_w^* - \frac{1}{2} K_a^{*2} C_{bf}\right) C_{H^+} - K_a^{*2} K_w^* \end{array}\right)}{\left(K_a^* C_{H^+} + {C_{H^+}}^2\right)^2}. \tag{2.559}$$

Condensation of terms alike simplifies Eq. (2.559) to

$$-\frac{\beta}{C_{H^+}\ln 10} = -\frac{C_{H^+}{}^4 + 2K_a^* C_{H^+}{}^3 + \left(K_w^* + K_a^{*2} + K_a^* C_{bf}\right) C_{H^+}{}^2 + 2K_a^* K_w^* C_{H^+} + K_a^{*2} K_w^*}{\left(K_a^* C_{H^+} + C_{H^+}{}^2\right)^2}, \tag{2.560}$$

where multiplication of both sides by $-C_{H^+}\ln 10$ unfolds

$$\beta = \frac{C_{H^+}{}^5 + 2K_a^* C_{H^+}{}^4 + \left(K_w^* + K_a^{*2} + K_a^* C_{bf}\right) C_{H^+}{}^3 + 2K_a^* K_w^* C_{H^+}{}^2 + K_a^{*2} K_w^* C_{H^+}}{\left(K_a^* C_{H^+} + C_{H^+}{}^2\right)^2} \ln 10; \tag{2.561}$$

after revisiting Eq. (2.274) as

$$C_{H^+} = 10^{-pH}, \tag{2.562}$$

Equation (2.561) may instead be coined as

$$\beta = \frac{\left(10^{-pH}\right)^5 + 2K_a^*\left(10^{-pH}\right)^4 + \left(K_w^* + K_a^{*2} + K_a^* C_{bf}\right)\left(10^{-pH}\right)^3 + 2K_a^* K_w^*\left(10^{-pH}\right)^2 + K_a^{*2} K_w^* 10^{-pH}}{\left(K_a^* 10^{-pH} + \left(10^{-pH}\right)^2\right)^2} \ln 10, \tag{2.563}$$

or else

$$\beta = \frac{10^{-5pH} + 2K_a^* 10^{-4pH} + \left(K_w^* + K_a^{*2} + K_a^* C_{bf}\right) 10^{-3pH} + 2K_a^* K_w^* 10^{-2pH} + K_a^{*2} K_w^* 10^{-pH}}{\left(K_a^* 10^{-pH} + 10^{-2pH}\right)^2} \ln 10 \tag{2.564}$$

upon lumping exponents in composite powers; multiplication of numerator and denominator by 10^{4pH} finally yields

$$\beta = \frac{10^{-pH} + 2K_a^* + \left(K_w^* + K_a^{*2} + K_a^* C_{bf}\right) 10^{pH} + 2K_a^* K_w^* 10^{2pH} + K_a^{*2} K_w^* 10^{3pH}}{\left(1 + K_a^* 10^{pH}\right)^2} \ln 10 \tag{2.565}$$

as depicted in Fig. 2.19. Note that β starts from $\ln 10 \approx 2.30$ at nil pH, as expected from

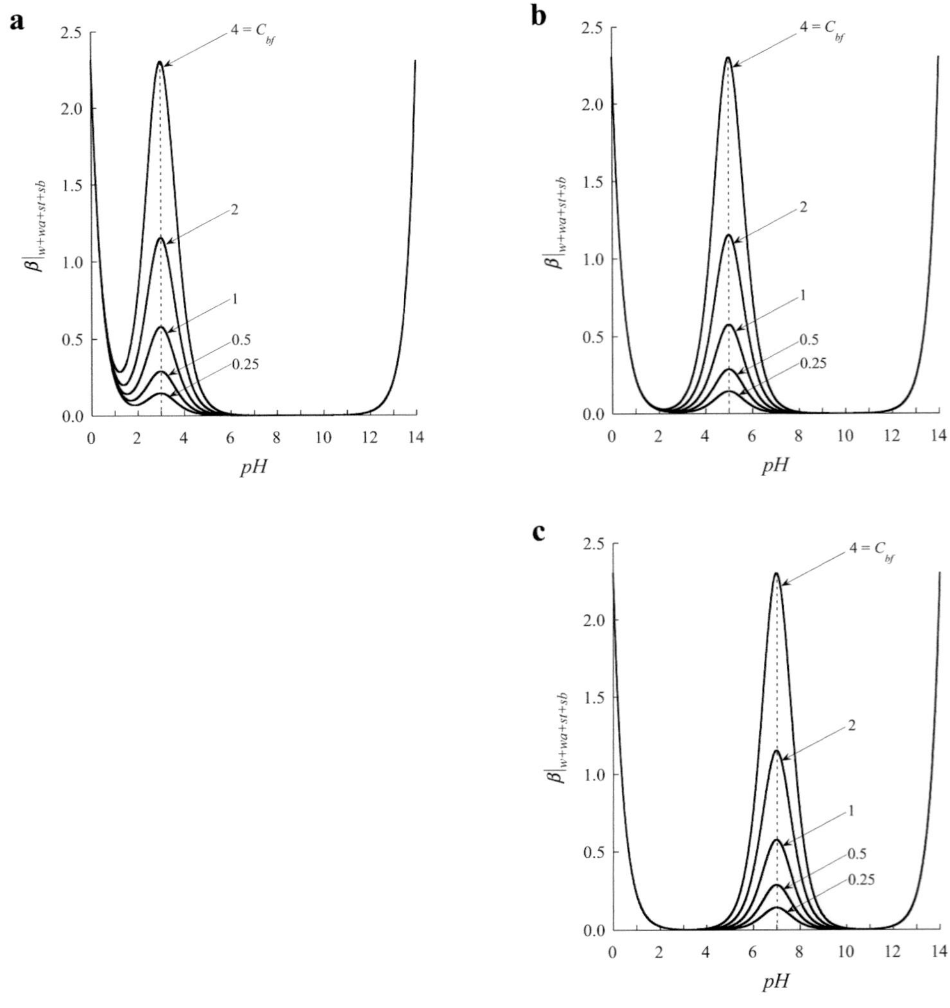

Fig. 2.19 Variation of buffering capacity, $\beta|_{w+wa+st+sb}$, of aqueous buffer solution of weak acid and its conjugate base in soluble salt form, characterized by selected values of protolysis constant, (**a**) $pK_a^*=3$, (**b**) $pK_a^*=5$, and (**c**) $pK_a^*=7$, in water, when in presence of strong base, versus pH for selected overall buffer concentrations, C_{bf}

$$\lim_{pH\to 0}\beta = \frac{10^0 + 2K_a^* + \left(10^{-14} + K_a^{*2} + K_a^* C_{bf}\right)10^0 + 2K_a^* 10^{-14}10^0 + K_a^{*2}10^{-14}10^0}{\left(1 + K_a^* 10^0\right)^2}\ln 10$$
$$\approx \frac{10^0}{1^2}\ln 10 = \ln 10, \tag{2.566}$$

because $K_a^{*2}K_w^* \ll K_a^*K_w^*$, $K_a^{*2} + K_w^* \ll K_a^* < K_a^*C_{bf} < 1$ and $K_a^*K_w^* \ll 1$ stemming from Eq. (2.565), and then decreases until a local minimum is reached. After this minimum, β increases toward the (intermediate) local maximum; it then decreases again to a flat minimum, and eventually rises again to ln 10 as pH 14 is approached—in agreement with

$$\lim_{pH \to 14} \beta = \frac{10^{-14} + 2K_a^* + \left(10^{-14} + K_a^{*2} + K_a^*C_{bf}\right)10^{14} + 2K_a^*10^{-14}10^{28} + K_a^{*2}\,10^{-14}10^{42}}{\left(1 + K_a^*10^{14}\right)^2} \ln 10$$
$$\approx \frac{K_a^{*2}10^{28}}{K_a^{*2}10^{28}} \ln 10 = \frac{1}{1} \ln 10 = \ln 10, \tag{2.567}$$

after taking into account that $10^{-14} + 2K_a^* \leq 2 \times 10^{-3}$ as well as $\left(K_w^* + K_a^{*2} + K_a^*C_{bf}\right)10^{pH} \leq \left(10^{-14} + \left(10^{-3}\right)^2 + 10^{-3}0.25\right)10^{14} \approx 2.5 \times 10^{10}$, and realizing that $2K_a^*K_w^*10^{2 \cdot 14} \leq 2 \times 10^{-3}10^{-14}10^{28} = 2 \times 10^{11}$ lie much below $(10^{-3})^2 10^{-14} 10^{3 \cdot 14} = 10^{22}$ in numerator, further to $K_a^*10^{14} \gg 10^{-3}10^{14} = 10^{11} \gg 1$ in denominator. The curves tend to be symmetrical relative to the vertical straight line with $pH = pK_a^*$ for horizontal intercept, with asymmetry becoming more and more notorious as pK_a^* is lowered; when this happens, the local minimum found at lower pH is no longer flat but is instead described by an ordinate above zero.

The maximum value for β occurs indeed at pH in the vicinity of pK_a^*, and the corresponding value of β increases with C_{bf}. In this region, one notices that ${C_{H^+}}^5 + 2K_a^*{C_{H^+}}^4 + 2K_a^*K_w^*{C_{H^+}}^2 + K_a^{*2}K_w^*C_{H^+}$ is of the order of magnitude of $K_a^{*5} + 2K_a^*K_a^{*4} + 2K_a^*10^{-14}K_a^{*2} + \quad K_a^{*2}10^{-14}K_a^* = 3K_a^{*5} + 3 \times 10^{-14}K_a^{*3} \approx 3K_a^{*5}$, and consequently much lower than $\left(K_w^* + K_a^{*2} + K_a^*C_{bf}\right){C_{H^+}}^3$ as it approaches $\left(10^{-14} + K_a^{*2} + K_a^*C_{bf}\right)K_a^{*3} \approx K_a^{*5} + C_{bf}K_a^{*4} \approx C_{bf}K_a^{*4}$. Under these circumstances, Eq. (2.561) reduces to approximately

$$\beta \approx \frac{\left(K_w^* + K_a^{*2} + K_a^*C_{bf}\right){C_{H^+}}^3}{\left(K_a^*C_{H^+} + {C_{H^+}}^2\right)^2} \ln 10; \tag{2.568}$$

the necessary condition for an optimum then reads

$$\frac{d\beta}{dC_{H^+}} = 0. \tag{2.569}$$

Equation (2.569) reduces, in the current problem, to

$$\frac{\left(\begin{array}{l}\left(K_w^*+K_w^{*2}+K_a^*C_{bf}\right)3C_{H^+}{}^2\left(K_a^*C_{H^+}+C_{H^+}{}^2\right)^2\\-\left(K_w^*+K_a^{*2}+K_a^*C_{bf}\right)C_{H^+}{}^3 2\left(K_a^*C_{H^+}+C_{H^+}{}^2\right)\left(K_a^*+2C_{H^+}\right)\end{array}\right)}{\left(\left(K_a^*C_{H^+}+C_{H^+}{}^2\right)^2\right)^2}\ln 10=0, \quad (2.570)$$

in view of Eq. (2.568); this is equivalent to writing

$$\frac{3\left(K_w^*+K_a^{*2}+K_a^*C_{bf}\right)C_{H^+}{}^2\left(K_a^*C_{H^+}+C_{H^+}{}^2\right)-2\left(K_w^*+K_a^{*2}+K_a^*C_{bf}\right)\left(K_a^*+2C_{H^+}\right)C_{H^+}{}^3}{\left(K_a^*C_{H^+}+C_{H^+}{}^2\right)^3}\ln 10=0 \quad (2.571)$$

upon multiplying exponents of composite powers in denominator and dropping $K_a^*C_{H^+}+C_{H^+}{}^2$ between numerator and denominator afterward. Since the denominator of Eq. (2.571) is finite and not nil, one can proceed with multiplication of both sides by $\left(K_a^*C_{H^+}+C_{H^+}{}^2\right)^3/\ln 10$ to get

$$3\left(K_w^*+K_a^{*2}+K_a^*C_{bf}\right)\left(K_a^*+C_{H^+}\right)C_{H^+}{}^3=2\left(K_w^*+K_a^{*2}+K_a^*C_{bf}\right)\left(K_a^*+2C_{H^+}\right)C_{H^+}{}^3, \quad (2.572)$$

along with factoring out of C_{H^+}, which prompts division of both sides by $\left(K_w^*+K_a^{*2}+K_a^*C_{bf}\right)C_{H^+}{}^3$ to obtain

$$3\left(K_a^*+C_{H^+}\right)=2\left(K_a^*+2C_{H^+}\right); \quad (2.573)$$

elimination of parentheses unfolds

$$3K_a^*+3C_{H^+}=2K_a^*+4C_{H^+}, \quad (2.574)$$

which is equivalent to merely

$$C_{H^+}=K_a^*. \quad (2.575)$$

After taking logarithms of both sides, and their negatives afterward, Eq. (2.575) becomes

$$-\log_{10}C_{H^+}=-\log_{10}K_a^*, \quad (2.576)$$

or else

$$pH_{opt} = pK_a^* \tag{2.577}$$

after recalling Eqs. (2.274) and (2.282); the local maxima in Fig. 2.19 are indeed described by Eq. (2.577) for abscissa. Insertion of Eq. (2.575) transforms Eq. (2.568) to

$$\beta_{max} \equiv \beta\big|_{C_{H^+}=K_a^*} = \frac{\left(K_w^* + K_a^{*2} + K_a^* C_{bf}\right) K_a^{*3}}{\left(K_a^* K_a^* + K_a^{*2}\right)^2} \ln 10, \tag{2.578}$$

which reduces to

$$\beta_{max} = \frac{\left(K_w^* + K_a^{*2} + K_a^* C_{bf}\right) K_a^{*3}}{\left(2K_a^{*2}\right)^2} \ln 10 = \frac{\left(K_w^* + K_a^{*2} + K_a^* C_{bf}\right) K_a^{*3}}{4K_a^{*4}} \ln 10 \tag{2.579}$$

once terms alike are collapsed and power of product taken; division of both numerator and denominator by K_a^{*4} ensues, namely,

$$\beta_{max} = \frac{\frac{K_w^*}{K_a^*} + K_a^* + C_{bf}}{4} \ln 10, \tag{2.580}$$

which simplifies further to

$$\beta_{max} \approx 0.576 C_{bf} \tag{2.581}$$

because $K_w^*/K_a^* + K_a^* \ll C_{bf}$ within the range of interest. In other words, a buffer is most effective when $C_{HA,0} = 0.5$ as per Eq. (2.522), and thus $C_{HA,0} = C_{MA,0}$ in view of Eq. (2.527); it keeps pH essentially unchanged in the vicinity of pK_a^* as conveyed by Eq. (2.577), and $C_{DOH,0}$ must vary by ca. $0.6C_{bf}$ in order for pH be displaced by one unit following Eqs. (2.538) and (2.581).

Problem 2.10

Consider again Eq. (2.565), which conveys $\beta \equiv \beta\{pH\}$ in general form.

(a) Find an exact expression for β_{max} as a function of pH.
(b) Obtain an approximate expression for β prevailing at low pH and then estimate $\beta\big|_{pH=0}$, as well as the first (local) minimum of β.

(continued)

Problem 2.10 (continued)

(c) Obtain an approximate expression for β prevailing at high pH and then estimate $\beta|_{pH=14}$, as well as the second (global) minimum of β.

Solution

(a) Calculation of the true optimum in terms of C_{H^+} (for convenience) requires differentiation of the whole expression labeled as Eq. (2.561) with regard to C_{H^+} rather than the approximate expression conveyed by Eq. (2.568), followed by combination with Eq. (2.569) according to

$$\frac{d}{dC_{H^+}}\left(\frac{{C_{H^+}}^5+2K_a^*{C_{H^+}}^4+\left(K_w^*+K_a^{*2}+K_a^*C_{bf}\right){C_{H^+}}^3+2K_a^*K_w^*{C_{H^+}}^2+K_a^{*2}K_w^*C_{H^+}}{\left(K_a^*C_{H^+}+{C_{H^+}}^2\right)^2}\ln 10\right)=0; \tag{2.582}$$

straight application of the classical rules of differentiation unfolds

$$\frac{\left(\begin{array}{l}\left(5{C_{H^+}}^4+2K_a^*4{C_{H^+}}^3+\left(K_w^*+K_a^{*2}+K_a^*C_{bf}\right)3{C_{H^+}}^2+2K_a^*K_w^*2C_{H^+}+K_a^{*2}K_w^*\right)\left(K_a^*C_{H^+}+{C_{H^+}}^2\right)^2\\-\left({C_{H^+}}^5+2K_a^*{C_{H^+}}^4+\left(K_w^*+K_a^{*2}+K_a^*C_{bf}\right){C_{H^+}}^3+2K_a^*K_w^*{C_{H^+}}^2+K_a^{*2}K_w^*C_{H^+}\right)2\left(K_a^*C_{H^+}+C_{H^+}^2\right)\left(K_a^*+2C_{H^+}\right)\end{array}\right)}{\left(\left(K_a^*C_{H^+}+{C_{H^+}}^2\right)^2\right)^2}=0, \tag{2.583}$$

which degenerates to

$$\begin{array}{l}\left(K_a^*C_{H^+}+{C_{H^+}}^2\right)^2\left(5{C_{H^+}}^4+8K_a^*{C_{H^+}}^3+3\left(K_w^*+K_a^{*2}+K_a^*C_{bf}\right){C_{H^+}}^2+4K_a^*K_w^*C_{H^+}+K_a^{*2}K_w^*\right)\\-2\left(K_a^*+2C_{H^+}\right)\left(K_a^*C_{H^+}+{C_{H^+}}^2\right)\left(\begin{array}{l}{C_{H^+}}^5+2K_a^*{C_{H^+}}^4+\left(K_w^*+K_a^{*2}+K_a^*C_{bf}\right){C_{H^+}}^3\\+2K_a^*K_w^*{C_{H^+}}^2+K_a^{*2}K_w^*C_{H^+}\end{array}\right)=0\end{array} \tag{2.584}$$

upon lumping of constant factors and elimination of denominators, because $\left(\left(K_a^*C_{H^+}+{C_{H^+}}^2\right)^2\right)^2=\left(K_a^*C_{H^+}+{C_{H^+}}^2\right)^4$ is not nil but finite. After having divided both sides by $K_a^*C_{H^+}+{C_{H^+}}^2\neq 0$, Eq. (2.584) reduces to

(continued)

Problem 2.10 (continued)

$$\left(K_a^*C_{H^+}+C_{H^+}{}^2\right)\left(5C_{H^+}{}^4+8K_a^*C_{H^+}{}^3+3\left(K_w^*+K_a^{*2}+K_a^*C_{bf}\right)C_{H^+}{}^2+4K_a^*K_w^*C_{H^+}+K_a^{*2}K_w^*\right)$$
$$-2\left(K_a^*+2C_{H^+}\right)\left(C_{H^+}{}^5+2K_a^*C_{H^+}{}^4+\left(K_w^*+K_a^{*2}+K_a^*C_{bf}\right)C_{H^+}{}^3+2K_a^*K_w^*C_{H^+}{}^2+K_a^{*2}K_w^*C_{H^+}\right)=0, \quad (2.585)$$

where elimination of outer parentheses yields

$$5K_a^*C_{H^+}{}^5+8K_a^{*2}C_{H^+}{}^4+3K_a^*\left(K_w^*+K_a^{*2}+K_a^*C_{bf}\right)C_{H^+}{}^3+4K_a^{*2}K_w^*C_{H^+}{}^2+K_a^{*3}K_w^*C_{H^+}$$
$$+5C_{H^+}{}^6+8K_a^*C_{H^+}{}^5+3\left(K_w^*+K_a^{*2}+K_a^*C_{bf}\right)C_{H^+}{}^4+4K_a^*K_w^*C_{H^+}{}^3+K_a^{*2}K_w^*C_{H^+}{}^2$$
$$-2K_a^*C_{H^+}{}^5-4K_a^{*2}C_{H^+}{}^4-2K_a^*\left(K_w^*+K_a^{*2}+K_a^*C_{bf}\right)C_{H^+}{}^3-4K_a^{*2}K_w^*C_{H^+}{}^2-2K_a^{*3}K_w^*C_{H^+}$$
$$-4C_{H^+}{}^6-8K_a^*C_{H^+}{}^5-4\left(K_w^*+K_a^{*2}+K_a^*C_{bf}\right)C_{H^+}{}^4-8K_a^*K_w^*C_{H^+}{}^3-4K_a^{*2}K_w^*C_{H^+}{}^2=0; \quad (2.586)$$

once alike terms have been collapsed, Eq. (2.586) becomes

$$(5-4)C_{H^+}{}^6+\left(5K_a^*+8K_a^*-2K_a^*-8K_a^*\right)C_{H^+}{}^5$$
$$+\left(8K_a^{*2}+3\left(K_w^*+K_a^{*2}+K_a^*C_{bf}\right)-4K_a^{*2}-4\left(K_w^*+K_a^{*2}+K_a^*C_{bf}\right)\right)C_{H^+}{}^4$$
$$+\left(3K_a^*\left(K_w^*+K_a^{*2}+K_a^*C_{bf}\right)+4K_a^*K_w^*-2K_a^*\left(K_w^*+K_a^{*2}+K_a^*C_{bf}\right)-8K_a^*K_w^*\right)C_{H^+}{}^3$$
$$+\left(4K_a^{*2}K_w^*+K_a^{*2}K_w^*-4K_a^{*2}K_w^*-4K_a^{*2}K_w^*\right)C_{H^+}{}^2+\left(K_a^{*3}K_w^*-2K_a^{*3}K_w^*\right)C_{H^+}=0, \quad (2.587)$$

where further lumping of constants leaves

$$C_{H^+}{}^6+3K_a^*C_{H^+}{}^5+\left(3K_a^{*2}-K_w^*-K_a^*C_{bf}\right)C_{H^+}{}^4$$
$$+\left(K_a^{*3}+K_a^{*2}C_{bf}-3K_a^*K_w^*\right)C_{H^+}{}^3-3K_a^{*2}K_w^*C_{H^+}{}^2-K_a^{*3}K_w^*C_{H^+}=0. \quad (2.588)$$

A final division of both sides of Eq. (2.588) by C_{H^+} is in order, namely,

$$C_{H^+}{}^5+3K_a^*C_{H^+}{}^4+\left(3K_a^{*2}-K_w^*-K_a^*C_{bf}\right)C_{H^+}{}^3$$
$$+K_a^*\left(K_a^{*2}+K_a^*C_{bf}-3K_w^*\right)C_{H^+}{}^2-3K_a^{*2}K_w^*C_{H^+}-K_a^{*3}K_w^*=0, \quad (2.589)$$

along with factoring out of K_a^* in the fourth term; since a fifth-order polynomial is found, solution must hereafter proceed numerically.

(continued)

Problem 2.10 (continued)

(b) Within the low pH range, one realizes that 10^{-pH} and $\left(K_w^* + K_a^{*2} + K_a^* C_{bf}\right)10^{pH}$ are the dominant terms in numerator, and 1 the dominant term in denominator; hence, Eq. (2.565) reduces to

$$\beta\big|_{pH \ll pK_a^*} \approx \frac{10^{-pH} + \left(K_w^* + K_a^{*2} + K_a^* C_{bf}\right)10^{pH}}{1^2} \ln 10, \tag{2.590}$$

while realization that $K_w^* + K_a^{*2} \ll K_a^* C_{bf}$ supports further simplification to

$$\beta\big|_{pH \ll pK_a} \approx \left(10^{-pH} + K_a^* C_{bf} 10^{pH}\right) \ln 10. \tag{2.591}$$

The necessary condition for a minimum of β, with regard to pH, reads

$$\frac{d\beta}{dpH} = 0, \tag{2.592}$$

where insertion of Eq. (2.591) unfolds

$$\frac{d}{dpH}\left(\left(10^{-pH} + K_a^* C_{bf} 10^{pH}\right) \ln 10\right) = 0; \tag{2.593}$$

application of the classical theorems on differentiation produces

$$\left(10^{-pH}(-\ln 10) + K_a^* C_{bf} 10^{pH} \ln 10\right) \ln 10 = 0, \tag{2.594}$$

or else

$$\left(K_a^* C_{bf} 10^{pH} - 10^{-pH}\right)\ln^2 10 = 0 \tag{2.595}$$

after ln 10 has been factored out. Following division of both sides by $\ln^2 10$, Eq. (2.595) reduces to

$$K_a^* C_{bf} 10^{pH} - 10^{-pH} = 0 \tag{2.596}$$

or, equivalently,

(continued)

Problem 2.10 (continued)

$$K_a^* C_{bf} 10^{pH} = 10^{-pH} \tag{2.597}$$

upon addition of 10^{-pH} to both sides; should decimal logarithms be taken of both sides, Eq. (2.597) yields

$$\log_{10} K_a^* C_{bf} + pH = -pH, \tag{2.598}$$

where terms alike can be lumped as

$$2pH = -\log_{10} K_a^* - \log_{10} C_{bf} \tag{2.599}$$

since the logarithm of a product coincides with the sum of logarithms of its factors. In view of Eq. (2.282), it is possible to rewrite Eq. (2.599) as

$$2pH = pK_a^* - \log_{10} C_{bf}, \tag{2.600}$$

with division of both sides by 2 leaving

$$pH_{min, pH \ll pK_a^*} = \frac{1}{2} pK_a^* - \frac{1}{2} \log_{10} C_{bf}; \tag{2.601}$$

Equation (2.601) describes indeed the abscissa of the first minima of the curves in Fig. 2.19, located exactly at $pK_a^*/2$ when $C_{bf} = 1$, and moving leftward or rightward as C_{bf} increases or decreases, respectively. Specifically at $pH = 0$, the only relevant term in Eq. (2.591) is the first, so one obtains

$$\beta\big|_{pH=0} \approx 10^{-pH}\big|_{pH=0} \ln 10, \tag{2.602}$$

which is the same to write

$$\beta\big|_{pH=0} = 10^0 \ln 10 = \ln 10 \approx 2.30 \tag{2.603}$$

as apparent from inspection of Fig. 2.19.

(c) Order of magnitude analysis of the terms in Eq. (2.565) within the high-pH range unfolds $\left(K_w^* + K_a^{*2} + K_a^* C_{bf}\right) 10^{pH}$ and $K_a^{*2} K_w^* 10^{3pH}$ as dominant ones in numerator, whereas $K_a^* 10^{pH}$ overrides in denominator; hence, one may resort to

(continued)

Problem 2.10 (continued)

$$\beta|_{pH\gg pK_a^*} \approx \frac{\left(K_w^* + K_a^{*2} + K_a^* C_{bf}\right)10^{pH} + K_a^{*2}K_w^* 10^{3pH}}{\left(K_a^* 10^{pH}\right)^2} \ln 10 \quad (2.604)$$

as simplified version of Eq. (2.565), where splitting of the power in denominator unfolds

$$\beta|_{pH\gg pK_a^*} = \frac{\left(K_w^* + K_a^{*2} + K_a^* C_{bf}\right)10^{pH} + K_a^{*2}K_w^* 10^{3pH}}{K_a^{*2} 10^{2pH}} \ln 10; \quad (2.605)$$

this prompts splitting of the fraction as

$$\beta|_{pH\gg pK_a^*} = \left(\frac{K_w^* + K_a^{*2} + K_a^* C_{bf}}{K_a^{*2}} 10^{-pH} + K_w^* 10^{pH}\right) \ln 10, \quad (2.606)$$

along with lumping of similar terms. A similar order of magnitude analysis further indicates that $K_w^* + K_a^{*2} \ll K_a^* C_{bf}$, so one may resort to

$$\beta|_{pH\gg pK_a^*} \approx \left(\frac{K_a^* C_{bf}}{K_a^{*2}} 10^{-pH} + K_w^* 10^{pH}\right) \ln 10 \quad (2.607)$$

in lieu of Eq. (2.606), or simply

$$\beta|_{pH\gg pK_a^*} = \left(\frac{C_{bf}}{K_a^*} 10^{-pH} + K_w^* 10^{pH}\right) \ln 10 \quad (2.608)$$

instead of Eq. (2.606). Since Eq. (2.592) remains valid as necessary condition for a minimum, one will obtain the second minimum via

$$\frac{d}{dpH}\left(\left(\frac{C_{bf}}{K_a^*} 10^{-pH} + K_w^* 10^{pH}\right) \ln 10\right) = 0 \quad (2.609)$$

stemming now from Eq. (2.608); differentiation as indicated gives rise to

(continued)

Problem 2.10 (continued)

$$\left(\frac{C_{bf}}{K_a^*}10^{-pH}(-\ln 10)+K_w^*10^{pH}\ln 10\right)\ln 10=0, \tag{2.610}$$

which breaks down to

$$\left(K_w^*10^{pH}-\frac{C_{bf}}{K_a^*}10^{-pH}\right)\ln^2 10=0 \tag{2.611}$$

once ln 10 is factored out and lumped with the identical factor already there. Since $\ln^2 10 \neq 0$, division of both sides of Eq. (2.611) thereby ensues as

$$K_w^*10^{pH}-\frac{C_{bf}}{K_a^*}10^{-pH}=0, \tag{2.612}$$

where additional division of both sides by K_w^* leaves

$$10^{pH}-\frac{C_{bf}}{K_a^*K_w^*}10^{-pH}=0; \tag{2.613}$$

multiplication of both sides by 10^{pH} then produces

$$10^{2pH}=\frac{C_{bf}}{K_a^*K_w^*}. \tag{2.614}$$

When decimal logarithms are taken of both sides, Eq. (2.614) becomes

$$2pH=\log_{10}\frac{C_{bf}}{K_a^*K_w^*}, \tag{2.615}$$

with the operational features of the logarithm function taken advantage of when writing

$$2pH=\log_{10}C_{bf}-\log_{10}K_a^*-\log_{10}K_w^*; \tag{2.616}$$

recalling Eq. (2.282), one may proceed to

$$2pH=\log_{10}C_{bf}+pK_a^*+pK_w^*, \tag{2.617}$$

and finally to

(continued)

Problem 2.10 (continued)

$$pH_{min,pH\gg pK_a^*} = \frac{1}{2}\left(pK_a^* + pK_w^*\right) + \frac{1}{2}\log_{10}C_{bf} \tag{2.618}$$

upon division of both sides by 2. Equation (2.618) describes the abscissae of the minima of the curves in Fig. 2.19 located after the maximum, which move rightward as C_{bf} increases. In the immediate neighborhood of the upper limit of the pH scale, only $K_w^* 10^{pH}$ plays a significant role in Eq. (2.608), so one is left with

$$\beta\big|_{pH=14} \approx K_w^* 10^{pH}\big|_{pH=14} \ln 10; \tag{2.619}$$

Equation (2.619) is equivalent to

$$\beta\big|_{pH=14} = 10^{-14} 10^{14} \ln 10 = 10^0 \ln 10 = \ln 10 \approx 2.30 \tag{2.620}$$

in view of $pK_w^* = 14$, thus justifying the ordinate where all curves in Fig. 2.19 cross the vertical straight line described by 14 for horizontal intercept.

2.4 Foaming

Liquid foams are ubiquitous in Nature, and a likely event whenever gas/liquid systems are processed from laboratory scale up to industrial level. In bioreactors, their presence is prone to reduce product yield and overall process performance, and thus seen as a nuisance; this justifies inclusion of a foam sensor and provision of a foam preventer device/additive in most apparatuses.

Foaming is promoted by larger flow rates of air through and smaller bubbles at the sparger tip, higher broth temperature, stronger mixing, and heavier mechanical stress during fermentation, as well as fast depressurization following sterilization. One major problem related thereto is (overhead) foam volume proper, which may account for a considerable portion of the bioreactor working volume; this is typically the case of beer fermentation, where it may occupy between one-fourth and one-third of the whole volume available. Remember that choice of impeller(s) for, and positioning of baffles in a given bioreactor normally aim at maximizing degree of mixing and associated rate of aeration—further to minimizing development of stagnant zones; plain water, with its intrinsic rheological features, is routinely used as model medium. However, water and foam in an aqueous medium exhibit disparate behaviors, so a lower final efficiency will often be attained, caused by poorer mixing quality and concomitantly higher power input required. In addition, foam may hinder or jam the control probes in the bioreactor, thus leading to false readings and eventually hampering adequate performance; besides leaving residues therein for the next batch, and constrain delivery and sampling. Serious clogging may also occur if foam reaches the air filters, with consequent pressure buildup eventually blocking the air inlet, while reducing efficiency of filtering and thus promoting possible contamination of culture. On the other hand, existence of foam in the top portion of a bioreactor creates a barrier to mass transfer of oxygen in its headspace; since the oxygen carried by the bubbles is rapidly taken up, potentially suboptimal aeration may arise, with detrimental effects upon many aerobic microorganisms.

However, the unfavorable impact of foam is not restricted to operation of the reacting vessel, as it often extends to broth and microorganism themselves. Molecules bearing amphiphilic character or hydrophobic nature will find a preferential energetic balance at the liquid/gas interface and so tend to concentrate there, thus stabilizing foams. Key nutrient molecules included in broth formulation for biomass expansion or such metabolites as exopolysaccharides, proteins, and lipoproteins (e.g., surfactin by *Bacillus subtilis*) synthesized and secreted by cells themselves can be trapped in the foam, thus becoming inaccessible to the microorganism cells or absent in the final fraction recovered; caramelized sugars and products of Maillard's reactions previously generated during broth sterilization also play a role in stabilizing foam. A high rate of microbial death correlates with enhanced foaming as well, owing to the amphiphilicity of the phospholipid constituents of cell membranes.

In addition, microorganisms trapped around rising bubbles will be dragged to the surface of the medium where bubble bursting is often detrimental thereto, especially in the case of animal cells behaving planktonically; if the bubble is added to an existing foam, no immediate damage results, while the strain produced by eventual collapse of bubbles and draining of lamellae will act otherwise. Shear protectants used to formulate the medium are meant to hamper cell attachment to bubbles and strengthen cell membrane, yet a few of them are hydrophobic; hence, they will inherently increase stability of foam, and thus lower its rate of draining. On the other hand, supply of CO_2 will be enhanced by continuous regeneration of foam—useful to grow microalgae autotrophically, while reducing the amount of available O_2.

While entrapment of microorganisms on bubbles can be used to isolate them when in the surface foam, fractionation of extracellular metabolites is also possible—should they concentrate preferentially in the foam; this is the case of surface-active agents, which play the role of either target products or contaminants.

Problem 2.11

Formation of foam on the surface of liquid media brings about a number of unwanted outcomes, including adsorption of extracellular enzymes and flotation of microorganisms; the latter has even been taken advantage of in separation of distinct species of bacteria or bacteria from yeasts, in concentration of bacterial spores and vegetative cells, and even in recovery of microalgae from their growth medium. The germane model considers flotation to result from a set of direct bubble–particle interactions, dependent on three subprocesses—collision, adhesion, and detachment; the probability, P, of occurrence of this phenomenon is accordingly given by

$$P = P_c P_a (1 - P_d), \tag{2.621}$$

where P_c denotes probability of collision, P_a denotes probability of adhesion, and P_d denotes probability of detachment. For microorganisms and spores, P_d is negligibly small owing to their low inertia, so one may safely take

$$P_d \approx 0; \tag{2.622}$$

P_c depends on particle and bubble size distributions, according to

$$P_c = \frac{2}{3}\left(\frac{D_c}{D_b}\right)^2, \tag{2.623}$$

with D_c denoting cell diameter and D_b denoting bubble diameter. Bubble-cell adhesion will occur when the sliding time of the cell on the bubble surface is longer than the induction time, t_i, defined as the minimum time taken to thin and rupture the bubble film; it is dependent on London's dispersion, van der Waals', and electrical double-layer forces, besides hydrophobic attraction between cell and bubble. Yoon and Luttrell's correlation for P_a indeed reads

$$P_a = \sin^2\left\{2\tan^{-1}\left\{\exp\left\{-\frac{\left(45 + 8\,{Re_b}^{0.72}\right)v_b t_i}{D_c + D_b}\right\}\right\}\right\}, \tag{2.624}$$

where Reynolds' number abides to

$$Re_b \equiv \frac{\rho_l v_b D_b}{\mu_l} \tag{2.625}$$

(continued)

Problem 2.11 (continued)

and pertains to bubble of diameter D_b, and liquid medium of mass density ρ_l and viscosity μ_l, with velocity of bubble, v_b, satisfying

$$v_b = \frac{1}{18}\frac{g(\rho_l - \rho_g){D_b}^2}{\mu_l}, \tag{2.626}$$

as long as g denotes acceleration of gravity and ρ_g denotes mass density of gas. Once P is known, the first-order kinetic constant of flotation, k_m, is obtainable via

$$k_m = \frac{1}{4}S_b P, \tag{2.627}$$

provided that S_b denotes bubble surface area flow rate—given, in turn, by

$$S_b = 6\frac{Q_g}{A_v D_b}, \tag{2.628}$$

with Q_g denoting velocity of gas supplied at bottom of vessel and A_v denoting cross-sectional area of vessel. For representative values of the germane physico-chemical properties, and bubbles of 2.0 mm for (average) diameter, discuss the discriminating potential of foaming using dimensionless parameters as far as possible.

Solution

Insertion of Eq. (2.628) converts Eq. (2.627) to

$$k_m = \frac{1}{4}6\frac{Q_g}{A_v D_b}P = \frac{3}{2}\frac{Q_g}{A_v D_b}P, \tag{2.629}$$

which prompts definition of a dimensionless kinetic constant as

$$k^*{}_m \equiv \frac{k_m}{\frac{Q_g}{A_v D_b}}; \tag{2.630}$$

hence, Eq. (2.629) will reduce to

$$k^*_m = \frac{3}{2}P. \tag{2.631}$$

(continued)

Problem 2.11 (continued)

Reynolds' number is obtainable from combination of Eqs. (2.625) and (2.626), namely,

$$Re_b = \frac{\rho_l \frac{1}{18} \frac{g(\rho_l - \rho_g) D_b{}^2}{\mu_l} D_b}{\mu_l} = \frac{1}{18} \frac{g\rho_l(\rho_l - \rho_g) D_b{}^3}{\mu_l{}^2} \equiv \frac{1}{18} Ga, \quad (2.632)$$

thus supporting emergence of Galileo's number, Ga; for $g = 9.8$ m s^{-2}, as well as $\rho_l = 1.0 \times 10^3$ kg m^{-3} (as for water), $\rho_g = 1.3$ kg m^{-3} (as for air), $D_b = 2.0 \times 10^{-3}$ m (as per hypothesis), and $\mu_l = 8.9 \times 10^{-4}$ Pa s (as for water), Eq. (2.632) becomes

$$Re_b = \frac{1}{18} \frac{9.8 \cdot 1.0 \times 10^3 (1.0 \times 10^3 - 1.3)(2.0 \times 10^{-3})^3}{(8.9 \times 10^{-4})^2} = 5.5 \times 10^3. \quad (2.633)$$

After rewriting Eq. (2.624) as

$$P_a = \sin^2 \left\{ 2\tan^{-1} \left\{ \exp \left\{ -\frac{45 + 8\,Re_b{}^{0.72}}{1 + \frac{D_c}{D_b}} \frac{v_b t_i}{D_b} \right\} \right\} \right\} \quad (2.634)$$

following division by D_b of numerator and denominator of the argument of the exponential function to allow explicit appearance of two dimensionless parameters (i.e., D_c/D_b and $v_b t_i/D_b$), insertion of Eq. (2.633) produces

$$P_a = \sin^2 \left\{ 2\tan^{-1} \left\{ \exp \left\{ -\frac{45 + 8(5.5 \times 10^3)^{0.72}}{1 + \frac{D_c}{D_b}} \frac{v_b t_i}{D_b} \right\} \right\} \right\}; \quad (2.635)$$

Equation (2.635) breaks down to

$$P_a = \sin^2 \left\{ 2\tan^{-1} \left\{ \exp \left\{ -4.0 \times 10^3 \frac{\frac{v_b t_i}{D_b}}{1 + \frac{D_c}{D_b}} \right\} \right\} \right\}. \quad (2.636)$$

Insertion of Eqs. (2.622), (2.623), and (2.636) turns Eq. (2.621) to

(continued)

Problem 2.11 (continued)

$$P = \frac{2}{3}\left(\frac{D_c}{D_b}\right)^2 \sin^2\left\{2\tan^{-1}\left\{\exp\left\{-4.0\times10^3\frac{\frac{v_b t_i}{D_b}}{1+\frac{D_c}{D_b}}\right\}\right\}\right\}(1-0), \tag{2.637}$$

which will, in turn, be inserted in Eq. (2.631) to give

$$k_m^* = \frac{3}{2}\frac{2}{3}\left(\frac{D_c}{D_b}\right)^2 \sin^2\left\{2\tan^{-1}\left\{\exp\left\{-4.0\times10^3\frac{\frac{v_b t_i}{D_b}}{1+\frac{D_c}{D_b}}\right\}\right\}\right\}, \tag{2.638}$$

or merely

$$k_m^* = \left(\frac{D_c}{D_b}\right)^2 \sin^2\left\{2\tan^{-1}\left\{\exp\left\{-4.0\times10^3\frac{\frac{v_b t_i}{D_b}}{1+\frac{D_c}{D_b}}\right\}\right\}\right\} \tag{2.639}$$

upon lumping of constants; a graphical interpretation is provided in Fig. 2.20, using D_c/D_b and $v_b t_i/D_b$ as continuous and discrete dimensionless independent variables, respectively. Inspection of this figure unfolds a noteworthy discriminatory pattern of flotation toward cell size, especially at lower values of $v_b t_i/D_b$, say below 0.0002; with change in k_m^* essentially coincident with change in D_c/D_b, when $v_b t_i/D_b$ is close to 0.00001.

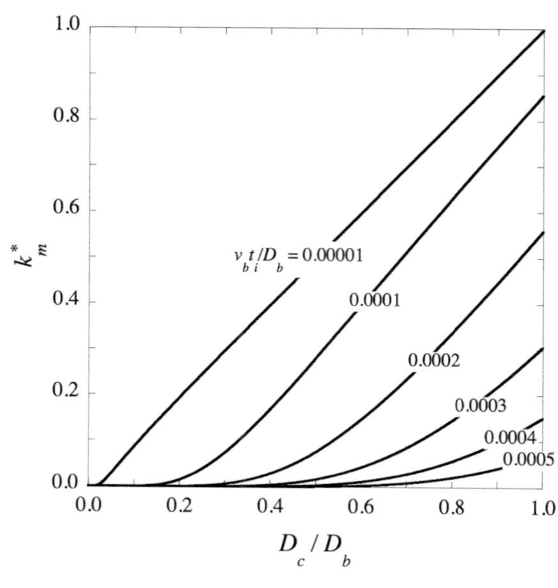

Fig. 2.20 Variation of dimensionless kinetic constant, k_m^*, versus normalized cell diameter, D_c/D_b, for selected values of dimensionless velocity, $v_b t_i/D_b$

2.4.1 Foam Buildup

A foam results from some process of shaking, stirring, or bubble nucleation; the excess liquid then drains out, and the bubble matrix becomes more and more compact—to eventually bring about foam instability that degenerates to film rupture. The time scale associated with liquid drainage is much shorter than that describing gas diffusion between bubbles, so these two phenomena can be decoupled for modeling purposes, and the latter discarded.

Foam may be defined as a complex structure, composed of gas pockets separated by liquid membranes, but exhibiting a number of unique features; such systems cannot, in particular, be regarded as simple discretized liquids or gases, yet they obey physics rules in a strict manner. When closely observed, a well-defined substructure is unveiled—strongly influenced by the liquid fraction; formation of a foam from rising bubbles is depicted in Fig. 2.21a, whereas actual patterns found in foam are provided in Fig. 2.21b–d. A variety of size distributions is normally found in foams, in which case they are termed polydispersed; although a truly monodispersed foam is essentially theoretical, it can be deliberately generated in practice and has accordingly served as background for mechanistic studies focused on foam structure.

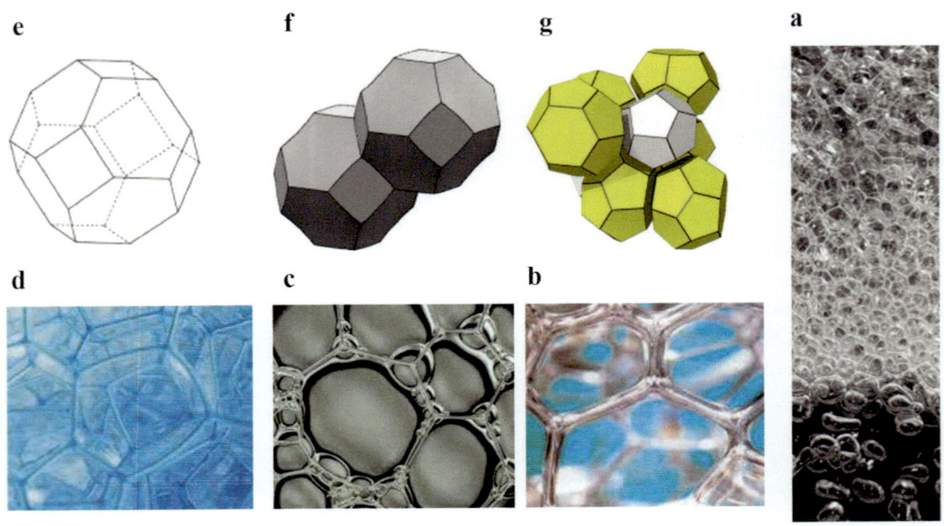

Fig. 2.21 Photos of (**a**) accumulation of gaseous bubbles at the top of a liquid in the form of a foam, with (**b**–**d**) details of geometrical features produced, and geometric interpretation of Kelvin's cell(s) as (**e**) single unit and (**f**) juxtaposed units, and of (**g**) Weaire and Phelan's cells in lattice

In its very essence, a liquid/gas foam is therefore a compact matrix of gaseous bubbles separated by liquid membranes, with buoyancy pushing the bubbles upward; hence, the spontaneous spherical shape of a bubble, when taken individually in the bulk of a liquid, is no longer suitable for the foam, because a relatively large volume of liquid would necessarily remain between adjacent bubbles and thus prevent their compactation toward an ultimate gaseous continuum. The associated distortion actually converts spheres to polyhedra, as long as they can stick perfectly to one another, and the remaining space in between is accounted for only by their faces constituted by (thin) liquid films. In a monodispersed i-th lattice, all cells are identical and characterized by f_i faces, e_i edges, and v_i vertices; however, which type will be spontaneously assumed by the gas pockets within a liquid foam remains to be found. Toward this goal, one should recall Euler's formula for polyhedra, namely,

$$f_i - e_i + v_i = 2, \tag{2.640}$$

which should be combined with the corresponding incidence relationship, that is,

$$2e_i - 3v_i = 0; \tag{2.641}$$

any one of f_i, e_i, or v_i will obviously (and uniquely) determine the other two, via the set of Eqs. (2.640) and (2.641). On the other hand, Kusner's theorem, pertaining to the minimum (average) number of faces in a minimal foam of a Riemannian compact, nonpositively curved, three-dimensional manifold, states that

$$\overline{f} \geq \frac{6\delta}{3\delta - \pi} \tag{2.642}$$

in general; here, $\overline{f}$ denotes the average number of faces in the collection of adjacent cells that constitutes the foam, and δ denotes (dihedral) center angle of a regular tetrahedron. It is known from analytical geometry that

$$\delta = \cos^{-1}\frac{1}{3} = 1.23\ \text{rad} = 70.5^\circ; \tag{2.643}$$

insertion of Eq. (2.643) converts Eq. (2.642) to

$$\bar{f} \geq \frac{6 \cdot 1.23}{3 \cdot 1.23 - \pi} = 13.46, \tag{2.644}$$

thus implying that the minimum (average) number of faces will read

$$f_i = 14 \tag{2.645}$$

for a periodic foam constituted by a single type of (repeating) polyhedron. Once in possession of Eq. (2.645), one will revisit Eq. (2.640) as

$$14 - e_i + v_i = 2, \tag{2.646}$$

or else

$$-2e_i + 2v_i = -24 \tag{2.647}$$

upon condensation of constants and multiplication of both sides by −2 afterward; ordered addition of Eqs. (2.641) and (2.647) unfolds

$$v_i = 24, \tag{2.648}$$

which may in turn be inserted in Eq. (2.641) to get

$$2e_i - 3 \cdot 24 = 0 \tag{2.649}$$

that breaks down to

$$e_i = 36. \tag{2.650}$$

The solution consubstantiated in Eqs. (2.645), (2.648), and (2.650) is known as Kelvin's conjecture for a periodic minimum foam in space; its fundamental cell is (combinatorially) a truncated octahedron with 14 faces (i.e., 6 flat quadrilaterals and 8 curved hexagons), 36 edges, and 24 vertices—also termed tetrakaidecahedron or Kelvin's cell, and depicted in Fig. 2.21e. Note its perfect fit to every other alike cell in space, without empty spaces in between, thus generating a three-dimensional network as per Fig. 2.21f; this lattice minimizes indeed the interfacial area of the polyhedron for a given volume while satisfying

Plateau's borders, and Occam's razor intrinsically justifies its adoption by Nature for being the simplest one bearing the required features. This point is emphasized by inspection of Table 2.5, where the first set of space-filling, convex polyhedra are described, together with the corresponding outer surface area per unit volume. More complex possibilities exist that resort to more than one type of polyhedron, such as tetrahedra and octahedra or octahedra, truncated octahedra, and cubes; pyritohedrons and truncated hexagonal trapezohedrons or Weaire and Phelan's cells, consisting of six 14-sides and two 12-sided polyhedra, deserve special mention (see Fig. 2.21g) because their average number of faces, that is, $\overline{f} = (6{\cdot}14 + 2{\cdot}12)/(6 + 2) = 13.50 > 13.46$, still lies above the threshold conveyed by Eq. (2.644). However, the improvement of the latter set of cells over Kelvin's cells is marginal, besides being bidispersed and thus much more complex to handle; hence, it will not deserve further discussion for modelling purposes.

It is useful to distinguish between the two extremes of foam features, that is, dry foam and wet foam, according to the value of the underlying liquid fraction, ϕ_l. The former is well defined and associated to a small ϕ_l, with thin membranes and dense liquid channels; the associated network of nodes follows a specific set of constraints. As the bubbles push against each other, the interfacial forces push the channels to adopt specific angles, which, in turn, shape the bubble edges into polygons. Although a large variety of shapes and sizes are possible, polygons possessing a small number of corners appear unfavorable in aging foams; in three-dimensional networks, polyhedral bubbles with fewer than 14 edges will eventually collapse. Conversely, the structure of wet foam is best viewed as a network of contacting soft spheres, and is found when ϕ_l approaches the value at which the bubbles separate from each other; the structure is poorly defined in this case, thus justifying why dry foam has classically been chosen for modelling purposes.

The elements (or cells) of the complex structure exhibited by foam are individual liquid lamellae (or faces) that meet at Plateau's borders, constituted by channels (or edges) that connect to each other at nodes (or intersections); the underlying topology is schematized in Fig. 2.22a, b. The cross sections of the aforementioned borders thicken as more liquid is incorporated into the foam; channels join at 109.4° in such four-channel nodes, as a reminiscent of the stereochemical molecular configuration of methane, because the associated configurations are a result of an energetic equilibrium. The underlying tetrahedron, centered at 0, is depicted as [*ABCD*] in Fig. 2.22c and is constituted by four triangular faces, that is, [*ABD*], [*BCD*], [*ACD*], and [*ABC*], and four corners, that is, *A*, *B*, *C*, and *D*.

Plateau's junctions may be simplified to the portion of equilateral triangle [*ABD*], of cross-sectional surface S, with base of length $2R$, and height of length H, besides (equal) inner angles ε, as sketched in Fig. 2.22d. Since the inner angles of the equilateral triangle abide to

$$3\varepsilon = \pi, \tag{2.651}$$

one promptly obtains

Table 2.5 List of first space-filling convex polyhedra sorted by decreasing number of outer surface area per unit volume, with corresponding sequential faces laid on a plane

Polyhedron			Side length = 1		
Type	Plannified layout	Number of sides	Volume (V)	Outer surface area (S)	S/V
Triangular prism	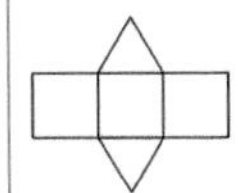	5	$\frac{\sqrt{3}}{4}$	$\frac{6+\sqrt{3}}{2}$	8.93
Gyrobifastigium	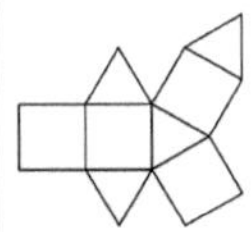	8	$\frac{\sqrt{3}}{2}$	$4+\sqrt{3}$	6.62
Cube	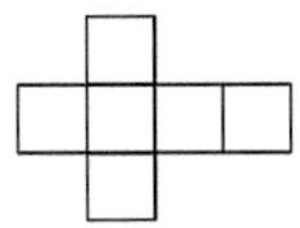	6	1	6	6.00
Hexagonal prism	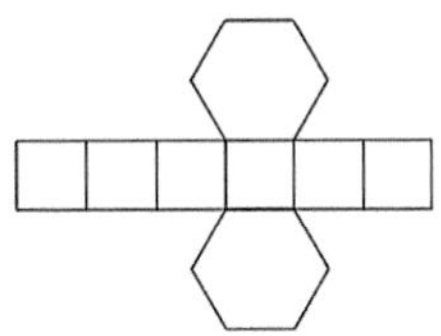	8	$\frac{3\sqrt{3}}{2}$	$3(2+\sqrt{3})$	4.31
Rhombic dodecahedron	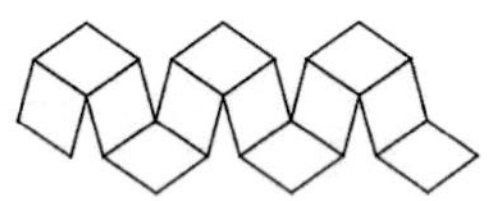	12	$\frac{16}{9}\sqrt{3}$	$8\sqrt{2}$	3.67

Elongated dodecahedron	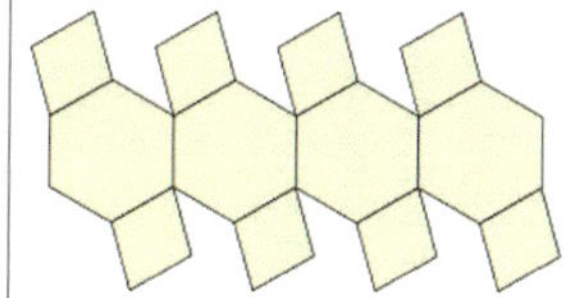	12	6	$2\sqrt{3}(3+\sqrt{5})$	3.02
Trapezo-rhombic dodecahedron	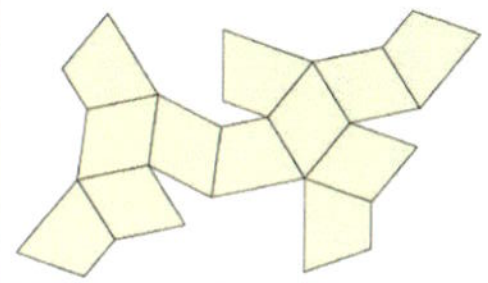	12	$6\sqrt{3}$	$18\sqrt{2}$	2.45
Truncated octahedron	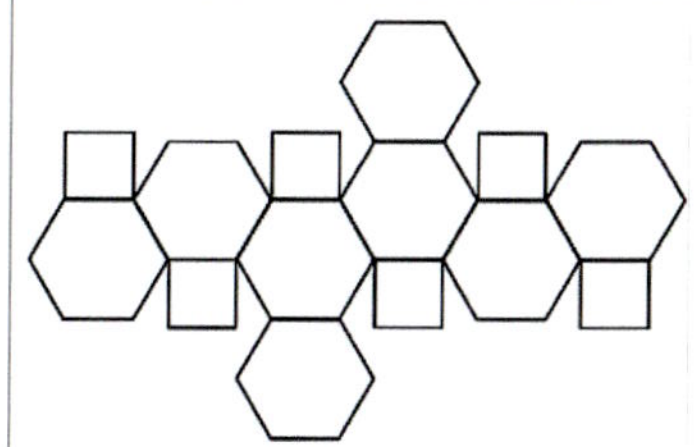	14	$8\sqrt{2}$	$6+12\sqrt{3}$	2.37

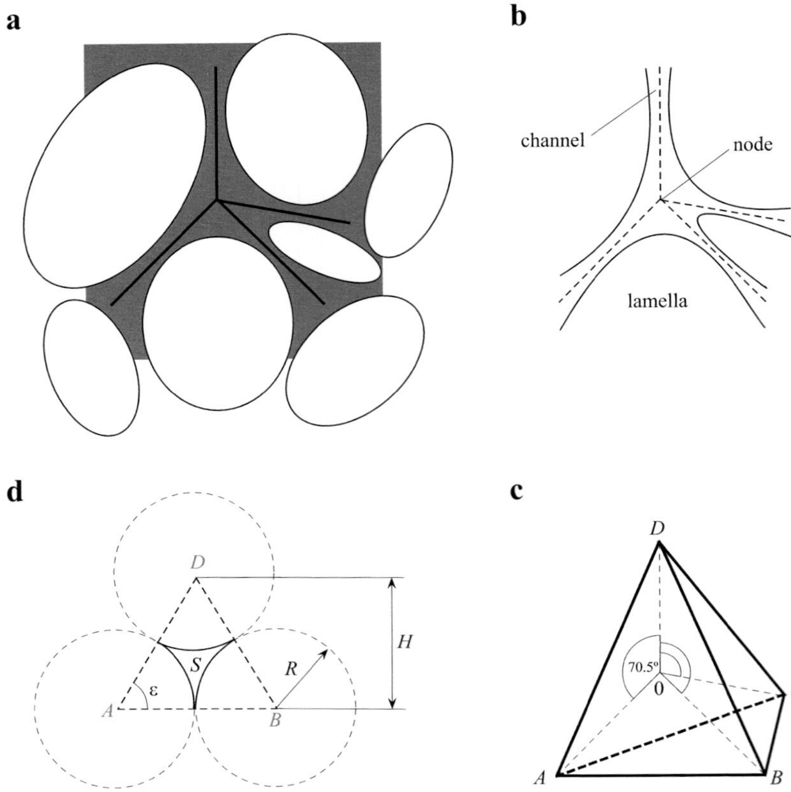

Fig. 2.22 Schematic representation of (**a**) distorted bubbles packed in foam; (**b**) Plateau's border framing the lamellae, in the form of channels joining at nodes; (**c**) tetrahedron-shaped cell, [*ABCD*], with node located at 0 and angles $\angle A0D = \angle B0D = \angle C0D = 70.5°$; and (**d**) plane projection of Plateau's border of cross-sectional surface *S*, inscribed in equilateral triangle of inner angle ε, base $2R$, and height *H*

$$\varepsilon = \frac{\pi}{3}; \tag{2.652}$$

the area S_{cs} of the subtended circular sector accordingly reads

$$S_{cs} = \frac{\varepsilon}{2\pi}\pi R^2, \tag{2.653}$$

where insertion of Eq. (2.652) leaves

$$S_{cs} = \frac{\frac{\pi}{3}}{2\pi} \pi R^2 = \frac{\pi}{6} R^2. \tag{2.654}$$

On the other hand, the height of said triangle abides to

$$\tan \varepsilon = \frac{H}{R}, \tag{2.655}$$

where isolation of H and combination with Eq. (2.652) give rise to

$$H = R \tan \varepsilon = R \tan \frac{\pi}{3} = \sqrt{3} R; \tag{2.656}$$

the surface area of the triangle, S_{tr}, will ensue as

$$S_{tr} = \frac{1}{2} H 2R = HR, \tag{2.657}$$

or else

$$S_{tr} = \sqrt{3} RR = \sqrt{3} R^2 \tag{2.658}$$

with the aid of Eq. (2.656). The area of Plateau's border proper will finally be given by

$$S = S_{tr} - 3S_{cs}, \tag{2.659}$$

where insertion of Eqs. (2.654) and (2.658) produces

$$S = \sqrt{3} R^2 - 3\frac{\pi}{6} R^2 = \left(\sqrt{3} - \frac{\pi}{2}\right) R^2 \tag{2.660}$$

with R^2 meanwhile factored out; isolation of R finally yields

$$R = \frac{\sqrt{S}}{\vartheta_1}, \tag{2.661}$$

with constant ϑ_1 abiding to

$$\vartheta_1 \equiv \sqrt{\sqrt{3} - \frac{\pi}{2}} \tag{2.662}$$

that supports a relationship of the radius of curvature of the lateral surface, R, to the cross-sectional area of Plateau's border, S.

The key assumption in Verbist, Weaire, and Kraynik's model of foam is that drainage occurs only along Plateau's borders, which is strictly valid only for dry foams but approximately valid otherwise; in the wet limit, flow can also occur through the side films, or lamellae—the thickness of which would no longer be negligible compared to that of Plateau's borders. Furthermore, a monomolecular layer of surfactant on the film surfaces reduces surface tension and retards liquid flow, thus producing sufficiently rigid and essentially nonslip conditions on the boundaries of the channels; hence, flow therethrough will resemble Poiseuille's flow. Unlike solid tubing (or porous media), the cross section of Plateau's borders is not fixed in time, and also depends on flow rate itself; in fact, the cross-sectional area is directly proportional to the liquid inventory, so liquid motion throughout the foam will be described by the evolution of S. Further simplifying hypotheses include random and homogeneous network of Plateau's borders, negligible contribution of shearing motion associated with flow through the nodes to overall viscous dissipation (expected in view of the borders being much longer than the nodes themselves), incompressibility and constant viscosity of liquid, constant surface tension at liquid/gas interface (and thus independent of interface elasticity), and constant gas pressure throughout the foam. Although validity of these hypotheses may be somehow questioned, usefulness of the model to be derived therefrom rests largely on the circumstancial evidence of its success in practice.

Assuming, as a first approach, that Plateau's channels are vertically oriented and possess a cylindrical shape, then a balance of momentum, under steady-state conditions, to an elementary annulus of inner radius r, thickness dr, and length dx can be put forward as

$$P_l|_x 2\pi r\,dr - \mu 2\pi r\,dx \left.\frac{\partial v_x}{\partial r}\right|_r = P_l|_{x+dx} 2\pi r\,dr - \mu 2\pi (r+dr)\,dx \left.\frac{\partial v_x}{\partial r}\right|_{r+dr} - \rho g 2\pi r\,dr\,dx; \tag{2.663}$$

the first terms in either side represent contact forces associated with (normal) pressure, the second terms in either side represent contact forces associated with (tangential) stress, and the last term in the right-hand side represents gravitational force active at a distance, with P_l denoting pressure on the liquid side, μ denoting liquid viscosity, and v_x denoting velocity in the vertical direction. Division of both sides by 2π reduces Eq. (2.663) to

$$P_l|_x r\,dr - \mu r \left.\frac{\partial v_x}{\partial r}\right|_r dx = P_l|_{x+dx} r\,dr - \mu(r+dr)\left.\frac{\partial v_x}{\partial r}\right|_{r+dr} dx - \rho g r\,dr\,dx, \tag{2.664}$$

thus paving the way to condensation of terms alike as

$$(P_l|_{x+dx} - P_l|_x) r\,dr - \mu\left((r+dr)\left.\frac{\partial v_x}{\partial r}\right|_{r+dr} - r\left.\frac{\partial v_x}{\partial r}\right|_r\right)dx - \rho g r\,dr\,dx = 0 \tag{2.665}$$

or, recalling the definition of differential,

$$r\,dr\,dP_l - \mu\,dx\,d\left(r\frac{\partial v_x}{\partial r}\right) - \rho g r\,dr\,dx = 0; \tag{2.666}$$

division of both sides by $r\ dr\ dx$ ensues as

$$\frac{\partial P_l}{\partial x} - \mu\frac{1}{r}\frac{\partial}{\partial r}\left(r\frac{\partial v_x}{\partial r}\right) - \rho g = 0, \tag{2.667}$$

thus leaving a combination of a first-order derivative on x as independent variable, a second-order derivative on r as independent variable, and a constant term. Upon isolation of the term containing the compounded differential operators, Eq. (2.667) will read

$$\mu\frac{1}{r}\frac{\partial}{\partial r}\left(r\frac{\partial v_x}{\partial r}\right) = \frac{\partial P_l}{\partial x} - \rho g, \tag{2.668}$$

while multiplication of both sides by r unfolds

$$\mu\frac{\partial}{\partial r}\left(r\frac{\partial v_x}{\partial r}\right) = \left(\frac{\partial P_l}{\partial x} - \rho g\right)r; \tag{2.669}$$

integration is now in order, via separation of variables, as

$$\mu\int d\left(r\frac{\partial v_x}{\partial r}\right) = \left(\frac{\partial P_l}{\partial x} - \rho g\right)\int r\,dr \tag{2.670}$$

on the hypothesis of a constant $\partial P_l/\partial x$, which leads to

$$\mu r\frac{\partial v_x}{\partial r} = \left(\frac{\partial P_l}{\partial x} - \rho g\right)\frac{r^2}{2} + k_1 \tag{2.671}$$

with k_1 denoting a constant or, after dividing both sides by r,

$$\mu \frac{\partial v_x}{\partial r} = \frac{1}{2}\left(\frac{\partial P_l}{\partial x} - \rho g\right) r + \frac{k_1}{r}. \tag{2.672}$$

The vertical velocity remains finite and so must remain its radial derivative everywhere within the liquid channel and, in particular,

$$\left.\frac{\partial v_x}{\partial r}\right|_{r=0} = \text{finite} \tag{2.673}$$

at the channel axis; along the very same location, Eq. (2.672) should read

$$\mu \left.\frac{\partial v_x}{\partial r}\right|_{r=0} = \frac{1}{2}\left(\frac{\partial P_l}{\partial x} - \rho g\right) 0 + \frac{k_1}{0} = 0 + \frac{k_1}{0} = \frac{k_1}{0}, \tag{2.674}$$

which is incompatible with Eq. (2.673) unless

$$k_1 = 0, \tag{2.675}$$

since 0/0 in the right-hand side would then accommodate (as unknown quantity) a finite left-hand side. Insertion of Eq. (2.675) transforms Eq. (2.672) to

$$\mu \frac{\partial v_x}{\partial r} = \frac{1}{2}\left(\frac{\partial P_l}{\partial x} - \rho g\right) r; \tag{2.676}$$

a second integration, via separation of variables, yields

$$\mu \int dv_x = \frac{1}{2}\left(\frac{\partial P_l}{\partial x} - \rho g\right) \int r\, dr, \tag{2.677}$$

equivalent to

$$\mu v_x = \frac{1}{2}\left(\frac{\partial P_l}{\partial x} - \rho g\right) \frac{r^2}{2} + k_2 \tag{2.678}$$

as per the fundamental theorem of integral calculus, with k_2 denoting another arbitrary constant. At the outer radius R_0 of the putatively cylindrical channel,

$$v_x|_{r=R_0} = 0 \tag{2.679}$$

is to be enforced, in view of the essentially static nature of the lamellae when some surfactant is present, otherwise no foam would form and remain in the first place; Eq. (2.679) will accordingly serve as boundary condition to Eq. (2.678), namely,

$$\mu 0 = \frac{1}{2}\left(\frac{\partial P_l}{\partial x} - \rho g\right)\frac{{R_0}^2}{2} + k_2 \tag{2.680}$$

which breaks down to

$$k_2 = -\frac{1}{4}\left(\frac{\partial P_l}{\partial x} - \rho g\right){R_0}^2 \tag{2.681}$$

upon isolation of k_2. Combination with Eq. (2.681) converts Eq. (2.678) to

$$\mu v_x = \frac{1}{4}\left(\frac{\partial P_l}{\partial x} - \rho g\right)r^2 - \frac{1}{4}\left(\frac{\partial P_l}{\partial x} - \rho g\right){R_0}^2, \tag{2.682}$$

where $(\partial P_l/\partial x - \rho g)/4$ will be factored out to get

$$v_x\{x, r\} = -\frac{1}{4\mu}\left(\frac{\partial P_l}{\partial x} - \rho g\right)\left({R_0}^2 - r^2\right) \tag{2.683}$$

along with division of both sides by μ. The average velocity, within [0, R_0] as radial amplitude, can be ascertained via

$$\overline{v}_x\{x\} \equiv \frac{\int\limits_0^{R_0} v_x\{r\} 2\pi r\, dr}{\int\limits_0^{R_0} 2\pi r\, dr}, \tag{2.684}$$

where insertion of Eq. (2.683) unfolds

$$\overline{v}_x = \frac{\int\limits_0^{R_0} \left(-\frac{1}{4\mu}\left(\frac{\partial P_l}{\partial x} - \rho g\right)\left(R_0{}^2 - r^2\right)\right) 2\pi r\, dr}{\int\limits_0^{R_0} 2\pi r\, dr}; \tag{2.685}$$

constant factors will then be taken off kernel as

$$\overline{v}_x = -\frac{\frac{2\pi}{4\mu}\left(\frac{\partial P_l}{\partial x} - \rho g\right)\int\limits_0^{R_0} \left(R_0{}^2 - r^2\right) r\, dr}{2\pi \int\limits_0^{R_0} r\, dr}. \tag{2.686}$$

After cancelling 2π between numerator and denominator, and factoring in r in the kernel, Eq. (2.686) becomes

$$\overline{v}_x = -\frac{1}{4\mu}\left(\frac{\partial P_l}{\partial x} - \rho g\right)\frac{\int\limits_0^{R_0} \left(R_0{}^2 r - r^3\right) dr}{\int\limits_0^{R_0} r\, dr}; \tag{2.687}$$

the fundamental theorem of integral calculus may again be invoked to write

$$\overline{v}_x = -\frac{1}{4\mu}\left(\frac{\partial P_l}{\partial x} - \rho g\right)\frac{\left.\left(R_0{}^2 \frac{r^2}{2} - \frac{r^4}{4}\right)\right|_0^{R_0}}{\left.\frac{r^2}{2}\right|_0^{R_0}}. \tag{2.688}$$

Replacement by upper and lower limits of integration in Eq. (2.688) leaves

$$\overline{v}_x = -\frac{1}{4\mu}\left(\frac{\partial P_l}{\partial x} - \rho g\right)\frac{R_0{}^2 \frac{R_0{}^2}{2} - \frac{R_0{}^4}{4}}{\frac{R_0{}^2}{2}} = -\frac{1}{4\mu}\left(\frac{\partial P_l}{\partial x} - \rho g\right)\frac{2}{R_0{}^2}\left(\frac{R_0{}^4}{2} - \frac{R_0{}^4}{4}\right), \tag{2.689}$$

along with reduction in number of fraction levels; Eq. (2.689) breaks down to

$$\overline{v}_x\{x\} = -\frac{1}{8\mu}\left(\frac{\partial P_l}{\partial x} - \rho g\right){R_0}^2, \tag{2.690}$$

after factoring in $2/{R_0}^2$ and removing the last parenthesis.

Equation (2.690) is strictly valid for a cylindrical channel; however, inspection of Fig. 2.22d indicates a disparate shape. Still under the assumption of creeping flow, one should therefore replace R_0 by the hydraulic radius, R_h, associated with a typical Plateau's channel, defined as

$$R_h \equiv 2\frac{S}{\mathcal{P}}, \tag{2.691}$$

where S denotes cross-sectional area and $\mathcal{P}$ denotes perimeter of liquid in contact with air. Said perimeter is given by

$$\mathcal{P} = 3\frac{\varepsilon}{2\pi}2\pi R, \tag{2.692}$$

corresponding to the three bounding arcs in Fig. 2.22d, each with angular amplitude ε, to be normalized by that of a full angle of 2π rad that would hold an outer perimeter of $2\pi R$; in view of Eq. (2.652), one obtains merely

$$\mathcal{P} = 3\frac{\frac{\pi}{3}}{2\pi}2\pi R = \pi R, \tag{2.693}$$

which may be inserted in Eq. (2.691) together with Eq. (2.660) to obtain

$$R_h = 2\frac{\left(\sqrt{3} - \frac{\pi}{2}\right)R^2}{\pi R}, \tag{2.694}$$

which simplifies to

$$R_h = \frac{2}{\pi}\left(\sqrt{3} - \frac{\pi}{2}\right)R \tag{2.695}$$

after droping R from both numerator and denominator. Replacement of R_0 in Eq. (2.690) by R_h as conveyed by Eq. (2.695) leads to

$$\overline{v}_x = -\frac{1}{8\mu}\left(\frac{\partial P_l}{\partial x} - \rho g\right)\left(\frac{2}{\pi}\left(\sqrt{3} - \frac{\pi}{2}\right)R\right)^2, \tag{2.696}$$

where straightforward algebraic manipulation produces

$$\overline{v}_x = -\frac{1}{8\mu}\left(\frac{\partial P_l}{\partial x} - \rho g\right)\frac{4}{\pi^2}\left(\sqrt{3} - \frac{\pi}{2}\right)\left(\sqrt{3} - \frac{\pi}{2}\right)R^2. \tag{2.697}$$

Equation (2.697) can be rewritten as

$$\overline{v}_x\{x\} = -\frac{1}{\vartheta_2\mu}\left(\frac{\partial P_l}{\partial x} - \rho g\right)S, \tag{2.698}$$

following comparative inspection with Eq. (2.660) and provided that a new constant ϑ_2 is defined as

$$\vartheta_2 \equiv \frac{2\pi^2}{\sqrt{3} - \frac{\pi}{2}} \equiv \frac{2\pi^2}{\vartheta_1{}^2} \tag{2.699}$$

with the aid of Eq. (2.662); hence, a relationship has been obtained for $\overline{v}_x$ as a function of x, indirectly since $P_l \equiv P_l\{x\}$ and $S \equiv S\{x\}$.

In view of the curved triangular sections of Plateau's border, as emphasized in Fig. 2.22d, the pressure across the interface, $P_l - P_g$, should obey Young and Laplace's equation, namely,

$$P_g - P_l = \frac{2\sigma\cos\theta}{R}, \tag{2.700}$$

with R serving as radius of curvature independent of orientation due to its underlying symmetry, coupled with (the hypothesized) uniformity of bubble size; here, σ denotes surface tension, θ denotes contact angle between liquid and gas, and P_l and P_g denote pressure on the liquid and gas side of a lamella, respectively. After inserting Eq. (2.661), one gets

$$P_g - P_l = \frac{2\vartheta_1 \sigma \cos\theta}{\sqrt{S}} \tag{2.701}$$

from Eq. (2.700), usually rewritten as

$$P_g - P_l = \frac{2\vartheta_1 \sigma}{\sqrt{S}}, \tag{2.702}$$

since θ is essentially nil when a surfactant is present (thus implying a unit value for its cosine). Differentiation of both sides of Eq. (2.702) with regard to S leads then to

$$-\frac{\partial P_l}{\partial S} = 2\vartheta_1 \sigma \left(-\frac{1}{2}\right) S^{-\frac{3}{2}}, \tag{2.703}$$

together with the assumption of homogeneous pressure in the gaseous phase; Eq. (2.703) breaks down to

$$\frac{\partial P_l}{\partial S} = \vartheta_1 \sigma S^{-\frac{3}{2}} \tag{2.704}$$

upon dropping of alike factors between numerator and denominator, and taking negatives of both sides afterward. Following multiplication of both sides by $\partial S/\partial x$, Eq. (2.704) turns to

$$\frac{\partial P_l}{\partial S}\frac{\partial S}{\partial x} = \vartheta_1 \sigma S^{-\frac{3}{2}} \frac{\partial S}{\partial x}, \tag{2.705}$$

which may be rephrased as

$$\frac{\partial P_l}{\partial x} = \vartheta_1 \sigma S^{-\frac{3}{2}} \frac{\partial S}{\partial x} \tag{2.706}$$

once dS is cancelled between numerator and denominator; insertion in Eq. (2.698) generates

$$\overline{v}_x\{x\} = -\frac{1}{\vartheta_2 \mu}\left(\vartheta_1 \sigma S^{-\frac{3}{2}} \frac{\partial S}{\partial x} - \rho g\right) S, \tag{2.707}$$

or else

$$\bar{v}_x\{x\} = \frac{S}{\vartheta_2\mu}\left(\rho g - \vartheta_1\sigma S^{-\frac{3}{2}}\frac{\partial S}{\partial x}\right) \tag{2.708}$$

after factoring in -1.

Remember that the continuity equation regarding (scalar) quantity χ takes the general form

$$\nabla\cdot(\chi\boldsymbol{v}) + \frac{\partial\chi}{\partial t} = 0, \tag{2.709}$$

where $\boldsymbol{v}$ denotes velocity vector; for instance, $\chi \equiv V$ converts Eq. (2.709) to

$$\nabla\cdot(V\boldsymbol{v}) + \frac{\partial V}{\partial t} = 0, \tag{2.710}$$

with expansion of the scalar product giving

$$V(\nabla\cdot\boldsymbol{v}) + \boldsymbol{v}\cdot\nabla V + \frac{\partial V}{\partial t} = 0 \tag{2.711}$$

that, for $\nabla V = \mathbf{0}$, simplifies as

$$V(\nabla\cdot\boldsymbol{v}) + \frac{\partial V}{\partial t} = 0. \tag{2.712}$$

If the quantity under scrutiny is cross-sectional area S, then Eq. (2.709) takes the form

$$\nabla\cdot(S\boldsymbol{v}) + \frac{\partial S}{\partial t} = 0; \tag{2.713}$$

since only the vertical direction is currently of interest,

$$\frac{\partial}{\partial x}(Sv_x) + \frac{\partial S}{\partial t} = 0 \tag{2.714}$$

may be used hereafter in lieu of Eq. (2.713). Insertion of Eq. (2.708) transforms Eq. (2.714) to

$$\frac{\partial}{\partial x}\left(S\frac{S}{\vartheta_2\mu}\left(\rho g - \vartheta_1\sigma S^{-\frac{3}{2}}\frac{\partial S}{\partial x}\right)\right) + \frac{\partial S}{\partial t} = 0, \tag{2.715}$$

where S^2 will be factored in and $1/\vartheta_2\mu$ factored out to obtain

$$\frac{1}{\vartheta_2\mu}\frac{\partial}{\partial x}\left(\rho g S^2 - \vartheta_1\sigma S^{\frac{1}{2}}\frac{\partial S}{\partial x}\right) + \frac{\partial S}{\partial t} = 0 \tag{2.716}$$

as underlying partial differential equation; solution of Eq. (2.716) will accordingly convey $S \equiv S\{x,t\}$ pertaining to the hypothesized single, vertical channel.

It should be emphasized that Plateau's borders are not (all) vertical, yet it may be assumed, to a good approximation, that the foam network consists of randomly oriented channels. If one such channel is tilted at a polar angle ω relative to the vertical, then longitudinal coordinate x should be replaced by its ratio to cos ω—because a longer path, x/cos ω (susceptible of a larger viscous drop of pressure), would have to be tracked for a given vertical displacement x; this rationale is illustrated in Fig. 2.23a. By the same token, the gravitational force per unit volume, acting along the direction of flow, will be lower than ρg—more specifically, equal to ρg cos ω; the corresponding sketch is provided in Fig. 2.23b.

Under the above circumstances, Eq. (2.716) should be corrected as

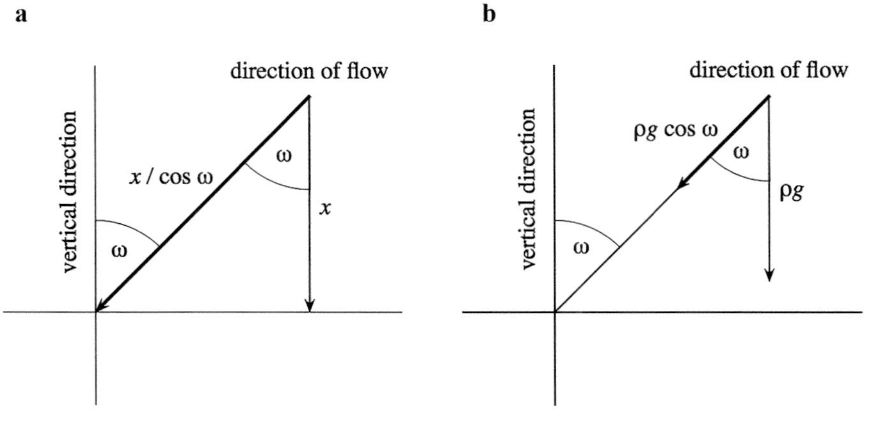

Fig. 2.23 Correction for (**a**) longer path, x/cos ω, than vertical path, x, and (**b**) lower driving force, ρg cos ω, than vertical weight per unit volume, ρg, associated with liquid draining in a direction tilted by angle ω relative to the vertical direction

$$\frac{1}{\vartheta_2\mu}\frac{\partial}{\partial\left(\frac{x}{\cos\omega}\right)}\left(S^2\rho g\cos\omega-\vartheta_1\sigma S^{\frac{1}{2}}\frac{\partial S}{\partial\left(\frac{x}{\cos\omega}\right)}\right)+\frac{\partial S}{\partial t}=0, \tag{2.717}$$

where cos ω—for being set in advance for each orientation, can be taken off the differential operators, namely,

$$\frac{\cos\omega}{\vartheta_2\mu}\frac{\partial}{\partial x}\left(\rho g S^2\cos\omega-\vartheta_1\sigma S^{\frac{1}{2}}\frac{\partial S}{\partial x}\cos\omega\right)+\frac{\partial S}{\partial t}=0; \tag{2.718}$$

the outstanding cosine function can as well be taken off the outer differential operator to get

$$\frac{\cos^2\omega}{\vartheta_2\mu}\frac{\partial}{\partial x}\left(\rho g S^2-\vartheta_1\sigma S^{\frac{1}{2}}\frac{\partial S}{\partial x}\right)+\frac{\partial S}{\partial t}=0. \tag{2.719}$$

The average of Eq. (2.719), taken over the whole range of values of ω, will accordingly ensue as

$$\overline{\frac{\cos^2\omega}{\vartheta_2\mu}\frac{\partial}{\partial x}\left(\rho g S^2-\vartheta_1\sigma S^{\frac{1}{2}}\frac{\partial S}{\partial x}\right)+\frac{\partial S}{\partial t}}=\overline{0}, \tag{2.720}$$

which degenerates to just

$$\frac{\overline{\cos^2\omega}}{\vartheta_2\mu}\frac{\partial}{\partial x}\left(\rho g S^2-\vartheta_1\sigma S^{\frac{1}{2}}\frac{\partial S}{\partial x}\right)+\frac{\partial S}{\partial t}=0 \tag{2.721}$$

in view of linearity of the average operator. Since every value of ω is equally probable, interval $[0, 2\pi]$ may be taken as a reference as it reflects the periodicity of the trigonometric functions; therefore,

$$\overline{\cos^2\omega}=\frac{\int_0^{2\pi}\cos^2\omega\,d\omega}{\int_0^{2\pi}d\omega}. \tag{2.722}$$

Recall the trigonometric formula of duplication of an angle, namely,

$$\cos 2\omega = \cos^2\omega - \sin^2\omega, \tag{2.723}$$

which can be combined with the fundamental formula of trigonometry, that is,

$$\cos^2\omega + \sin^2\omega = 1, \tag{2.724}$$

to eventually produce

$$\cos 2\omega = \cos^2\omega - \left(1 - \cos^2\omega\right) = 2\cos^2\omega - 1; \tag{2.725}$$

isolation of $\cos^2\omega$ prompts

$$\cos^2\omega = \frac{1 + \cos 2\omega}{2}. \tag{2.726}$$

Insertion of Eq. (2.726) converts Eq. (2.722) to

$$\overline{\cos^2\omega} = \frac{\int_0^{2\pi} \frac{1 + \cos 2\omega}{2} d\omega}{\int_0^{2\pi} d\omega} = \frac{\int_0^{2\pi} (1 + \cos 2\omega) d\omega}{2\int_0^{2\pi} d\omega}, \tag{2.727}$$

with 1/2 already taken off the kernel for its constancy; the fundamental theorem of integral calculus has it that

$$\overline{\cos^2\omega} = \frac{\left(\omega + \frac{\sin 2\omega}{2}\right)\Big|_0^{2\pi}}{2\omega\Big|_0^{2\pi}}, \tag{2.728}$$

which degenerates to

$$\overline{\cos^2\omega} = \frac{2\pi + \frac{\sin 2\cdot 2\pi}{2} - 0 - \frac{\sin 2\cdot 0}{2}}{2(2\pi - 0)} = \frac{2\pi + \frac{\sin 4\pi}{2} - \frac{\sin 0}{2}}{4\pi} = \frac{2\pi + \frac{0}{2} - \frac{0}{2}}{4\pi} = \frac{2\pi}{4\pi} = \frac{1}{2}. \tag{2.729}$$

Insertion of Eq. (2.729) supports conversion of Eq. (2.721) to

$$\frac{\frac{1}{2}}{\vartheta_2 \mu} \frac{\partial}{\partial x}\left(\rho g S^2 - \vartheta_1 \sigma S^{\frac{1}{2}} \frac{\partial S}{\partial x}\right) + \frac{\partial S}{\partial t} = 0; \tag{2.730}$$

or else

$$\frac{1}{\vartheta_3 \mu} \frac{\partial}{\partial x}\left(\rho g S^2 - \vartheta_1 \sigma S^{\frac{1}{2}} \frac{\partial S}{\partial x}\right) + \frac{\partial S}{\partial t} = 0, \tag{2.731}$$

as long as parameter ϑ_3 abides to

$$\vartheta_3 \equiv \frac{4\pi^2}{\sqrt{3} - \frac{\pi}{2}} \equiv 2\vartheta_2, \tag{2.732}$$

written with the aid of Eq. (2.699). At first sight, it is not obvious that the foregoing adjustment would correctly account for the effect of the nodes where the channels merge; nevertheless, it does indeed and with remarkable accuracy. As a matter of fact, the conservation rule on each junction is satisfied by gravity-driven flow at the symmetrical tetrahedral nodes, for which one realizes that

$$-\cos\angle A0D - \cos\angle B0D - \cos\angle C0D + \cos\angle D0D = -3\cos 70.5^\circ + \cos 0^\circ = -3\frac{1}{3} + 1 = 0, \tag{2.733}$$

with reference to Fig. 2.22c; in other words, $\cos\omega$ adds up to 0 over those four channels, whenever they meet.

Inspection of the functional form of Eq. (2.731) prompts definition of auxiliary independent variables

$$\xi \equiv \sqrt{\frac{\rho g}{\vartheta_1 \sigma}}\, x \tag{2.734}$$

and

$$\tau \equiv \frac{\sqrt{\vartheta_1 \sigma \rho g}}{\vartheta_3 \mu}\, t, \tag{2.735}$$

for vertical location and elapsed time, respectively; further to auxiliary surface area as

$$\alpha \equiv \frac{\rho g}{\vartheta_1 \sigma} S, \tag{2.736}$$

all of which are dimensionless. Differentiation of Eq. (2.734) unfolds

$$dx = \sqrt{\frac{\vartheta_1 \sigma}{\rho g}} d\xi, \tag{2.737}$$

and likewise

$$dt = \frac{\vartheta_3 \mu}{\sqrt{\vartheta_1 \sigma \rho g}} d\tau \tag{2.738}$$

arises from Eq. (2.735); while

$$dS = \frac{\vartheta_1 \sigma}{\rho g} d\alpha \tag{2.739}$$

stems from Eq. (2.736). Combination with Eqs. (2.736)–(2.739) converts Eq. (2.731) to

$$\frac{1}{\vartheta_3 \mu} \frac{\partial}{\sqrt{\frac{\vartheta_1 \sigma}{\rho g}} \partial \xi} \left(\rho g \left(\frac{\vartheta_1 \sigma}{\rho g} \alpha \right)^2 - \vartheta_1 \sigma \left(\frac{\vartheta_1 \sigma}{\rho g} \alpha \right)^{\frac{1}{2}} \frac{\frac{\vartheta_1 \sigma}{\rho g} \partial \alpha}{\sqrt{\frac{\vartheta_1 \sigma}{\rho g}} \partial \xi} \right) + \frac{\frac{\vartheta_1 \sigma}{\rho g} \partial \alpha}{\frac{\vartheta_3 \mu}{\sqrt{\vartheta_1 \sigma \rho g}} \partial \tau} = 0, \tag{2.740}$$

where constant factors will be taken off the differential operators as

$$\frac{1}{\vartheta_3 \mu \sqrt{\frac{\vartheta_1 \sigma}{\rho g}}} \frac{\partial}{\partial \xi} \left(\rho g \left(\frac{\vartheta_1 \sigma}{\rho g} \right)^2 \alpha^2 - \vartheta_1 \sigma \left(\frac{\vartheta_1 \sigma}{\rho g} \right)^{\frac{1}{2}} \alpha^{\frac{1}{2}} \frac{\frac{\vartheta_1 \sigma}{\rho g}}{\sqrt{\frac{\vartheta_1 \sigma}{\rho g}}} \frac{\partial \alpha}{\partial \xi} \right) + \frac{\frac{\vartheta_1 \sigma}{\rho g}}{\frac{\vartheta_3 \mu}{\sqrt{\vartheta_1 \sigma \rho g}}} \frac{\partial \alpha}{\partial \tau} = 0; \tag{2.741}$$

condensation of factors alike and reduction in number of fraction levels produce

$$\frac{\rho^{\frac{1}{2}} g^{\frac{1}{2}}}{\vartheta_3 \mu \vartheta_1^{\frac{1}{2}} \sigma^{\frac{1}{2}}} \frac{\partial}{\partial \xi} \left(\frac{\vartheta_1^2 \sigma^2}{\rho g} \alpha^2 - \frac{\vartheta_1^2 \sigma^2}{\rho g} \alpha^{\frac{1}{2}} \frac{\partial \alpha}{\partial \xi} \right) + \frac{\vartheta_1^{\frac{3}{2}} \sigma^{\frac{3}{2}}}{\vartheta_3 \mu \rho^{\frac{1}{2}} g^{\frac{1}{2}}} \frac{\partial \alpha}{\partial \tau} = 0, \tag{2.742}$$

while factoring out of $\vartheta_1^2 \sigma^2 / \rho g$ leaves

$$\frac{\rho^{\frac{1}{2}} g^{\frac{1}{2}}}{\vartheta_3 \mu \vartheta_1^{\frac{1}{2}} \sigma^{\frac{1}{2}}} \frac{\vartheta_1^2 \sigma^2}{\rho g} \frac{\partial}{\partial \xi} \left(\alpha^2 - \alpha^{\frac{1}{2}} \frac{\partial \alpha}{\partial \xi} \right) + \frac{\vartheta_1^{\frac{3}{2}} \sigma^{\frac{3}{2}}}{\vartheta_3 \mu \rho^{\frac{1}{2}} g^{\frac{1}{2}}} \frac{\partial \alpha}{\partial \tau} = 0. \tag{2.743}$$

Condensation of similar factors in Eq. (2.743) gives rise to

$$\frac{\vartheta_1^{\frac{3}{2}}\sigma^{\frac{3}{2}}}{\vartheta_3\mu\rho^{\frac{1}{2}}g^{\frac{1}{2}}}\frac{\partial}{\partial\xi}\left(\alpha^2-\alpha^{\frac{1}{2}}\frac{\partial\alpha}{\partial\xi}\right)+\frac{\vartheta_1^{\frac{3}{2}}\sigma^{\frac{3}{2}}}{\vartheta_3\mu\rho^{\frac{1}{2}}g^{\frac{1}{2}}}\frac{\partial\alpha}{\partial\tau}=0 \tag{2.744}$$

or, after dividing both sides by $\vartheta_1^{\frac{3}{2}}\sigma^{\frac{3}{2}}/\vartheta_3\mu\rho^{\frac{1}{2}}g^{\frac{1}{2}}$,

$$\frac{\partial}{\partial\xi}\left(\alpha^2-\alpha^{\frac{1}{2}}\frac{\partial\alpha}{\partial\xi}\right)+\frac{\partial\alpha}{\partial\tau}=0; \tag{2.745}$$

Equation (2.745), as nonlinear, partial differential equation, conveys $\alpha \equiv \alpha\{\xi,\tau\}$ as solution. As previously outlined, the model of foam drainage developed above is quite accurate relative to experimental evidence, yet it fails at longer times when foam coarsening (or disproportionation), due to interbubble gas diffusion, occurs to a considerable extent; this is a form of Ostwald's ripening, with larger bubbles growing at the expense of smaller ones.

Despite its usefulness in attempts to ascertain the rate of spontaneous foam draining, remember that S in Eq. (2.731) represents cross-sectional surface of Plateau's channels; liquid fraction, ϕ_l, would indeed be more informative. Should V_b denote volume of each bubble, N_b denote number of bubbles per unit volume of foam, N_p denote number of Plateau's channels per bubble, and L denote channel length, then ϕ_l would obey

$$\phi_l=\frac{N_bN_pSL}{N_bV_b+N_bN_pSL}, \tag{2.746}$$

where division of both numerator and denominator by N_bN_pL unfolds

$$\phi_l=\frac{S}{\frac{V_b}{N_pL}+S}. \tag{2.747}$$

Hence, one obtains $\phi_l \equiv \phi_l\{S, V_b/N_pL\}$, where only two geometric parameters are germane: cross-sectional surface (i.e., S), obtainable from solution of Eq. (2.716), and ratio of bubble volume (i.e., V_b) to full length of Plateau's channel per bubble (i.e., N_pL), characteristic of the foam itself. In view of the functional form of Eq. (2.745) vis-à-vis with Eq. (2.716), one should again revisit Eq. (2.736) as

$$S=\frac{\vartheta_1\sigma}{\rho g}\alpha; \tag{2.748}$$

insertion of Eq. (2.748) converts Eq. (2.747) to

$$\phi_l = \frac{\frac{\vartheta_1 \sigma}{\rho g}\alpha}{\frac{V_b}{N_p L} + \frac{\vartheta_1 \sigma}{\rho g}\alpha}, \tag{2.749}$$

where multiplication of both numerator and denominator by $\rho g / \vartheta_1 \sigma$ yields

$$\phi_l = \frac{\alpha}{\frac{\rho g V_b}{\vartheta_1 \sigma N_p L} + \alpha} \tag{2.750}$$

that unfolds a rectangular hyperbola on α. An increase in liquid fraction will relax the well-defined angle of Plateau's border present in dry foams; said liquid concentrates in the channels, thus alleviating strain on the surfaces and allowing them to adopt a tendentially spherical configuration, after a critical threshold of liquid fraction, $\phi_{l,crt}$, is reached. In a true wet foam, almost no interactions occur between surfaces, so the whole network behaves like a suspension of nearly undistorted bubbles.

In a bioreactor where gas is continuously sparged at the bottom of the liquid medium, and rises to the surface carrying some liquid therewith, a steady state will eventually be attained; it should be defined by

$$\frac{\partial \alpha}{\partial \tau} = 0, \tag{2.751}$$

regarding the dimensionless form of the cross-sectional area of foam, α. Equation (2.751) allows simplification of Eq. (2.745) to

$$\frac{d}{d\xi}\left(\alpha^2 - \alpha^{\frac{1}{2}} \frac{d\alpha}{d\xi}\right) = 0, \tag{2.752}$$

which entails an ordinary, nonlinear differential equation of α on ξ. At this stage, it is convenient to revisit Eq. (2.708) as

$$Q = \frac{S^2}{\vartheta_2 \mu}\left(\rho g - \vartheta_1 \sigma S^{-\frac{3}{2}} \frac{dS}{dx}\right), \tag{2.753}$$

where Q, given by

$$Q \equiv S\overline{v}_x, \tag{2.754}$$

denotes volumetric flow rate of liquid being drained downward from the foam network or, owing to the aforementioned steady state, of liquid brought to the base of the foam by the ascending gas bubbles; upon combination with Eqs. (2.737), (2.739), (2.748), and (2.754), one obtains

$$S\overline{v}_x = \frac{\left(\frac{\vartheta_1\sigma}{\rho g}\alpha\right)^2}{\vartheta_2\mu}\left(\rho g - \vartheta_1\sigma\left(\frac{\vartheta_1\sigma}{\rho g}\alpha\right)^{-\frac{3}{2}}\frac{\frac{\vartheta_1\sigma}{\rho g}d\alpha}{\sqrt{\frac{\vartheta_1\sigma}{\rho g}}d\xi}\right) \tag{2.755}$$

from Eq. (2.753), where constant factors will be taken off the differential operators to get

$$S\overline{v}_x = \frac{\left(\frac{\vartheta_1\sigma}{\rho g}\right)^2}{\vartheta_2\mu}\alpha^2\left(\rho g - \frac{\vartheta_1\sigma}{\left(\frac{\vartheta_1\sigma}{\rho g}\right)^{\frac{3}{2}}}\alpha^{-\frac{3}{2}}\frac{\frac{\vartheta_1\sigma}{\rho g}}{\sqrt{\frac{\vartheta_1\sigma}{\rho g}}}\frac{d\alpha}{d\xi}\right). \tag{2.756}$$

After splitting or lumping powers of $\vartheta_1\sigma/\rho g$ (as appropriate), Eq. (2.756) reduces to

$$S\overline{v}_x = \frac{(\vartheta_1\sigma)^2}{\vartheta_2\mu\rho g}\frac{\alpha^2}{\rho g}\left(\rho g - \frac{\vartheta_1\sigma}{\frac{\vartheta_1\sigma}{\rho g}}\alpha^{-\frac{3}{2}}\frac{d\alpha}{d\xi}\right), \tag{2.757}$$

while the coefficient preceding $\alpha^{-\frac{3}{2}}$ will be algebraically rearranged to read

$$S\overline{v}_x = \frac{{\vartheta_1}^2\sigma^2}{\vartheta_2\mu\rho g}\frac{\alpha^2}{\rho g}\left(\rho g - \rho g\,\alpha^{-\frac{3}{2}}\frac{d\alpha}{d\xi}\right); \tag{2.758}$$

once $\alpha^2/\rho g$ is factored in, one gets

$$S\overline{v}_x = \frac{{\vartheta_1}^2\sigma^2}{\vartheta_2\mu\rho g}\left(\alpha^2 - \alpha^{\frac{1}{2}}\frac{d\alpha}{d\xi}\right). \tag{2.759}$$

Upon multiplication of both sides by $\vartheta_2\mu\rho g/{\vartheta_1}^2\sigma^2$, Eq. (2.759) becomes

$$\frac{\vartheta_2\mu\rho g}{{\vartheta_1}^2\sigma^2}S\overline{v}_x = \alpha^2 - \alpha^{\frac{1}{2}}\frac{d\alpha}{d\xi}, \tag{2.760}$$

where notation simplifies to

$$\alpha^2 - \alpha^{\frac{1}{2}}\frac{d\alpha}{d\xi} = Q^* \tag{2.761}$$

provided that the dimensionless volumetric flow rate, Q^*, satisfies

$$Q^* \equiv \frac{\vartheta_2 \mu \rho g}{{\vartheta_1}^2 \sigma^2} S\overline{v}_x; \tag{2.762}$$

since Q^* must be constant through the foam for the sake of steady-state operation, one realizes that

$$Q^* = {\alpha_0}^2 \tag{2.763}$$

on account of $d\alpha/d\xi = 0$ under those conditions, so Eq. (2.763) may serve as boundary condition for Eq. (2.752), with ${\alpha_0}^2$ representing dimensionless volumetric flow rate of supply of liquid to the wet foam. Integration of Eq. (2.752) ensues as

$$\int d\left(\alpha^2 - \alpha^{\frac{1}{2}}\frac{d\alpha}{d\xi}\right) = \int 0 d\xi, \tag{2.764}$$

which is the same as writing

$$\alpha^2 - \alpha^{\frac{1}{2}}\frac{d\alpha}{d\xi} = k_1; \tag{2.765}$$

arbitrary constant k_1 can then be calculated upon combination of Eqs. (2.761), (2.763), and (2.765), that is,

$$k_1 = {\alpha_0}^2, \tag{2.766}$$

which converts Eq. (2.765) to

$$\alpha^2 - \alpha^{\frac{1}{2}}\frac{d\alpha}{d\xi} = {\alpha_0}^2. \tag{2.767}$$

Note that Q^* being constant everywhere throughout the foam does not preclude $\alpha \equiv \alpha\{\xi\}$; in fact, volumetric flow rate is equal to product of cross-sectional surface of flow by (average) velocity of flow as per Eq. (2.754), so $\overline{v}_x$ will actually increase as S reduces (and vice versa) along the vertical coordinate. Isolation of the differential term in Eq. (2.767) unfolds

$$\frac{d\alpha}{d\xi}=\frac{\alpha^2-{\alpha_0}^2}{\alpha^{\frac{1}{2}}}, \tag{2.768}$$

which suggests definition of an auxiliary variable χ as

$$\chi \equiv \alpha^{\frac{1}{2}} \tag{2.769}$$

or, equivalently,

$$\alpha=\chi^2 \tag{2.770}$$

upon squaring both sides; differentiation then generates

$$d\alpha=2\chi\, d\chi. \tag{2.771}$$

Insertion of Eqs. (2.769)–(2.771) converts Eq. (2.768) to

$$\frac{2\chi\, d\chi}{d\xi}=\frac{\left(\chi^2\right)^2-{\alpha_0}^2}{\chi}, \tag{2.772}$$

so integration via separation of variables is in order, namely,

$$2\int\frac{\chi^2}{\left(\chi^2\right)^2-{\alpha_0}^2}d\chi=\int d\xi; \tag{2.773}$$

after splitting the denominator in the kernel of the first integral as product of two conjugated binomials on χ^2, Eq. (2.773) gives rise to

$$2\int\frac{\chi^2}{\left(\chi^2-\alpha_0\right)\left(\chi^2+\alpha_0\right)}d\chi=\int d\xi \tag{2.774}$$

and a similar splitting of $\chi^2-\alpha_0$ leaves, in turn,

$$2\int \frac{\chi^2}{\left(\chi-\sqrt{\alpha_0}\right)\left(\chi+\sqrt{\alpha_0}\right)\left(\chi^2+\alpha_0\right)}d\chi=\int d\xi. \tag{2.775}$$

The poles of the kernel in the left-hand side of Eq. (2.775) become accessible via solution of

$$\left(\chi-\sqrt{\alpha_0}\right)\left(\chi+\sqrt{\alpha_0}\right)\left(\chi^2+\alpha_0\right)=0, \tag{2.776}$$

which degenerates to

$$\chi=\sqrt{\alpha_0}\vee\chi=-\sqrt{\alpha_0}\vee\chi=\pm\sqrt{-\alpha_0} \tag{2.777}$$

or, in a more condensed fashion,

$$\chi=\pm\sqrt{\alpha_0}\vee\chi=\pm\imath\sqrt{\alpha_0} \tag{2.778}$$

with $\imath\equiv\sqrt{-1}$ denoting imaginary unit; Heaviside's decomposition of the kernel of Eq. (2.775) will accordingly take the form

$$\frac{\chi^2}{\left(\chi-\sqrt{\alpha_0}\right)\left(\chi+\sqrt{\alpha_0}\right)\left(\chi^2+\alpha_0\right)}=\frac{k_1}{\chi-\sqrt{\alpha_0}}+\frac{k_2}{\chi+\sqrt{\alpha_0}}+\frac{k_3+k_4\chi}{\chi^2+\alpha_0} \tag{2.779}$$

where k_1, k_2, k_3, and k_4 denote constants (to be determined), with elimination of denominators producing

$$\begin{aligned}\chi^2=&\,k_1\left(\chi+\sqrt{\alpha_0}\right)\left(\chi^2+\alpha_0\right)+k_2\left(\chi-\sqrt{\alpha_0}\right)\left(\chi^2+\alpha_0\right)\\&+\left(k_3+k_4\chi\right)\left(\chi-\sqrt{\alpha_0}\right)\left(\chi+\sqrt{\alpha_0}\right).\end{aligned} \tag{2.780}$$

Since the equality labelled as Eq. (2.780) must be obeyed for every value of χ, it should, in particular, guarantee

$$\begin{aligned}\left(-\sqrt{\alpha_0}\right)^2=&\,k_1\left(-\sqrt{\alpha_0}+\sqrt{\alpha_0}\right)\left(\left(-\sqrt{\alpha_0}\right)^2+\alpha_0\right)+k_2\left(-\sqrt{\alpha_0}-\sqrt{\alpha_0}\right)\left(\left(-\sqrt{\alpha_0}\right)^2+\alpha_0\right)\\&+\left(k_3+k_4\left(-\sqrt{\alpha_0}\right)\right)\left(-\sqrt{\alpha_0}-\sqrt{\alpha_0}\right)\left(-\sqrt{\alpha_0}+\sqrt{\alpha_0}\right)\end{aligned} \tag{2.781}$$

at $\chi=-\sqrt{\alpha_0}$; this simplifies to

$$\alpha_0 = k_1 0 \cdot 2\alpha_0 + k_2(-2\sqrt{\alpha_0})2\alpha_0 + (k_3 - k_4\sqrt{\alpha_0})(-2\sqrt{\alpha_0})0 = -4\alpha_0\sqrt{\alpha_0}k_2, \quad (2.782)$$

after collapsing terms alike and composing square root with square power or, upon isolation of k_2,

$$k_2 = -\frac{1}{4\sqrt{\alpha_0}}. \quad (2.783)$$

By the same token, $\chi = \sqrt{\alpha_0}$ converts Eq. (2.780) to

$$\begin{aligned}(\sqrt{\alpha_0})^2 &= k_1(\sqrt{\alpha_0} + \sqrt{\alpha_0})\left((\sqrt{\alpha_0})^2 + \alpha_0\right) + k_2(\sqrt{\alpha_0} - \sqrt{\alpha_0})\left((\sqrt{\alpha_0})^2 + \alpha_0\right) \\ &+ (k_3 + k_4\sqrt{\alpha_0})(\sqrt{\alpha_0} - \sqrt{\alpha_0})(\sqrt{\alpha_0} + \sqrt{\alpha_0})\end{aligned}, \quad (2.784)$$

where similar terms can be lumped and squares cancelled with roots as

$$\alpha_0 = k_1 2\sqrt{\alpha_0}\, 2\alpha_0 + k_2 0 \cdot 2\alpha_0 + (k_3 + k_4\sqrt{\alpha_0})0 \cdot 2\sqrt{\alpha_0} = 4\alpha_0\sqrt{\alpha_0}k_1; \quad (2.785)$$

after solving for k_1, Eq. (2.785) yields

$$k_1 = \frac{1}{4\sqrt{\alpha_0}}. \quad (2.786)$$

Another two (independent) equations are required to calculate k_3 and k_4, which may similarly be generated via setting α_0 in Eq. (2.780) equal to, say, 0 and 1; this leads to

$$0^2 = k_1(0 + \sqrt{\alpha_0})(0^2 + \alpha_0) + k_2(0 - \sqrt{\alpha_0})(0^2 + \alpha_0) + (k_3 + k_4 0)(0 - \sqrt{\alpha_0})(0 + \sqrt{\alpha_0}) \quad (2.787)$$

and

$$1^2 = k_1(1 + \sqrt{\alpha_0})(1^2 + \alpha_0) + k_2(1 - \sqrt{\alpha_0})(1^2 + \alpha_0) + (k_3 + k_4 1)(1 - \sqrt{\alpha_0})(1 + \sqrt{\alpha_0}), \quad (2.788)$$

respectively. Equation (2.787) gives rise to

$$k_1\sqrt{\alpha_0}\,\alpha_0 + k_2(-\sqrt{\alpha_0})\alpha_0 + k_3(-\sqrt{\alpha_0})\sqrt{\alpha_0} = 0, \tag{2.789}$$

which breaks down to

$$k_1\alpha_0\sqrt{\alpha_0} - k_2\alpha_0\sqrt{\alpha_0} - k_3\alpha_0 = 0 \tag{2.790}$$

upon lumping factors alike, or else

$$k_1\sqrt{\alpha_0} - k_2\sqrt{\alpha_0} - k_3 = 0 \tag{2.791}$$

once both sides are divided by α_0; isolation of k_3 then unfolds

$$k_3 = (k_1 - k_2)\sqrt{\alpha_0}, \tag{2.792}$$

where $\sqrt{\alpha_0}$ was duly factored out. Insertion of Eqs. (2.783) and (2.786) transforms Eq. (2.792) to

$$k_3 = \left(\frac{1}{4\sqrt{\alpha_0}} - \left(-\frac{1}{4\sqrt{\alpha_0}}\right)\right)\sqrt{\alpha_0}, \tag{2.793}$$

with elimination of parentheses leaving

$$k_3 = \frac{2}{4\sqrt{\alpha_0}}\sqrt{\alpha_0}; \tag{2.794}$$

lumping of factors alike will leave merely

$$k_3 = \frac{1}{2}. \tag{2.795}$$

Equation (2.788) can similarly be redone to

$$k_1(1+\sqrt{\alpha_0})(1+\alpha_0) + k_2(1-\sqrt{\alpha_0})(1+\alpha_0) + (k_3+k_4)(1-\sqrt{\alpha_0})(1+\sqrt{\alpha_0}) = 1, \tag{2.796}$$

where $1 + \alpha_0$ will be factored out, inner parentheses removed, and $1 - \sqrt{\alpha_0}$ and $1 + \sqrt{\alpha_0}$ lumped with each other to get

$$\left(k_1 + k_1\sqrt{\alpha_0} + k_2 - k_2\sqrt{\alpha_0}\right)(1+\alpha_0) + (k_3+k_4)\left(1^2 - \left(\sqrt{\alpha_0}\right)^2\right) = 1; \tag{2.797}$$

after factoring out $\sqrt{\alpha_0}$, Eq. (2.797) becomes

$$\left(k_1 + k_2 + (k_1 - k_2)\sqrt{\alpha_0}\right)(1+\alpha_0) + (k_3+k_4)(1-\alpha_0) = 1. \tag{2.798}$$

Combination with Eqs. (2.783), (2.786), and (2.795) converts Eq. (2.798) to

$$\left(\frac{1}{4\sqrt{\alpha_0}} - \frac{1}{4\sqrt{\alpha_0}} + \left(\frac{1}{4\sqrt{\alpha_0}} - \left(-\frac{1}{4\sqrt{\alpha_0}}\right)\right)\sqrt{\alpha_0}\right)(1+\alpha_0) + \left(\frac{1}{2} + k_4\right)(1-\alpha_0) = 1, \tag{2.799}$$

where symmetrical terms will be dropped off and similar terms condensed to obtain

$$\frac{2}{4\sqrt{\alpha_0}}\sqrt{\alpha_0}(1+\alpha_0) + \left(\frac{1}{2} + k_4\right)(1-\alpha_0) = 1; \tag{2.800}$$

$2\sqrt{\alpha_0}$ then cancels out between numerator and denominator, namely,

$$\frac{1}{2}(1+\alpha_0) + \left(\frac{1}{2} + k_4\right)(1-\alpha_0) = 1. \tag{2.801}$$

After subtracting $(1 + \alpha_0)/2$ from both sides, Eq. (2.801) yields

$$\left(\frac{1}{2} + k_4\right)(1-\alpha_0) = 1 - \frac{1}{2} - \frac{1}{2}\alpha_0 = \frac{1}{2} - \frac{1}{2}\alpha_0 = \frac{1}{2}(1-\alpha_0), \tag{2.802}$$

along with lumping of terms alike and eventual factoring out of 1/2; division of both sides by $1 - \alpha_0$ is now in order, according to

$$\frac{1}{2} + k_4 = \frac{1}{2}, \tag{2.803}$$

thus implying

$$k_4 = 0. \tag{2.804}$$

In view of Eqs. (2.783), (2.786), (2.795), and (2.804), one will redo Eq. (2.779) as

$$\frac{\chi^2}{\left(\chi-\sqrt{\alpha_0}\right)\left(\chi+\sqrt{\alpha_0}\right)\left(\chi^2+\alpha_0\right)}=\frac{\frac{1}{4\sqrt{\alpha_0}}}{\chi-\sqrt{\alpha_0}}+\frac{-\frac{1}{4\sqrt{\alpha_0}}}{\chi+\sqrt{\alpha_0}}+\frac{\frac{1}{2}+0\chi}{\chi^2+\alpha_0}, \tag{2.805}$$

which simplifies to

$$\frac{\chi^2}{\left(\chi-\sqrt{\alpha_0}\right)\left(\chi+\sqrt{\alpha_0}\right)\left(\chi^2+\alpha_0\right)}=\frac{1}{4\sqrt{\alpha_0}}\left(\frac{1}{\chi-\sqrt{\alpha_0}}-\frac{1}{\chi+\sqrt{\alpha_0}}\right)+\frac{1}{2}\frac{1}{\chi^2+\alpha_0} \tag{2.806}$$

once $1/4\sqrt{\alpha_0}$ is factored out and the nil term dropped; after dividing both sides by 2, combination with Eq. (2.806) converts Eq. (2.775) to

$$\int\frac{1}{4\sqrt{\alpha_0}}\left(\frac{1}{\chi-\sqrt{\alpha_0}}-\frac{1}{\chi+\sqrt{\alpha_0}}\right)d\chi+\int\frac{1}{2}\frac{1}{\chi^2+\alpha_0}d\chi=\frac{1}{2}\int d\xi, \tag{2.807}$$

where the splitting theorem for integrals was revisited. Equation (2.807) should, to advantage, be algebraically rearranged as

$$\frac{1}{4\sqrt{\alpha_0}}\int\left(\frac{1}{\chi-\sqrt{\alpha_0}}-\frac{1}{\chi+\sqrt{\alpha_0}}\right)d\chi+\frac{1}{2\sqrt{\alpha_0}}\int\frac{1}{1+\left(\frac{\chi}{\sqrt{\alpha_0}}\right)^2}d\left(\frac{\chi}{\sqrt{\alpha_0}}\right)=\frac{1}{2}\int d\xi \tag{2.808}$$

upon taking constant factors off the integrals, and dividing both numerator and denominator of the kernel in the second integral by $\alpha_0=\sqrt{\alpha_0}\sqrt{\alpha_0}$; following multiplication of both sides by $2\sqrt{\alpha_0}$, Eq. (2.808) becomes

$$\frac{1}{2}\int\left(\frac{1}{\chi-\sqrt{\alpha_0}}-\frac{1}{\chi+\sqrt{\alpha_0}}\right)d\chi+\int\frac{1}{1+\left(\frac{\chi}{\sqrt{\alpha_0}}\right)^2}d\left(\frac{\chi}{\sqrt{\alpha_0}}\right)=\sqrt{\alpha_0}\int d\xi. \tag{2.809}$$

For physical significance, $\chi<\sqrt{\alpha_0}$ (see Eq. (2.769)), so one should rephrase Eq. (2.809) as

$$-\frac{1}{2}\int\left(\frac{1}{\sqrt{\alpha_0}-\chi}+\frac{1}{\sqrt{\alpha_0}+\chi}\right)d\chi+\int\frac{1}{1+\left(\frac{\chi}{\sqrt{\alpha_0}}\right)^2}d\left(\frac{\chi}{\sqrt{\alpha_0}}\right)=\sqrt{\alpha_0}\int d\xi, \tag{2.810}$$

thus assuring that the denominators of both terms in the first kernel are positive. One may eventually proceed to

$$-\frac{1}{2}\left(\frac{\ln\{\sqrt{\alpha_0}-\chi\}}{-1}+\ln\{\sqrt{\alpha_0}+\chi\}\right)+\tan^{-1}\frac{\chi}{\sqrt{\alpha_0}}=\sqrt{\alpha_0}\,\xi+k_2 \tag{2.811}$$

with k_2 denoting a constant, as a result of direct application of the fundamental theorem of integral calculus to Eq. (2.810), or else

$$-\frac{1}{2}(\ln\{\sqrt{\alpha_0}+\chi\}-\ln\{\sqrt{\alpha_0}-\chi\})+\tan^{-1}\frac{\chi}{\sqrt{\alpha_0}}=\sqrt{\alpha_0}\,\xi+k_2; \tag{2.812}$$

the two logarithmic functions are now to be collapsed as

$$-\frac{1}{2}\ln\frac{\sqrt{\alpha_0}+\chi}{\sqrt{\alpha_0}-\chi}+\tan^{-1}\frac{\chi}{\sqrt{\alpha_0}}=\sqrt{\alpha_0}\,\xi+k_2. \tag{2.813}$$

Both numerator and denominator of the argument of the outstanding logarithm in Eq. (2.813) can be divided by $\sqrt{\alpha_0}$ to get

$$-\frac{1}{2}\ln\frac{1+\frac{\chi}{\sqrt{\alpha_0}}}{1-\frac{\chi}{\sqrt{\alpha_0}}}+\tan^{-1}\frac{\chi}{\sqrt{\alpha_0}}=\sqrt{\alpha_0}\,\xi+k_2; \tag{2.814}$$

the first term coincides with the (negative of the) inverse hyperbolic tangent, thus allowing notation be simplified as

$$-\tanh^{-1}\frac{\chi}{\sqrt{\alpha_0}}+\tan^{-1}\frac{\chi}{\sqrt{\alpha_0}}=\sqrt{\alpha_0}\,\xi+k_2. \tag{2.815}$$

The original notation, as per Eq. (2.769), will finally be retrieved to rephrase Eq. (2.815) to

$$\tan^{-1}\frac{\sqrt{\alpha}}{\sqrt{\alpha_0}}-\tanh^{-1}\frac{\sqrt{\alpha}}{\sqrt{\alpha_0}}=\sqrt{\alpha_0}\,\xi+k_2; \tag{2.816}$$

condensation of square powers is finally in order, namely,

$$\tan^{-1}\sqrt{\frac{\alpha}{\alpha_0}}-\tanh^{-1}\sqrt{\frac{\alpha}{\alpha_0}}=\sqrt{\alpha_0}\,\xi+k_2 \tag{2.817}$$

that entails $\alpha\equiv\alpha\{\xi\}$ in implicit form, yet still containing an arbitrary constant. At this stage, one may resort to

$$\alpha\big|_{\xi=0}=\alpha\big|_{\phi_l=\phi_{l,crt}}\equiv\alpha_{crt} \tag{2.818}$$

for boundary condition, as long as bulk liquid remains in contact with foam (as is normally the case); here, α_{crt} denotes the critical value of α, corresponding to ϕ_l equal to $\phi_{l,crt}$ (known as rigidity-loss transition) when bubbles start coming apart, while $\xi = 0$ corresponds to the bottom of foam (and thus to the top of bulk liquid). Upon elimination of denominators, Eq. (2.750) takes the form

$$\frac{\rho g V_b}{\vartheta_1 \sigma N_p L}\phi_l+\phi_l\alpha=\alpha, \tag{2.819}$$

which can be solved for α as

$$\alpha=\frac{\rho g V_b}{\vartheta_1 \sigma N_p L}\frac{\phi_l}{1-\phi_l}; \tag{2.820}$$

in particular, Eq. (2.820) justifies

$$\alpha_{crt}=\frac{\rho g V_b}{\vartheta_1 \sigma N_p L}\frac{\phi_{l,crt}}{1-\phi_{l,crt}}, \tag{2.821}$$

meant to be inserted in Eq. (2.818). Compatibility between Eqs. (2.817) and (2.818) requires

$$\tan^{-1}\sqrt{\frac{\alpha_{crt}}{\alpha_0}}-\tanh^{-1}\sqrt{\frac{\alpha_{crt}}{\alpha_0}}=\sqrt{\alpha_0}\,0+k_2, \tag{2.822}$$

or just

$$k_2=\tan^{-1}\sqrt{\frac{\alpha_{crt}}{\alpha_0}}-\tanh^{-1}\sqrt{\frac{\alpha_{crt}}{\alpha_0}} \tag{2.823}$$

upon removal of the nil term; the arbitrary constant may thus be taken off Eq. (2.817) to get

$$\tan^{-1}\sqrt{\frac{\alpha}{\alpha_0}} - \tanh^{-1}\sqrt{\frac{\alpha}{\alpha_0}} = \sqrt{\alpha_0}\,\xi + \tan^{-1}\sqrt{\frac{\alpha_{crt}}{\alpha_0}} - \tanh^{-1}\sqrt{\frac{\alpha_{crt}}{\alpha_0}}, \tag{2.824}$$

which will finally appear as

$$\sqrt{\alpha_0}\,\xi = \tan^{-1}\sqrt{\frac{\alpha}{\alpha_0}} - \tan^{-1}\sqrt{\frac{\alpha_{crt}}{\alpha_0}} - \tanh^{-1}\sqrt{\frac{\alpha}{\alpha_0}} + \tanh^{-1}\sqrt{\frac{\alpha_{crt}}{\alpha_0}} \tag{2.825}$$

upon isolation of $\sqrt{\alpha_0}\,\xi$. A graphical interpretation of Eq. (2.825) is provided in Fig. 2.24, as α/α_0-vs.-$\sqrt{\alpha_0}\,\xi$, knowing that $\alpha < \alpha_{crt} < \alpha_0$. In the bulk liquid, sufficiently below the foam line (described by $\sqrt{\alpha_0}\,\xi = 0$), $\alpha/\alpha_0 \to 1$; this is consistent with Eq. (2.825), that is,

$$\begin{aligned} \lim_{\xi \to -\infty} \sqrt{\alpha_0}\,\xi = -\infty &= \frac{\pi}{4} - \tan^{-1}\sqrt{\frac{\alpha_{crt}}{\alpha_0}} - \infty + \tanh^{-1}\sqrt{\frac{\alpha_{crt}}{\alpha_0}} \\ = \tan^{-1}\sqrt{1} - \tan^{-1}\sqrt{\frac{\alpha_{crt}}{\alpha_0}} &- \tanh^{-1}\sqrt{1} + \tanh^{-1}\sqrt{\frac{\alpha_{crt}}{\alpha_0}} = \lim_{\frac{\alpha}{\alpha_0} \to 1} \sqrt{\alpha_0}\,\xi \end{aligned} \tag{2.826}$$

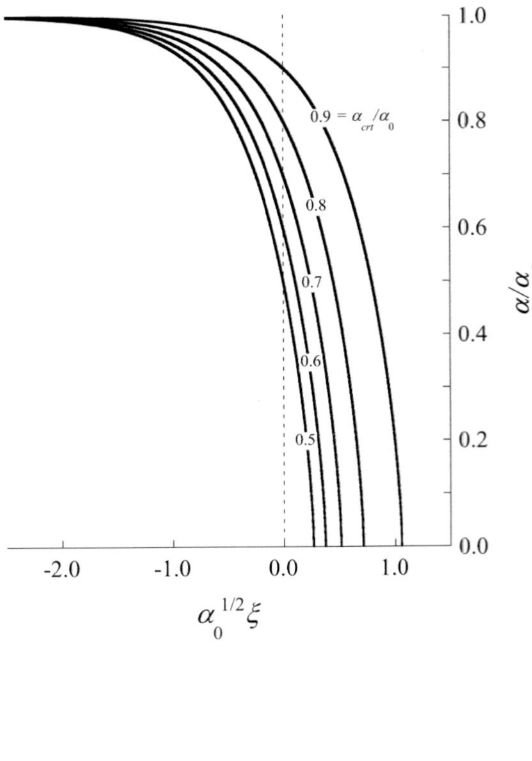

Fig. 2.24 Variation of normalized cross-sectional area of Plateau's channels in foam, α/α_0, versus dimensionless vertical coordinate, $\sqrt{\alpha_0}\,\xi$ (—), for selected values of α_{crt}/α_0, with indication of interface (- - -)

serves as horizontal asymptote to the lines in Fig. 2.24. The value of the relative cross-sectional surface of liquid then decreases gradually until reaching α_{crt}, which sets the interface between bubbled liquid and foam; decrease thereafter becomes more abrupt, and reflects thinner and thinner Plateau's borders in the foam as one climbs it up, until α becomes nil at some finite value of $\sqrt{\alpha_0}\,\xi$ corresponding to foam collapse. The loci of the horizontal intercepts in Fig. 2.24, or coordinates $(\sqrt{\alpha_0}\,\xi_{max}, 0)$, are described by

$$\sqrt{\alpha_0}\,\xi_{max} = \tan^{-1}\sqrt{0} - \tan^{-1}\sqrt{\frac{\alpha_{crt}}{\alpha_0}} - \tanh^{-1}\sqrt{0} + \tanh^{-1}\sqrt{\frac{\alpha_{crt}}{\alpha_0}} \tag{2.827}$$

stemming from Eq. (2.825); after realizing that $\tan^{-1}\sqrt{0} = \tan^{-1}0 = 0$ and $\tanh^{-1}\sqrt{0} = \tanh^{-1}0 = 0$, Eq. (2.827) reduces to

$$\sqrt{\alpha_0}\,\xi_{max} = \tanh^{-1}\sqrt{\frac{\alpha_{crt}}{\alpha_0}} - \tan^{-1}\sqrt{\frac{\alpha_{crt}}{\alpha_0}} \tag{2.828}$$

that conveys $\xi_{max} \equiv \xi_{max}\{\alpha_{crt}\}$, as illustrated in Fig. 2.25. The maximum height attained by foam, under steady-state operation, grows smoothly as α_{crt}/α_0 increases; however, it undergoes a sudden increase in the vicinity of unity. Once in possession of α via solution of

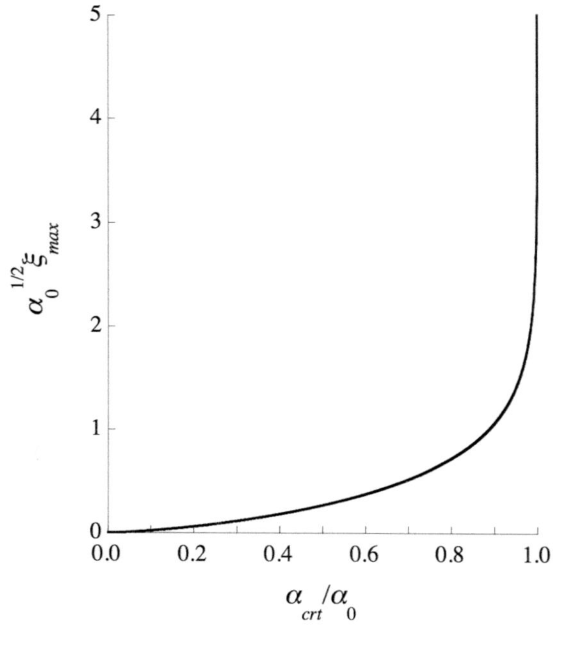

Fig. 2.25 Variation of dimensionless maximum height of foam, $\sqrt{\alpha_0}\,\xi_{max}$, versus normalized critical cross-sectional area of Plateau's channels in foam, α_{crt}/α_0

Eq. (2.825), one may calculate the steady-state profile of liquid fraction, ϕ_l, in gas/liquid system by resorting to Eq. (2.750).

Despite the general validity, and remarkable accuracy of Eq. (2.828), its application in practice suffers from the requirement of ${\alpha_0}^2$, as alias to volumetric flow rate of liquid dragged by the gas bubbles, Q^* (see Eq. (2.763)); the volumetric flow rate of gas is indeed more readily accessible. However, in a typical bioreactor, bulk liquid is stirred and either recycled (in batch operation) or supplied/withdrawn (in continuous operation) perpendicularly to the foam interface; whereas gas is bubbled at the bottom of the liquid, spans the liquid height until reaching the foam, remains temporarily therein, and eventually leaves the bioreactor when foam breaks apart, thus calling for a wavelike solution to Eq. (2.745). If a solitary wave moves upward with velocity v_w, then modeling should preferentially resort to a frame moving with said velocity, thus implying $\alpha \equiv \alpha\{\zeta - \tau\}$ as basic postulate coupled with definition of location variable z as

$$z \equiv x - v_w t; \tag{2.829}$$

insertion of Eqs. (2.734) and (2.735) then generates

$$z = \sqrt{\frac{\vartheta_1 \sigma}{\rho g}}\xi - v_w \frac{\vartheta_3 \mu}{\sqrt{\vartheta_1 \sigma \rho g}}\tau. \tag{2.830}$$

Multiplication of both sides of Eq. (2.830) by $\sqrt{\rho g / \vartheta_1 \sigma}$ unfolds

$$\sqrt{\frac{\rho g}{\vartheta_1 \sigma}} z = \sqrt{\frac{\rho g}{\vartheta_1 \sigma}}\sqrt{\frac{\vartheta_1 \sigma}{\rho g}}\xi - v_w \sqrt{\frac{\rho g}{\vartheta_1 \sigma}}\frac{\vartheta_3 \mu}{\sqrt{\vartheta_1 \sigma \rho g}}\tau, \tag{2.831}$$

with lumping of similar factors leaving

$$\sqrt{\frac{\rho g}{\vartheta_1 \sigma}} z = \xi - \frac{\vartheta_3 \mu}{\vartheta_1 \sigma} v_w \tau; \tag{2.832}$$

Equation (2.832) may be redone as

$$\zeta \equiv \xi - v_w^* \tau, \tag{2.833}$$

provided that (dimensionless) auxiliary variable ξ and parameter v_w^* satisfy

$$\zeta = \sqrt{\frac{\rho g}{\vartheta_1 \sigma}}\, z \tag{2.834}$$

and

$$v_w^* \equiv \frac{\vartheta_3 \mu}{\vartheta_1 \sigma} v_w. \tag{2.835}$$

Under such circumstances, the partial derivatives of α with regard to ξ and τ will look like

$$\frac{\partial \alpha}{\partial \xi} = \frac{d\alpha}{d\zeta}\frac{\partial \zeta}{\partial \xi} = \frac{d\alpha}{d\zeta} \tag{2.836}$$

and

$$\frac{\partial \alpha}{\partial \tau} = \frac{d\alpha}{d\zeta}\frac{\partial \zeta}{\partial \tau} = -v_w^* \frac{d\alpha}{d\zeta}, \tag{2.837}$$

respectively, based on Eq. (2.833). In view of Eqs. (2.836) and (2.837), one can redo Eq. (2.745) as

$$\frac{d}{d\zeta}\left(\alpha^2 - \alpha^{\frac{1}{2}}\frac{d\alpha}{d\zeta}\right) - v_w^* \frac{d\alpha}{d\zeta} = 0, \tag{2.838}$$

where the two derivatives can be joined as

$$\frac{d}{d\zeta}\left(\alpha^2 - v_w^* \alpha - \alpha^{\frac{1}{2}}\frac{d\alpha}{d\zeta}\right) = 0; \tag{2.839}$$

Equation (2.839) should thus be used in lieu of Eq. (2.745) in the problem under scrutiny, with the major advantage residing in its being an ordinary rather than partial differential equation. After realizing that

$$\alpha^2 - v_w^* \alpha - \alpha^{\frac{1}{2}}\frac{d\alpha}{d\zeta} = \kappa = 0 \tag{2.840}$$

is equivalent to Eq. (2.839), where the arbitrary constant κ will hereafter be considered nil for simplicity, one should revisit the change of dependent variable labelled as

Eq. (2.769), together with (equivalent) Eqs. (2.770) and (2.771), to reformulate Eq. (2.840) as

$$\chi^4 - v_w^* \chi^2 - \chi \frac{2\chi \, d\chi}{d\zeta} = 0. \tag{2.841}$$

Division of both sides by χ^2 simplifies Eq. (2.841) to

$$\chi^2 - v_w^* - \frac{2 \, d\chi}{d\zeta} = 0, \tag{2.842}$$

or else

$$2\frac{d\chi}{d\zeta} = \chi^2 - v_w^* \tag{2.843}$$

upon isolation of the differential term. Integration of Eq. (2.843) ensues via separation of variables, namely,

$$2\int \frac{d\chi}{\chi^2 - v_w^*} = \int d\zeta, \tag{2.844}$$

where the difference in squares in denominator will be converted to the product of two conjugated binomials as

$$2\int \frac{d\chi}{\left(\chi - \sqrt{v_w^*}\right)\left(\chi + \sqrt{v_w^*}\right)} = \int d\zeta. \tag{2.845}$$

The kernel of the first integral in Eq. (2.845) is thus to be splitted as

$$\frac{1}{\left(\chi - \sqrt{v_w^*}\right)\left(\chi + \sqrt{v_w^*}\right)} = \frac{k_5}{\chi - \sqrt{v_w^*}} + \frac{k_6}{\chi + \sqrt{v_w^*}} \tag{2.846}$$

with k_5 and k_6 denoting constants (to be determined), while elimination of denominators leaves

$$1 = k_5\left(\chi + \sqrt{v_w^*}\right) + k_6\left(\chi - \sqrt{v_w^*}\right); \tag{2.847}$$

for simplicity, one may start by setting $\chi = -\sqrt{v_w}$ to obtain

$$1 = k_5\left(-\sqrt{v_w^*} + \sqrt{v_w^*}\right) + k_6\left(-\sqrt{v_w^*} - \sqrt{v_w^*}\right) = -2\sqrt{v_w^*}\,k_6, \tag{2.848}$$

which can be solved for k_6 as

$$k_6 = -\frac{1}{2\sqrt{v_w^*}}. \tag{2.849}$$

By the same token, $\chi = \sqrt{v_w^*}$ converts Eq. (2.847) to

$$1 = k_5\left(\sqrt{v_w^*} + \sqrt{v_w^*}\right) + k_6\left(\sqrt{v_w^*} - \sqrt{v_w^*}\right) = 2\sqrt{v_w^*}\,k_5, \tag{2.850}$$

where division of both sides by $2\sqrt{v_w^*}$ unfolds

$$k_5 = \frac{1}{2\sqrt{v_w^*}}; \tag{2.851}$$

insertion of Eqs. (2.849) and (2.851) transforms Eq. (2.846) to

$$\frac{1}{\left(\chi - \sqrt{v_w^*}\right)\left(\chi + \sqrt{v_w^*}\right)} = \frac{\frac{1}{2\sqrt{v_w^*}}}{\chi - \sqrt{v_w^*}} + \frac{-\frac{1}{2\sqrt{v_w^*}}}{\chi + \sqrt{v_w^*}} = \frac{1}{2\sqrt{v_w^*}}\left(\frac{1}{\chi - \sqrt{v_w^*}} - \frac{1}{\chi + \sqrt{v_w^*}}\right), \tag{2.852}$$

along with factoring out of $1/2\sqrt{v_w^*}$. In view of Eq. (2.852), it becomes possible to rephrase Eq. (2.845) as

$$2\int \frac{1}{2\sqrt{v_w^*}}\left(\frac{1}{\chi - \sqrt{v_w^*}} - \frac{1}{\chi + \sqrt{v_w^*}}\right)d\chi = \int d\zeta \tag{2.853}$$

or, after taking $-1/2\sqrt{v_w^*}$ off the integral for its constancy,

$$-\frac{1}{\sqrt{v_w^*}}\int\left(\frac{1}{\chi + \sqrt{v_w^*}} - \frac{1}{\chi - \sqrt{v_w^*}}\right)d\chi = \int d\zeta; \tag{2.854}$$

both sides will now be multiplied by $-\sqrt{v_w^*}/2$ to get

$$\frac{1}{2}\int\left(\frac{1}{\chi+\sqrt{v_w^*}}-\frac{1}{\chi-\sqrt{v_w^*}}\right)d\chi=-\frac{\sqrt{v_w^*}}{2}\int d\zeta. \tag{2.855}$$

After noting that $\chi > \sqrt{v_w^*}$ in practice, the theorem of integral calculus converts Eq. (2.855) to

$$\frac{1}{2}\left(\ln\{\chi+\sqrt{v_w^*}\}-\ln\{\chi-\sqrt{v_w^*}\}\right)=k_7-\frac{\sqrt{v_w^*}}{2}\zeta \tag{2.856}$$

with positive (and thus meaningful) arguments for both logarithmic functions and where k_7 denotes some arbitrary constant; after collapsing the two logarithmic functions, Eq. (2.856) will read

$$\frac{1}{2}\ln\frac{\chi+\sqrt{v_w^*}}{\chi-\sqrt{v_w^*}}=k_7-\frac{\sqrt{v_w^*}}{2}\zeta \tag{2.857}$$

with division of both numerator and denominator of the argument of the logarithmic function yielding

$$\frac{1}{2}\ln\frac{\frac{\chi}{\sqrt{v_w^*}}+1}{\frac{\chi}{\sqrt{v_w^*}}-1}=k_7-\frac{\sqrt{v_w^*}}{2}\zeta. \tag{2.858}$$

Notation in Eq. (2.858) can be condensed to advantage as

$$\coth^{-1}\frac{\chi}{\sqrt{v_w^*}}=k_7-\frac{\sqrt{v_w^*}}{2}\zeta; \tag{2.859}$$

the starting notation, as conveyed by Eq. (2.769), will now be retrieved as

$$\coth^{-1}\frac{\sqrt{\alpha}}{\sqrt{v_w^*}}=k_7-\frac{\sqrt{v_w^*}}{2}\zeta. \tag{2.860}$$

After taking hyperbolic cotangent of both sides, Eq. (2.860) gives rise to

$$\frac{\sqrt{\alpha}}{\sqrt{v_w^*}}=\coth\left\{k_7-\frac{\sqrt{v_w^*}}{2}\zeta\right\}, \tag{2.861}$$

where multiplication of both sides by $\sqrt{v_w^*}$ unfolds

$$\sqrt{\alpha} = \sqrt{v_w} \coth\left\{ k_7 - \frac{\sqrt{v_w^*}}{2}\zeta \right\}; \tag{2.862}$$

squares are finally to be taken of both sides as

$$\alpha = v_w^* \coth^2\left\{ k_7 - \frac{\sqrt{v_w^*}}{2}\zeta \right\}. \tag{2.863}$$

In view of Eq. (2.833), one may reformulate Eq. (2.863) as

$$\alpha\{\xi, \tau; v_w^*\} = v_w^* \coth^2\left\{ k_7 - \frac{\sqrt{v_w^*}(\xi - v_w^*\tau)}{2} \right\} \quad ; \quad \xi \leq v_w^*\tau \tag{2.864}$$

as general solution to Eq. (2.865); whereas the complementary solution looks trivially like

$$\alpha = 0 \quad ; \quad \xi > v_w^*\tau, \tag{2.865}$$

because foam cannot develop anywhere vertically along the liquid before the gas wave reaches that location. When v_w^* approaches zero, Eq. (2.864) supports

$$\lim_{v_w^* \to 0} \alpha = 0 \coth^2\left\{ k_7 - \frac{\sqrt{0}(\xi - 0\tau)}{2} \right\} = 0 \coth^2\left\{ k_7 - \frac{0\xi}{2} \right\} = 0 \coth^2 k_7. \tag{2.866}$$

On the other hand, if no flow (or $Q^* = 0$) took place through the foam, then Eq. (2.761) would degenerate to

$$\alpha^2 - \alpha^{\frac{1}{2}} \frac{d\alpha}{d\xi} = 0, \tag{2.867}$$

which simplifies to

$$\frac{d\alpha}{d\xi} = \alpha^{\frac{3}{2}} \tag{2.868}$$

upon isolation of the differential term, complemented by division of both sides by $\alpha^{\frac{1}{2}} \neq 0$; here $\alpha \equiv \alpha\{\xi\}$, for being an equilibrium (and thus stationary) solution, was meanwhile taken into acount. Integration of Eq. (2.868) ensues, via separation of variables, as

$$\int \alpha^{-\frac{3}{2}}\, d\alpha = \int d\xi \tag{2.869}$$

or, after bringing the fundamental theorem of integral calculus on board,

$$\frac{\alpha^{-\frac{1}{2}}}{-\frac{1}{2}} = \xi + k_8; \tag{2.870}$$

k_8 denotes a (necessarily finite) integration constant. Equation (2.870) may be rewritten as

$$\alpha^{-\frac{1}{2}} = -\frac{1}{2}(\xi + k_8) \tag{2.871}$$

or, after taking reciprocals,

$$\alpha^{\frac{1}{2}} = -\frac{1}{\frac{1}{2}(\xi + k_8)} = -\frac{2}{\xi + k_8}; \tag{2.872}$$

once squares are taken of both sides, one obtains

$$\alpha = \frac{4}{(\xi + k_8)^2}. \tag{2.873}$$

Upon retrieval of Eq. (2.818) as

$$\alpha\big|_{\xi=0} = \alpha_{crt}, \tag{2.874}$$

it may be used as boundary condition for Eq. (2.867), thus enforcing

$$\alpha_{crt} = \frac{4}{(0 + k_8)^2} = \frac{4}{k_8{}^2} \tag{2.875}$$

when combined with Eq. (2.873); isolation of k_8 gives rise to

$$k_8 = \frac{2}{\sqrt{\alpha_{crt}}}, \tag{2.876}$$

so Eq. (2.873) will finally look like

$$\lim_{v_w^* \to 0} \alpha = \frac{4}{\left(\xi + \frac{2}{\sqrt{\alpha_{cri}}}\right)^2}. \tag{2.877}$$

Inspection of the right-hand side of Eq. (2.877) indicates that α always takes a finite value for $\xi \geq 0$, that is,

$$\lim_{v_w^* \to 0} \alpha = \text{finite}; \tag{2.878}$$

elimination of $\lim_{v_w^* \to 0} \alpha$ between Eqs. (2.866) and (2.878) yields

$$0 \coth^2 k_7 = \text{finite} \tag{2.879}$$

which would not be feasible unless

$$\coth^2 k_7 = \infty, \tag{2.880}$$

since $0 \cdot \infty$ for left-hand side of Eq. (2.879), as unknown quantity, is the only option that accommodates Eq. (2.878). After taking square roots of Eq. (2.880), one gets

$$\coth k_7 = \infty, \tag{2.881}$$

which is equivalent to

$$k_7 = 0; \tag{2.882}$$

insertion of Eq. (2.882) converts Eq. (2.864) to

$$\alpha = v_w^* \coth^2 \left\{ 0 - \frac{\sqrt{v_w^*}\left(\xi - v_w^* \tau\right)}{2} \right\}, \tag{2.883}$$

or merely

$$\alpha = v_w^* \coth^2\left\{-\frac{\sqrt{v_w^*}\left(\xi - v_w^*\tau\right)}{2}\right\}; \tag{2.884}$$

since the odd nature of hyperbolic cotangent enforces an even behavior for its square,

$$\alpha\left\{\xi, \tau, v_w^*\right\} = v_w^* \coth^2 \frac{\sqrt{v_w^*}\left(\xi - v_w^*\tau\right)}{2} \tag{2.885}$$

may be utilized in lieu of Eq. (2.884).

A steady state will eventually be attained, where rate of supply of gas as bubbles at the bottom of liquid in the bioreactor balances rate of disappearance thereof at the top of foam as bubbles break out; this obviously assumes no rupture, and even maintenance of size of bubbles along the path in between, which also implies constancy of v irrespective of x. The frame of reference in which the foam network develops is static, for moving upward with the bubbles; consequently, liquid will be drained at velocity v everywhere, so as to assure steady state. Under these conditions, $v_w^*\tau$ in Eq. (2.885) should be replaced by some constant ξ^*, namely,

$$\alpha = v_w^* \coth^2 \frac{\sqrt{v_w^*}\left(\xi - \xi^*\right)}{2} = v_w^* \coth^2\left\{\frac{\sqrt{v_w^*}\,\xi}{2} - \frac{\sqrt{v_w^*}\,\xi^*}{2}\right\}, \tag{2.886}$$

subjected to

$$\alpha\big|_{\xi=0} = \alpha_{crt} \tag{2.887}$$

as boundary condition in parallel to Eq. (2.818); combination with Eq. (2.887) transforms Eq. (2.886) to

$$\alpha_{crt} = v_w^* \coth^2\left\{\frac{\sqrt{v_w^*}\,0}{2} - \frac{\sqrt{v_w^*}\,\xi^*}{2}\right\}, \tag{2.888}$$

where nil terms can be dropped to advantage to get

$$\alpha_{crt} = v_w^* \coth^2\left\{-\frac{\sqrt{v_w^*}\,\xi^*}{2}\right\}. \tag{2.889}$$

Upon dividing both sides by v_w^* and taking square roots afterward, Eq. (2.889) becomes

$$\sqrt{\frac{\alpha_{crt}}{v_w^*}} = \coth\left\{ -\frac{\sqrt{v_w^*}\,\xi^*}{2} \right\}, \tag{2.890}$$

thus paving the way to take inverse hyperbolic cotangent of both sides as

$$\coth^{-1}\sqrt{\frac{\alpha_{crt}}{v_w^*}} = -\frac{\sqrt{v_w^*}\,\xi^*}{2}; \tag{2.891}$$

insertion of Eq. (2.891) permits elimination of $-\sqrt{v_w^*}\,\xi^*/2$ from Eq. (2.886) as

$$\alpha = v_w^* \coth^2\left\{ \frac{\sqrt{v_w^*}\,\xi}{2} + \coth^{-1}\sqrt{\frac{\alpha_{crt}}{v_w^*}} \right\}. \tag{2.892}$$

By the same token, one realizes that

$$\alpha\Big|_{\xi=\xi_{max}} = \alpha_{min}, \tag{2.893}$$

where ξ_{max} denotes the maximum value taken by ξ, or foam height, and α_{min} denotes the minimum value of α compatible with a foam, corresponding to attainment of a critical pressure difference between Plateau's borders and thin liquid films—above which the latter breaks; Eq. (2.892) becomes, in this case,

$$\alpha_{min} = v_w^* \coth^2\left\{ \frac{\sqrt{v_w^*}\,\xi_{max}}{2} + \coth^{-1}\sqrt{\frac{\alpha_{crt}}{v_w^*}} \right\}. \tag{2.894}$$

Once again, one should divide both sides of Eq. (2.894) by v_w^* and then take square roots of both sides to get

$$\sqrt{\frac{\alpha_{min}}{v_w^*}} = \coth\left\{ \frac{\sqrt{v_w^*}\,\xi_{max}}{2} + \coth^{-1}\sqrt{\frac{\alpha_{crt}}{v_w^*}} \right\}, \tag{2.895}$$

thus allowing application of (the inverse) algebraic operator $\coth^{-1}$ as

$$\coth^{-1}\sqrt{\frac{\alpha_{min}}{v_w^*}} = \frac{\sqrt{v_w^*}\,\xi_{max}}{2} + \coth^{-1}\sqrt{\frac{\alpha_{crt}}{v_w^*}}; \tag{2.896}$$

isolation of $\sqrt{v_w^*}\xi_{max}/2$ then leaves

$$\frac{\sqrt{v_w^*}\,\xi_{max}}{2} = \coth^{-1}\sqrt{\frac{\alpha_{min}}{v_w^*}} - \coth^{-1}\sqrt{\frac{\alpha_{crt}}{v_w^*}}. \tag{2.897}$$

Equation (2.897) may be readily solved for ξ_{max} as

$$\xi_{max} = \frac{2}{\sqrt{v_w^*}}\left(\coth^{-1}\sqrt{\frac{\alpha_{min}}{v_w^*}} - \coth^{-1}\sqrt{\frac{\alpha_{crt}}{v_w^*}}\right), \tag{2.898}$$

which will be reformulated to

$$\xi_{max} = \frac{2}{\sqrt{v_w^*}}\left(\frac{1}{2}\ln\frac{\sqrt{\frac{\alpha_{min}}{v_w^*}}+1}{\sqrt{\frac{\alpha_{min}}{v_w^*}}-1} - \frac{1}{2}\ln\frac{\sqrt{\frac{\alpha_{crt}}{v_w^*}}+1}{\sqrt{\frac{\alpha_{crt}}{v_w^*}}-1}\right); \tag{2.899}$$

after factoring out ½ and collapsing the logarithmic functions, one gets

$$\xi_{max} = \frac{1}{\sqrt{v_w^*}}\ln\frac{\left(\sqrt{\frac{\alpha_{min}}{v_w^*}}+1\right)\left(\sqrt{\frac{\alpha_{crt}}{v_w^*}}-1\right)}{\left(\sqrt{\frac{\alpha_{min}}{v_w^*}}-1\right)\left(\sqrt{\frac{\alpha_{crt}}{v_w^*}}+1\right)}, \tag{2.900}$$

where elimination of parentheses is in order, namely,

$$\xi_{max} = \frac{1}{\sqrt{v_w^*}}\ln\frac{\sqrt{\frac{\alpha_{min}}{v_w^*}}\sqrt{\frac{\alpha_{crt}}{v_w^*}} - \sqrt{\frac{\alpha_{min}}{v_w^*}} + \sqrt{\frac{\alpha_{crt}}{v_w^*}} - 1}{\sqrt{\frac{\alpha_{min}}{v_w^*}}\sqrt{\frac{\alpha_{crt}}{v_w^*}} + \sqrt{\frac{\alpha_{min}}{v_w^*}} - \sqrt{\frac{\alpha_{crt}}{v_w^*}} - 1}. \tag{2.901}$$

After reshuffling both numerator and denominator, Eq. (2.901) yields

$$\xi_{max} = \frac{1}{\sqrt{v_w^*}}\ln\frac{\frac{\sqrt{\alpha_{crt}}}{\sqrt{v_w^*}}\frac{\sqrt{\alpha_{min}}}{\sqrt{v_w^*}} - 1 + \frac{\sqrt{\alpha_{crt}}}{\sqrt{v_w^*}} - \frac{\sqrt{\alpha_{min}}}{\sqrt{v_w^*}}}{\frac{\sqrt{\alpha_{crt}}}{\sqrt{v_w^*}}\frac{\sqrt{\alpha_{min}}}{\sqrt{v_w^*}} - 1 - \frac{\sqrt{\alpha_{crt}}}{\sqrt{v_w^*}} + \frac{\sqrt{\alpha_{min}}}{\sqrt{v_w^*}}}, \tag{2.902}$$

where multiplication of both numerator and denominator of the argument of the logarithm by v_w^* unfolds

$$\xi_{max} = \frac{1}{\sqrt{v_w^*}} \ln \frac{\sqrt{\alpha_{crt}}\sqrt{\alpha_{min}} - v_w^* + \sqrt{v_w^*}\sqrt{\alpha_{crt}} - \sqrt{v_w^*}\sqrt{\alpha_{min}}}{\sqrt{\alpha_{crt}}\sqrt{\alpha_{min}} - v_w^* - \sqrt{v_w^*}\sqrt{\alpha_{crt}} + \sqrt{v_w^*}\sqrt{\alpha_{min}}}; \tag{2.903}$$

division of both numerator and denominator, in turn, by $\sqrt{v_w^*}\left(\sqrt{\alpha_{crt}} - \sqrt{\alpha_{min}}\right)$, complemented by selective merging of square roots and factoring out of $\sqrt{v_w^*}$ are in order, namely,

$$\xi_{max} = \frac{1}{\sqrt{v_w^*}} \ln \frac{\dfrac{\sqrt{\alpha_{crt}\alpha_{min}} - v_w^* + \sqrt{v_w^*}\left(\sqrt{\alpha_{crt}} - \sqrt{\alpha_{min}}\right)}{\sqrt{v_w^*}\left(\sqrt{\alpha_{crt}} - \sqrt{\alpha_{min}}\right)}}{\dfrac{\sqrt{\alpha_{crt}\alpha_{min}} - v_w^* - \sqrt{v_w^*}\left(\sqrt{\alpha_{crt}} + \sqrt{\alpha_{min}}\right)}{\sqrt{v_w^*}\left(\sqrt{\alpha_{crt}} - \sqrt{\alpha_{min}}\right)}}, \tag{2.904}$$

which may be algebraically rearranged to read

$$\xi_{max} = \frac{2}{\sqrt{v_w^*}} \frac{1}{2} \ln \frac{\dfrac{1}{\sqrt{v_w^*}} \dfrac{\sqrt{\alpha_{crt}\alpha_{min}} - v_w^*}{\sqrt{\alpha_{crt}} - \sqrt{\alpha_{min}}} + 1}{\dfrac{1}{\sqrt{v_w^*}} \dfrac{\sqrt{\alpha_{crt}\alpha_{min}} - v_w^*}{\sqrt{\alpha_{crt}} - \sqrt{\alpha_{min}}} - 1} \tag{2.905}$$

or, using a more condensed notation,

$$\xi_{max} = \frac{2}{\sqrt{v_w^*}} \coth^{-1} \frac{1}{\sqrt{v_w^*}} \frac{\sqrt{\alpha_{crt}\alpha_{min}} - v_w^*}{\sqrt{\alpha_{crt}} - \sqrt{\alpha_{min}}}. \tag{2.906}$$

A graphical interpretation of Eq. (2.906) is provided in Fig. 2.26. As expected, the maximum foam height increases with velocity of the gas phase; note, however, that the dimensionless height of foam at nil gas velocity tends to a finite, non-nil value, while foam tends to run away at some finite velocity (in view of existence of vertical asymptotes) that proves higher at larger α_{min}.

In attempts to find the mathematical description of the aforementioned minimum value attained by the maximum foam height, one should proceed to

$$\begin{aligned} \lim_{v_w^* \to 0} \xi_{max} &= \frac{2}{\sqrt{0}} \left(\coth^{-1} \sqrt{\frac{\alpha_{min}}{0}} - \coth^{-1} \sqrt{\frac{\alpha_{crt}}{0}} \right) = \frac{2}{0} \left(\coth^{-1} \sqrt{\infty} - \coth^{-1} \sqrt{\infty} \right) \\ &= \frac{2\left(\coth^{-1} \infty - \coth^{-1} \infty\right)}{0} = \frac{2(0-0)}{0} = \frac{0}{0}, \end{aligned} \tag{2.907}$$

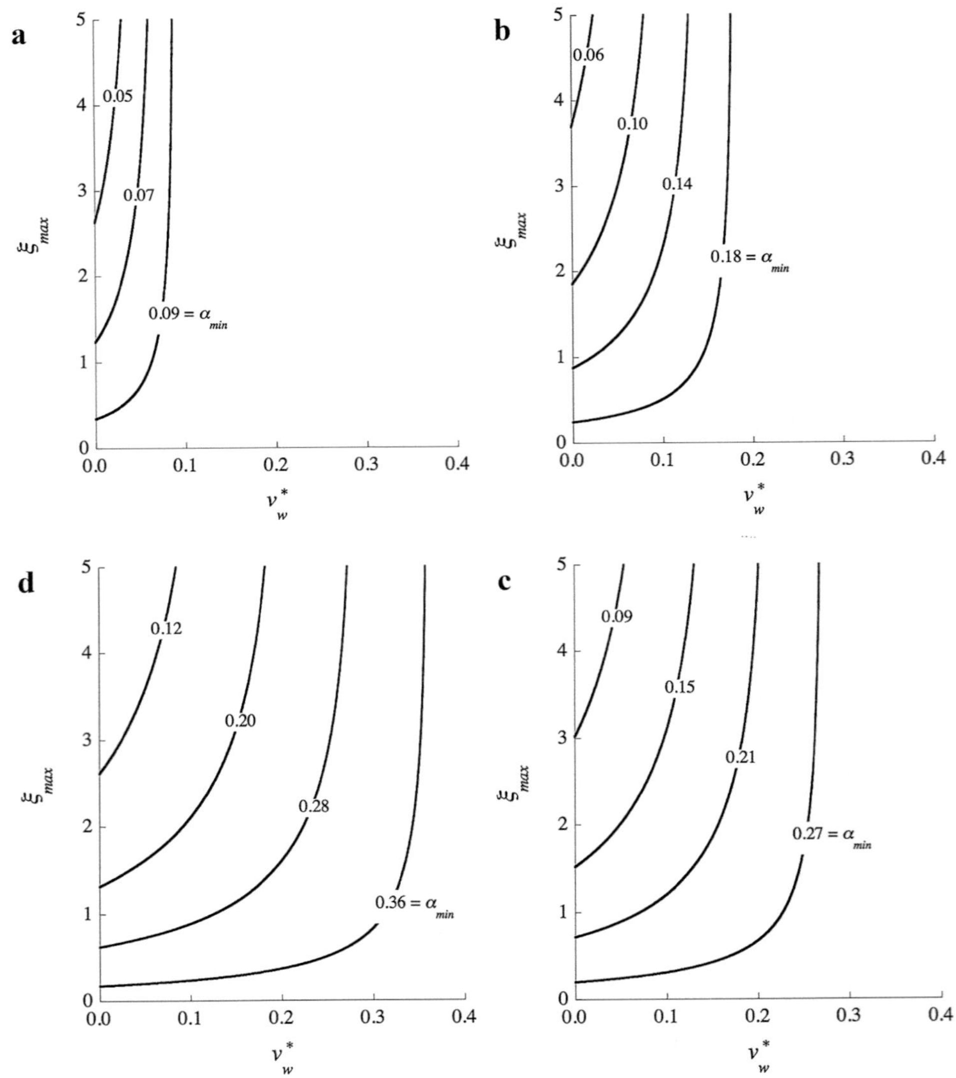

Fig. 2.26 Variation of dimensionless maximum height of foam, ξ_{max}, versus dimensionless velocity of gas bubbles, v_w^*, for selected values of α_{min} and (**a**) $\alpha_{crt} = 0.1$, (**b**) $\alpha_{crt} = 0.2$, (**c**) $\alpha_{crt} = 0.3$, and (**d**) $\alpha_{crt} = 0.4$

stemming from Eq. (2.898), and upon plain application of the classical theorems on limits. To circumvent the unknown quantity produced, one should resort to L'Hôpital's rule applied directly on Eq. (2.898), according to

$$\lim_{v_w^* \to 0} \xi_{max} = 2 \lim_{v_w^* \to 0} \frac{\frac{d}{dv_w^*}\left(\coth^{-1}\sqrt{\frac{\alpha_{min}}{v_w^*}} - \coth^{-1}\sqrt{\frac{\alpha_{crt}}{v_w^*}}\right)}{\frac{d\sqrt{v_w^*}}{dv_w^*}}; \tag{2.908}$$

the chain differentiation rule then supports

$$\lim_{v_w^* \to 0} \xi_{max} = 2 \lim_{v_w^* \to 0} \frac{\frac{d\left(\coth^{-1}\sqrt{\frac{\alpha_{min}}{v_w^*}}\right)}{d\sqrt{\frac{\alpha_{min}}{v_w^*}}} \frac{d\sqrt{\frac{\alpha_{min}}{v_w^*}}}{dv_w^*} - \frac{d\left(\coth^{-1}\sqrt{\frac{\alpha_{crt}}{v_w^*}}\right)}{d\sqrt{\frac{\alpha_{crt}}{v_w^*}}} \frac{d\sqrt{\frac{\alpha_{crt}}{v_w^*}}}{dv_w^*}}{\frac{d\sqrt{v^*}}{dv_w^*}}. \tag{2.909}$$

Remember, in this regard, that

$$\frac{d\coth^{-1}\chi}{d\chi} = \frac{d}{d\chi}\left(\frac{1}{2}\ln\frac{\chi+1}{\chi-1}\right) = \frac{1}{2}\frac{d\left(\ln\frac{\chi+1}{\chi-1}\right)}{d\left(\frac{\chi+1}{\chi-1}\right)}\frac{d\left(\frac{\chi+1}{\chi-1}\right)}{d\chi} \tag{2.910}$$

at the expense again of the chain differentiation rule, which gives rise to

$$\frac{d\coth^{-1}\chi}{d\chi} = \frac{1}{2}\frac{1}{\frac{\chi+1}{\chi-1}}\frac{1(\chi-1)-(\chi+1)1}{(\chi-1)^2} = \frac{1}{2}\frac{\chi-1}{\chi+1}\frac{\chi-1-\chi-1}{(\chi-1)^2}; \tag{2.911}$$

upon dropping symmetrical terms in numerator, splitting the square power, and cancelling common factors between numerator and denominator at the end, Eq. (2.911) becomes

$$\frac{d\coth^{-1}\chi}{d\chi} = \frac{1}{2}\frac{\chi-1}{\chi+1}\frac{-2}{(\chi-1)(\chi-1)} = -\frac{1}{(\chi+1)(\chi-1)}, \tag{2.912}$$

where elimination of parentheses leaves finally

$$\frac{d\coth^{-1}\chi}{d\chi} = -\frac{1}{\chi^2-1} = \frac{1}{1-\chi^2}. \tag{2.913}$$

In view of Eq. (2.913), one can redo Eq. (2.909) as

$$\lim_{v_w^*\to 0}\xi_{max}=2\lim_{v_w^*\to 0}\frac{\frac{1}{1-\left(\sqrt{\frac{\alpha_{min}}{v_w^*}}\right)^2}\frac{-\frac{\alpha_{min}}{v_w^{*2}}}{2\sqrt{\frac{\alpha_{min}}{v_w^*}}}-\frac{1}{1-\left(\sqrt{\frac{\alpha_{crt}}{v_w^*}}\right)^2}\frac{-\frac{\alpha_{crt}}{v_w^{*2}}}{2\sqrt{\frac{\alpha_{crt}}{v_w^*}}}}{\frac{1}{2\sqrt{v_w^*}}}, \tag{2.914}$$

where composition of square power with square root, followed by reduction in number of fraction levels permit simplification to

$$\begin{aligned}\lim_{v_w^*\to 0}\xi_{max}&=2\lim_{v_w^*\to 0}\frac{-\frac{1}{1-\frac{\alpha_{min}}{v_w^*}}\frac{\frac{\alpha_{min}}{v_w^{*2}}}{2\sqrt{\frac{\alpha_{min}}{v_w^*}}}+\frac{1}{1-\frac{\alpha_{crt}}{v_w^*}}\frac{\frac{\alpha_{crt}}{v_w^{*2}}}{2\sqrt{\frac{\alpha_{crt}}{v_w^*}}}}{\frac{1}{2\sqrt{v_w^*}}}\\&=2\lim_{v_w^*\to 0}2\sqrt{v_w^*}\left(-\frac{\alpha_{min}}{2v_w^{*2}\left(1-\frac{\alpha_{min}}{v_w^*}\right)\sqrt{\frac{\alpha_{min}}{v_w^*}}}+\frac{\alpha_{crt}}{2v_w^{*2}\left(1-\frac{\alpha_{crt}}{v_w^*}\right)\sqrt{\frac{\alpha_{crt}}{v_w^*}}}\right);\end{aligned} \tag{2.915}$$

after condensing constant factors, and splitting square roots and powers, Eq. (2.915) becomes

$$\begin{aligned}\lim_{v_w^*\to 0}\xi_{max}&=2\lim_{v_w^*\to 0}\sqrt{v_w^*}\left(\frac{\alpha_{crt}}{v_w^{*2}\left(1-\frac{\alpha_{crt}}{v_w^*}\right)\frac{\sqrt{\alpha_{crt}}}{\sqrt{v_w^*}}}-\frac{\alpha_{min}}{v_w^{*2}\left(1-\frac{\alpha_{min}}{v_w^*}\right)\frac{\sqrt{\alpha_{min}}}{\sqrt{v_w^*}}}\right)\\&=2\lim_{v_w^*\to 0}\sqrt{v_w^*}\left(\frac{\alpha_{crt}\sqrt{v_w^*}}{v_w^*\sqrt{\alpha_{crt}}\,v_w^*\left(1-\frac{\alpha_{crt}}{v_w^*}\right)}-\frac{\alpha_{min}\sqrt{v_w^*}}{v_w^*\sqrt{\alpha_{min}}\,v_w^*\left(1-\frac{\alpha_{min}}{v_w^*}\right)}\right).\end{aligned} \tag{2.916}$$

When $\sqrt{v_w^*}$ and v_w^* (as appropriate) are factored in, and v_w^* dropped from numerator and denominator afterward, Eq. (2.916) turns to

$$\begin{aligned}\lim_{v_w^*\to 0}\xi_{max}&=2\lim_{v_w^*\to 0}\left(\frac{\alpha_{crt}v_w^*}{v_w^*\sqrt{\alpha_{crt}}\left(v_w^*-\alpha_{crt}\right)}-\frac{\alpha_{min}v_w^*}{v_w^*\sqrt{\alpha_{min}}\left(v_w^*-\alpha_{min}\right)}\right)\\&=2\lim_{v_w^*\to 0}\left(\frac{\alpha_{crt}}{\sqrt{\alpha_{crt}}\left(v_w^*-\alpha_{crt}\right)}-\frac{\alpha_{min}}{\sqrt{\alpha_{min}}\left(v_w^*-\alpha_{min}\right)}\right);\end{aligned} \tag{2.917}$$

$\sqrt{\alpha_{crt}}$ or $\sqrt{\alpha_{min}}$ will now cancel out between numerator and denominator as

$$\lim_{v_w^* \to 0} \xi_{max} = 2 \lim_{v_w^* \to 0} \left(\frac{\sqrt{\alpha_{crt}}}{v_w^* - \alpha_{crt}} - \frac{\sqrt{\alpha_{min}}}{v_w^* - \alpha_{min}} \right). \tag{2.918}$$

Replacement of v_w^* by 0 in Eq. (2.918) ensues, namely,

$$\lim_{v_w^* \to 0} \xi_{max} = 2\left(\frac{\sqrt{\alpha_{crt}}}{0 - \alpha_{crt}} - \frac{\sqrt{\alpha_{min}}}{0 - \alpha_{min}} \right) = 2\left(\frac{\sqrt{\alpha_{crt}}}{-\alpha_{crt}} - \frac{\sqrt{\alpha_{min}}}{-\alpha_{min}} \right) = 2\left(\frac{\sqrt{\alpha_{min}}}{\alpha_{min}} - \frac{\sqrt{\alpha_{crt}}}{\alpha_{crt}} \right) \tag{2.919}$$

that finally degenerates to

$$\lim_{v_w^* \to 0} \xi_{max} = 2\left(\frac{1}{\sqrt{\alpha_{min}}} - \frac{1}{\sqrt{\alpha_{crt}}} \right) \tag{2.920}$$

as common factors are lumped between numerator and denominator. Equation (2.920) accordingly conveys the descriptor of the vertical intercepts of the curves in Fig. 2.26; the corresponding foam height will spontaneously be attained (speeded up by gentle stirring of the liquid surface) from the overhead gas, should some surfactant be already present in the medium.

As emphasized above, a maximum foam height will be asymptotically reached when v_w^* increases; inspection of the functional form of Eq. (2.898), coupled with

$$\lim_{\eta \to 1^+} \coth^{-1} \eta = \infty \tag{2.921}$$

and

$$\lim_{\eta \to -1^-} \coth^{-1} \eta = -\infty \tag{2.922}$$

as mathematical features of the inverse hyperbolic cotangent, indicates that a singularity will be attained when either $\sqrt{\alpha_{min}/v_w^*} \to 1^+$ or $\sqrt{\alpha_{crt}/v_w^*} \to -1^-$, which reduces to

$$v_w^* \to \alpha_{min}^{\;\;-} \tag{2.923}$$

as sole physically feasible possibility. Therefore,

$$\lim_{v_w^* \to \alpha_{min}^-} \xi_{max} = \frac{2}{\sqrt{\alpha_{min}}}\left(\coth^{-1}\sqrt{\frac{\alpha_{min}}{\alpha_{min}^-}} - \coth^{-1}\sqrt{\frac{\alpha_{crt}}{\alpha_{min}^-}}\right) = \frac{2}{\sqrt{\alpha_{min}}}\left(\coth^{-1}\sqrt{1^+} - \coth^{-1}\sqrt{\frac{\alpha_{crt}}{\alpha_{min}}}\right)$$
$$= \frac{2}{\sqrt{\alpha_{min}}}\left(\coth^{-1}1^+ - \coth^{-1}\sqrt{\frac{\alpha_{crt}}{\alpha_{min}}}\right) = \frac{2}{\sqrt{\alpha_{min}}}\left(\infty - \coth^{-1}\sqrt{\frac{\alpha_{crt}}{\alpha_{min}}}\right) \tag{2.924}$$

stemming from Eqs. (2.898), (2.921), and (2.923), or, equivalently,

$$\lim_{v_w^* \to \alpha_{min}^-} \xi_{max} = \infty \tag{2.925}$$

describes the vertical asymptotes to the curves in Fig. 2.26, characterized by (finite) α_{min} for horizontal intercept.

The structural components of foam can slide, change conformation, and even collapse; in rheological terms, foam accordingly exhibits a viscoelastic behavior. High shearing breaks the channels, so the foam becomes more viscous and less plastic; hence, foam reacts as a shear-thinning fluid, and Herschel and Bulkley's model, namely,

$$\tau^n = \tau_0{}^n + \kappa\dot{\gamma}^n \;\; ; \quad 0 < n \leq 1, \tag{2.926}$$

has been successfully applied. Furthermore, the liquid fraction of foam changes as time elapses, which also brings about changes in rheological behavior; therefore, foam can be considered as thixotropic, suggesting that a structural decay parameter is to be included in Eq. (2.926) as

$$\tau^n = \lambda(\tau_0{}^n + \kappa\dot{\gamma}^n). \tag{2.927}$$

Oscillatory shear tests have been widely applied to characterize the rheological behavior of foam, which also hold the advantage of being nondestructive.

In dry foam, bubbles push against each other, so the channels in between are thin and Plateau's borders are well defined; the interacting surfaces become harder to deform since the stiff channels transmit a major portion of any prevailing force. The force transmitted through the foam is elastic – and if large enough, the channels just break up and reorganization of the network ensues; this threshold is known as yield stress. Conversely, the bubbles inside the foam are rounder, and Plateau's borders exhibit a poor definition in wet foam; the channels are larger, and the bubbles also slide more freely. Less force is transmitted through the channels, as most of it is used up to bring about bubble motion; this constitutes the so-called viscous component of force. Transition between wet and dry foam can be assessed by observing decay in elastic behavior; once ϕ_{crt} is reached, interaction between bubbles ceases and the foam becomes unordered, so the yielding phenomenon gradually vanishes.

Estimation of the rate of draining of a foam is quite useful, yet a few simplifying assumptions were needed that consubstantiate the concept of dry foam; therefore, the results are strictly valid for this asymptotic form, but only approximately valid otherwise. Remember that the contribution to drainage by liquid flow along the lamellae has accordingly been neglected, meaning that liquid flow occurs solely along Plateau's borders; and that this flow has been taken as laminar and thus consistent with Poiseuille's model, coupled with nil velocity at its boundaries as expected from sufficiently rigid channels. The contribution of shearing motion associated to flow through the nodes to the overall viscous dissipation has also been neglected; this postulate is reasonable since borders are much longer that nodes themselves. Finally, a constant surface tension σ has been considered, and thus independent of surface elasticity effects; as well as a constant viscosity μ of bulk liquid.

Problem 2.12

Verbist, Weaire, and Kraynik's model of foam provides a coherent picture for the mechanism behind foam formation—in terms of cross-sectional area of Plateau's borders or, equivalently, liquid fraction; however, it also sheds light onto the phenomenon of foam draining.

(a) Based on the relationship between cross-sectional surface of Plateau's border and liquid fraction of a foam, obtain the partial differential equation that allows direct calculation of the latter.

(b) Investigate how an existing (dry) foam will behave if liquid is continuously fed in at its top.

(c) When surface tension effects are not relevant, prove that the simplified form of the foam building/draining equation looks like

$$\alpha = \frac{\xi - \xi^*}{2\tau} \tag{2.928}$$

where ξ^* denotes a constant; and specify it for the no-flow condition at the top of the foam, taken for origin of the spatial coordinate.

(d) In order to ensure a stable foam, surface tension effects should be relevant on the surface of the foam (even if not everywhere else); prove that $\alpha \equiv \alpha\{\xi, \tau\}$ obeys

(continued)

Problem 2.12 (continued)

$$\alpha^{\frac{3}{2}} - \frac{\xi}{2\tau}\alpha^{\frac{1}{2}} - \frac{1}{3\tau} = 0 \tag{2.929}$$

in this case, and plot the evolution of α on the top of the foam.

Solution

(a) After revisiting Eq. (2.747) as

$$\phi_l = \frac{S}{\beta + S}, \tag{2.930}$$

with the aid of

$$\beta \equiv \frac{V_b}{N_p L} \tag{2.931}$$

as definition of the surface-type geometrical parameter of relevance, one may proceed to elimination of denominators in Eq. (2.930) as

$$\beta\phi_l + \phi_l S = S; \tag{2.932}$$

after factoring out S, Eq. (2.932) becomes

$$(1 - \phi_l)S = \beta\phi_l, \tag{2.933}$$

which degenerates to

$$S = \beta\frac{\phi_l}{1 - \phi_l} \tag{2.934}$$

upon isolation of S. Partial differentiation of S with regard to x converts Eq. (2.934) to

$$\frac{\partial S}{\partial x} = \frac{dS}{d\phi_l}\frac{\partial \phi_l}{\partial x} \tag{2.935}$$

with the aid of the chain differentiation rule, which may be rewritten as

(continued)

Problem 2.12 (continued)

$$\frac{\partial S}{\partial x} = \beta \frac{1(1-\phi_l) - \phi_l(-1)}{(1-\phi_l)^2} \frac{\partial \phi_l}{\partial x} = \frac{\beta}{(1-\phi_l)^2} \frac{\partial \phi_l}{\partial x} \tag{2.936}$$

in view of Eq. (2.934); by the same token, partial differentiation of S with regard to t unfolds

$$\frac{\partial S}{\partial t} = \frac{dS}{d\phi_l} \frac{\partial \phi_l}{\partial t}, \tag{2.937}$$

which turns to

$$\frac{\partial S}{\partial t} = \beta \frac{1(1-\phi_l) - \phi_l(-1)}{(1-\phi_l)^2} \frac{\partial \phi_l}{\partial t} = \frac{\beta}{(1-\phi_l)^2} \frac{\partial \phi_l}{\partial t} \tag{2.938}$$

again after taking Eq. (2.934) in account. Insertion of Eqs. (2.934), (2.936), and (2.938) converts Eq. (2.716) to

$$\frac{1}{\vartheta_2 \mu} \frac{\partial}{\partial x} \left(\rho g \left(\beta \frac{\phi_l}{1-\phi_l} \right)^2 - \vartheta_1 \sigma \left(\beta \frac{\phi_l}{1-\phi_l} \right)^{\frac{1}{2}} \frac{\beta}{(1-\phi_l)^2} \frac{\partial \phi_l}{\partial x} \right) + \frac{\beta}{(1-\phi_l)^2} \frac{\partial \phi_l}{\partial t} = 0, \tag{2.939}$$

where lumping of factors alike coupled to multiplication of both sides by $(1 - \phi_l)^2/\beta$ leave

$$\frac{(1-\phi_l)^2}{\vartheta_2 \beta \mu} \frac{\partial}{\partial x} \left(\beta^2 \rho g \frac{\phi_l^2}{(1-\phi_l)^2} - \vartheta_1 \beta^{\frac{3}{2}} \sigma \frac{\phi_l^{\frac{1}{2}}}{(1-\phi_l)^{\frac{5}{2}}} \frac{\partial \phi_l}{\partial x} \right) + \frac{\partial \phi_l}{\partial t} = 0; \tag{2.940}$$

a final factoring out of the square of $\phi_l/(1 - \phi_l)$ leaves

$$\frac{(1-\phi_l)^2}{\vartheta_2 \beta \mu} \frac{\partial}{\partial x} \left(\beta^2 \frac{\phi_l^2}{(1-\phi_l)^2} \left(\rho g - \frac{\vartheta_1 \sigma}{\beta^{\frac{1}{2}}} \frac{\phi_l^{-\frac{3}{2}}}{(1-\phi_l)^{\frac{1}{2}}} \frac{\partial \phi_l}{\partial x} \right) \right) + \frac{\partial \phi_l}{\partial t} = 0 \tag{2.941}$$

as the partial differential equation sought for calculation of $\phi_l \equiv \phi_l\{x, t\}$.

(b) The problem of liquid fed at the top of an existing foam consubstantiates a solitary wave of liquid moving downward. Therefore, the derivation

(continued)

Problem 2.12 (continued)

presented in the case of a solitary wave of gas moving up can be retrieved in full, that is, Eq. (2.839) that eventually led to Eq. (2.855), as long as the origin of the vertical coordinate x is set as the top of the initial (dry) foam; on the other hand, $\chi < \sqrt{v_w^*}$ in this case, which prompts (convenient) algebraic rearrangement of Eq. (2.855) to

$$\frac{1}{2}\int\left(\frac{1}{\sqrt{v_w^*}+\chi}+\frac{1}{\sqrt{v_w^*}-\chi}\right)d\chi=-\frac{\sqrt{v_w^*}}{2}\int d\zeta. \tag{2.942}$$

The fundamental theorem of integral calculus supports conversion of Eq. (2.942) to

$$\frac{1}{2}\left(\ln\left\{\sqrt{v_w^*}+\chi\right\}+\frac{\ln\left\{\sqrt{v_w^*}-\chi\right\}}{-1}\right)=k_7-\frac{\sqrt{v_w^*}}{2}\zeta, \tag{2.943}$$

where the logarithmic functions will be collapsed as

$$\frac{1}{2}\ln\frac{\sqrt{v_w^*}+\chi}{\sqrt{v_w^*}-\chi}=k_7-\frac{\sqrt{v_w^*}}{2}\zeta; \tag{2.944}$$

division by $\sqrt{v_w^*}$ of both numerator and denominator of the argument of the outstanding logarithmic function leads then to

$$\frac{1}{2}\ln\frac{1+\frac{\chi}{\sqrt{v_w^*}}}{1-\frac{\chi}{\sqrt{v_w^*}}}=k_7-\frac{\sqrt{v_w^*}}{2}\zeta, \tag{2.945}$$

fully equivalent to

$$\tanh^{-1}\frac{\chi}{\sqrt{v_w^*}}=k_7-\frac{\sqrt{v_w^*}}{2}\zeta \tag{2.946}$$

as per a more concise notation. Similar argumentation as developed previously, in attempts to find the value of arbitrary constant k_7, supports Eq. (2.882), so Eq. (2.946) will break down to

(continued)

Problem 2.12 (continued)

$$\tanh^{-1}\frac{\chi}{\sqrt{v_w^*}}=-\frac{\sqrt{v_w^*}}{2}\zeta; \tag{2.947}$$

after taking hyperbolic tangent of both sides, one gets

$$\frac{\chi}{\sqrt{v_w^*}}=\tanh\left\{-\frac{\sqrt{v_w^*}}{2}\zeta\right\}, \tag{2.948}$$

or else

$$\frac{\chi}{\sqrt{v_w^*}}=-\tanh\frac{\sqrt{v_w^*}}{2}\zeta \tag{2.949}$$

in view of the odd nature of hyperbolic tangent. Insertion of Eq. (2.769) converts Eq. (2.949) to

$$\frac{\sqrt{\alpha}}{\sqrt{v_w^*}}=-\tanh\frac{\sqrt{v_w^*}}{2}\zeta, \tag{2.950}$$

where multiplication of both sides by $\sqrt{v_w^*}$ unfolds

$$\sqrt{\alpha}=-\sqrt{v_w^*}\tanh\frac{\sqrt{v_w^*}}{2}\zeta; \tag{2.951}$$

when squares are taken of both sides and Eq. (2.833) is plugged in, Eq. (2.951) eventually takes the form

$$\alpha=v_w^*\tanh^2\frac{\sqrt{v_w^*}}{2}\left(\xi-v_w^*\tau\right)\ ;\quad \xi\leq v_w^*\tau, \tag{2.952}$$

obviously complemented by

$$\alpha=0\ ;\quad \xi>v_w^*\tau \tag{2.953}$$

as alias to Eq. (2.865). To minimize overparameterization, it proves useful to divide both sides of Eq. (2.952) by v_w^*, and factor in $\sqrt{v_w^*}$ in the argument of the hyberbolic function, namely,

(continued)

Problem 2.12 (continued)

$$\frac{\alpha}{v_w^*} = \tanh^2 \frac{\sqrt{v_w^*}\,\xi - v_w^* \sqrt{v_w^*}\,\tau}{2}; \tag{2.954}$$

a sketch of α/v_w^*-vs.-$\sqrt{v_w^*}\,\xi$ is provided in Fig. 2.27, for selected values of $v_w^* \sqrt{v_w^*}\,\tau \equiv v_w^{*\frac{3}{2}}\tau$. The wetted region accordingly travels downward and is characterized by a sharply defined profile, as the dry foam reduces its thickness toward zero; obviously that only the portion of the curves to the left of the vertical line with horizontal intercept equal to the initial thickness of the dry foam is physically significant. The distance of the interface between wet and dry foam to the origin of coordinate, $\Delta\xi$, is given by

$$\Delta\xi \equiv \xi|_{\alpha=0} \tag{2.955}$$

to be estimated from Eq. (2.952) as

$$0 = v_w^* \tanh^2 \frac{\sqrt{v_w^*}}{2}\left(\Delta\xi - v_w^* \tau\right); \tag{2.956}$$

after dividing both sides by v_w^*, Eq. (2.956) becomes

(continued)

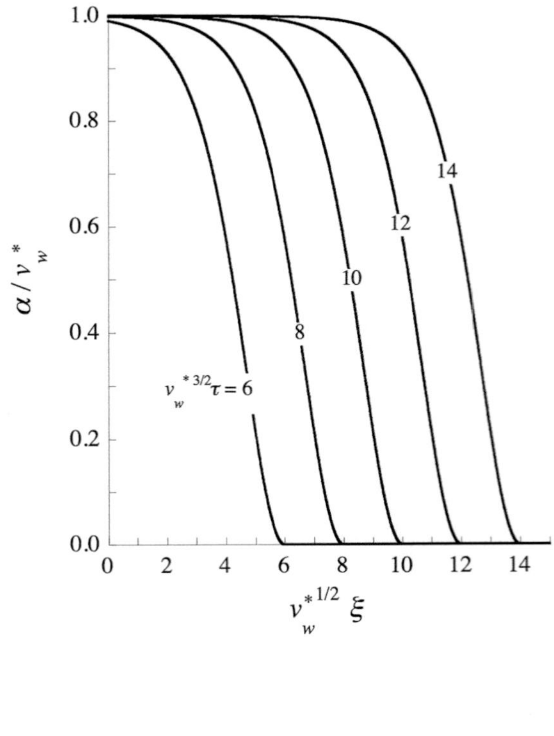

Fig. 2.27 Variation of ratio of normalized cross-sectional area of Plateau's channels in foam, α, to v_w^*, versus dimensionless vertical coordinate, $\sqrt{v_w^*}\,\xi$, for selected values of dimensionless time, $v_w^{*\frac{3}{2}}\tau$, under liquid supply on top of foam

Problem 2.12 (continued)

$$\tanh^2 \frac{\sqrt{v_w^*}}{2}\left(\Delta\xi - v_w^*\tau\right) = 0 \tag{2.957}$$

or, equivalently,

$$\tanh \frac{\sqrt{v_w^*}}{2}\left(\Delta\xi - v_w^*\tau\right) = 0 \tag{2.958}$$

once square roots are taken of both sides. After operator inverse hyperbolic tangent is applied to both sides, Eq. (2.958) becomes

$$\frac{\sqrt{v_w^*}}{2}\left(\Delta\xi - v_w^*\tau\right) = \tanh^{-1} 0 = 0, \tag{2.959}$$

where multiplication of both sides by $2/\sqrt{v_w^*}$ unfolds

$$\Delta\xi - v_w^*\tau = 0; \tag{2.960}$$

upon isolation of $\Delta\xi$ as

$$\Delta\xi = v_w^*\tau, \tag{2.961}$$

one readily concludes that

$$\frac{d\Delta\xi}{d\tau} = v_w^* \tag{2.962}$$

so the velocity of progression of the interface coincides with the velocity hypothesized for the liquid wave.

(c) Comparative inspection with Eqs. (2.704) and (2.708) indicates that $\sigma\sqrt{S}(\partial S/\partial x)$ in Eq. (2.716) is the term associated to surface tension; if such an effect is neglected throughout the liquid that makes up the foam, then Eq. (2.745) reduces to

$$\frac{\partial \alpha^2}{\partial \xi} + \frac{\partial \alpha}{\partial \tau} = 0 \tag{2.963}$$

or, after recalling the chain differentiation rule,

(continued)

Problem 2.12 (continued)

$$2\alpha\frac{\partial\alpha}{\partial\xi}+\frac{\partial\alpha}{\partial\tau}=0. \tag{2.964}$$

Taking Eq. (2.928) for tentative solution, one finds that

$$\frac{\partial\alpha}{\partial\xi}=\frac{1}{2\tau} \tag{2.965}$$

and

$$\frac{\partial\alpha}{\partial\tau}=-\frac{\xi-\xi^*}{2\tau^2}; \tag{2.966}$$

insertion of Eqs. (2.928), (2.965), and (2.966) transforms Eq. (2.964) to

$$2\frac{\xi-\xi^*}{2\tau}\frac{1}{2\tau}-\frac{\xi-\xi^*}{2\tau^2}=0, \tag{2.967}$$

where factors alike will be lumped to get

$$\frac{\xi-\xi^*}{2\tau^2}-\frac{\xi-\xi^*}{2\tau^2}=0. \tag{2.968}$$

Since Eq. (2.968) is a universal condition, one concludes that Eq. (2.928) is indeed a solution to Eq. (2.964). One should now revisit Eq. (2.761) as

$$\alpha^2-\sqrt{\alpha}\frac{d\alpha}{d\xi}=Q^* \tag{2.969}$$

with ξ measured upside down as per hypothesis, where Q^* denotes dimensionless volumetric flow rate of liquid; in the specific case of absence of surface tension, Eq. (2.969) reduces to

$$\alpha^2=Q^*, \tag{2.970}$$

or else

(continued)

Problem 2.12 (continued)

$$\alpha = \sqrt{Q^*} \tag{2.971}$$

after taking square roots of both sides. In view of Eq. (2.971), one realizes that the suggested no-flow conditions (or $Q^* = 0$) is described by

$$\alpha = 0; \tag{2.972}$$

hence,

$$\alpha\big|_{\xi=0} = 0 \tag{2.973}$$

should serve as boundary condition, as long as $\xi = 0$ is taken for descriptor of the top of the foam. Compatibility between Eqs. (2.928) and (2.973) requires

$$0 = \frac{0 - \xi^*}{2\tau}, \tag{2.974}$$

which enforces

$$\xi^* = 0 \tag{2.975}$$

upon multiplication of both sides by 2τ—irrespective of τ; insertion of Eq. (2.975) back in Eq. (2.928) leaves merely

$$\alpha = \frac{\xi}{2\tau} \tag{2.976}$$

for $\alpha \equiv \alpha\{\xi, \tau\}$, as depicted in Fig. 2.28. As apparent from inspection of the curves in this figure, Eq. (2.976) entails a linear variation of cross section with vertical position ξ within the foam (starting from its top) at any time, and a hyberbolically decreasing cross section with time τ at any vertical position.

(d) Since no change was introduced on the behavior of the bulk liquid, Eq. (2.964) remains valid; however,

(continued)

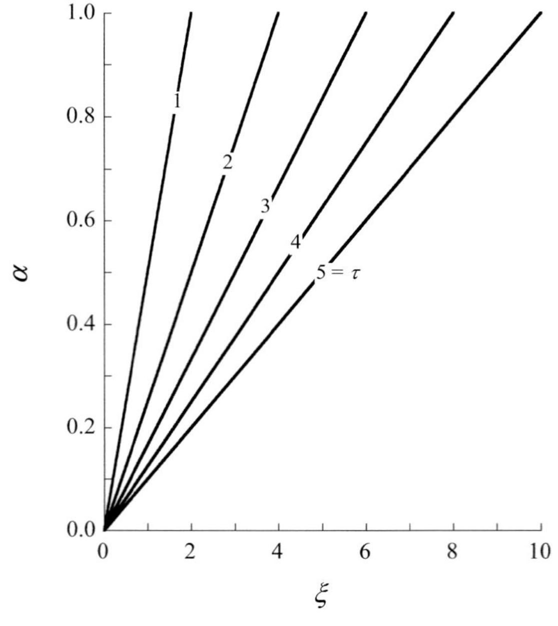

Fig. 2.28 Variation of normalized cross-sectional area of Plateau's channels in foam, α, versus dimensionless vertical coordinate, ξ, for selected values of dimensionless time, τ, under negligible surface tension

Problem 2.12 (continued)

$$\left(\alpha^2 - \sqrt{\alpha}\frac{d\alpha}{d\xi} = 0\right)\Bigg|_{\xi=0}, \tag{2.977}$$

which mimics Eq. (2.969) for $Q^* = 0$, should be used in lieu of Eq. (2.973) for boundary condition. After rewriting Eq. (2.929) as

$$\varphi\{\xi, \tau, \alpha\} = 0 \tag{2.978}$$

with auxiliary (multivariate) function φ defined as

$$\varphi \equiv \alpha^{\frac{3}{2}} - \frac{\xi}{2\tau}\alpha^{\frac{1}{2}} - \frac{1}{3\tau}, \tag{2.979}$$

one may proceed to implicit differentiation to obtain

$$\frac{\partial \alpha}{\partial \xi} = -\frac{\frac{\partial \varphi}{\partial \xi}}{\frac{\partial \varphi}{\partial \alpha}} \tag{2.980}$$

and

(continued)

Problem 2.12 (continued)

$$\frac{\partial \alpha}{\partial \tau} = -\frac{\frac{\partial \varphi}{\partial \tau}}{\frac{\partial \varphi}{\partial \alpha}}; \tag{2.981}$$

partial differentiation of Eq. (2.979) as indicated then transforms Eqs. (2.980) and (2.981) to

$$\frac{\partial \alpha}{\partial \xi} = -\frac{-\frac{1}{2\tau}\alpha^{\frac{1}{2}}}{\frac{3}{2}\alpha^{\frac{1}{2}} - \frac{\xi}{2\tau}\frac{1}{2}\alpha^{-\frac{1}{2}}} \tag{2.982}$$

and

$$\frac{\partial \alpha}{\partial \tau} = -\frac{\frac{\xi}{2\tau^2}\alpha^{\frac{1}{2}} + \frac{1}{3\tau^2}}{\frac{3}{2}\alpha^{\frac{1}{2}} - \frac{\xi}{2\tau}\frac{1}{2}\alpha^{-\frac{1}{2}}}, \tag{2.983}$$

respectively. After multiplying numerator and denominator of the right-hand side by $4\tau\sqrt{\alpha}$, Eq. (2.982) reduces to

$$\frac{\partial \alpha}{\partial \xi} = \frac{2\alpha}{6\tau\alpha - \xi}, \tag{2.984}$$

and Eq. (2.983) breaks down to

$$\frac{\partial \alpha}{\partial \tau} = -\frac{6\xi\alpha + 4\sqrt{\alpha}}{3\tau(6\tau\alpha - \xi)} \tag{2.985}$$

following multiplication of numerator and denominator of its right-hand side by $12\tau^2\sqrt{\alpha}$; Eq. (2.964) will accordingly become

$$2\alpha\frac{2\alpha}{6\tau\alpha - \xi} - \frac{6\xi\alpha + 4\sqrt{\alpha}}{3\tau(6\tau\alpha - \xi)} = 0 \tag{2.986}$$

upon insertion of Eqs. (2.984) and (2.985). After multiplying both sides by $6\tau\alpha - \xi$ and then splitting the last term, Eq. (2.986) takes the form

(continued)

Problem 2.12 (continued)

$$4\alpha^2 - \frac{2\xi}{\tau}\alpha - \frac{4}{3\tau}\sqrt{\alpha} = 0, \tag{2.987}$$

where division of both sides by $4\sqrt{\alpha}$ unfolds

$$\alpha^{\frac{3}{2}} - \frac{\xi}{2\tau}\alpha^{\frac{1}{2}} - \frac{1}{3\tau} = 0 \tag{2.988}$$

that retrieves Eq. (2.929); therefore, the equation provided satisfies the germane mass balance. On the other hand, Eq. (2.984) converts the left-hand side of Eq. (2.977) to

$$\alpha^2 - \sqrt{\alpha}\frac{d\alpha}{d\xi} = \alpha^2 - \sqrt{\alpha}\frac{2\alpha}{6\tau\alpha - \xi}, \tag{2.989}$$

where lumping of terms in the right-hand side produces

$$\alpha^2 - \sqrt{\alpha}\frac{d\alpha}{d\xi} = \frac{6\tau\alpha^3 - \xi\alpha^2 - 2\alpha\sqrt{\alpha}}{6\tau\alpha - \xi}; \tag{2.990}$$

after factoring out $6\tau\alpha^{\frac{3}{2}}$, Eq. (2.990) becomes

$$\alpha^2 - \sqrt{\alpha}\frac{d\alpha}{d\xi} = \frac{6\tau\alpha^{\frac{3}{2}}}{6\tau\alpha - \xi}\left(\alpha^{\frac{3}{2}} - \frac{\xi\alpha^{\frac{1}{2}}}{6\tau} - \frac{1}{3\tau}\right). \tag{2.991}$$

Specifically at $\xi = 0$, Eq. (2.991) yields

$$\left.\left(\alpha^2 - \sqrt{\alpha}\frac{d\alpha}{d\xi}\right)\right|_{\xi=0} = \frac{6\tau\alpha^{\frac{3}{2}}}{6\tau\alpha}\left(\alpha^{\frac{3}{2}} - \frac{1}{3\tau}\right), \tag{2.992}$$

which can be rewritten as

$$\left.\left(\alpha^2 - \sqrt{\alpha}\frac{d\alpha}{d\xi}\right)\right|_{\xi=0} = \alpha^{\frac{1}{2}}\left.\left(\alpha^{\frac{3}{2}} - \frac{\xi}{2\tau}\alpha^{\frac{1}{2}} - \frac{1}{3\tau}\right)\right|_{\xi=0}. \tag{2.993}$$

Since Eq. (2.929) is, by hypothesis, valid irrespective of the value taken by ξ, one can replace Eq. (2.993) by

(continued)

Problem 2.12 (continued)

$$\left(\alpha^2 - \sqrt{\alpha}\frac{d\alpha}{d\xi}\right)\Bigg|_{\xi=0} = \alpha^{\frac{1}{2}}0 = 0, \tag{2.994}$$

so Eq. (2.977) is indeed obeyed; one therefore concludes also on the suitability of Eq. (2.929) to satisfy the imposed boundary condition. The evolution of α in time, at various distances to the top of the foam, is depicted in Fig. 2.29. The curves depart from a finite value of α for $\xi = 0$ rather than 0 as in *c*), as expected because of the disparate boundary conditions thereon that affect the behavior in its vicinity; the curves then increase with ξ, but less notoriously (and less linearly) than when surface effects are completely neglected (see Fig. 2.29 vis-à-vis with Fig. 2.28). When $\xi = 0$, one obtains

$$\left(\alpha^{\frac{3}{2}} - \frac{1}{3\tau} = 0\right)\Bigg|_{\xi=0} \tag{2.995}$$

from Eq. (2.929), or else

(continued)

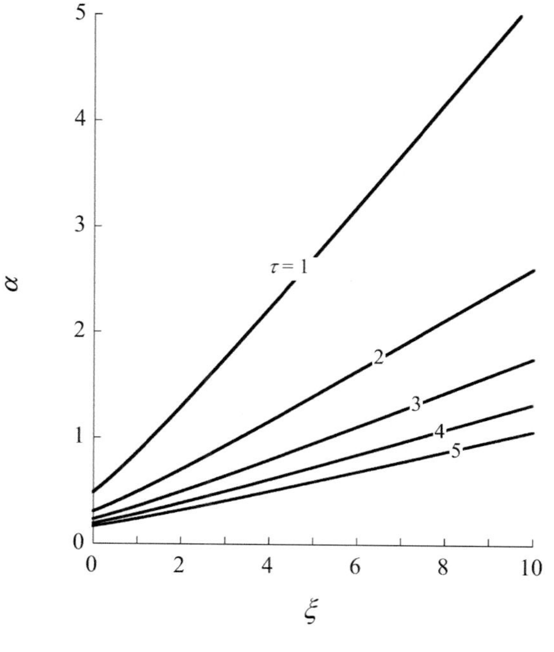

Fig. 2.29 Variation of normalized cross-sectional area of Plateau's channels in foam, α, versus dimensionless vertical coordinate, ξ, for selected values of dimensionless time, τ

Problem 2.12 (continued)

$$\left.\left(\alpha^{\frac{3}{2}} = \frac{1}{3\tau}\right)\right|_{\xi=0}; \tag{2.996}$$

after solving for α, Eq. (2.996) becomes

$$\alpha|_{\xi=0} = \frac{1}{(3\tau)^{\frac{2}{3}}}. \tag{2.997}$$

Equation (2.997) conveys the evolution of cross section of Plateau's borders at the top of the foam with dimensionless time, and predicts a steady decrease in time as liquid drains away; it likewise describes the vertical intercepts of the curves in Fig. 2.29.

2.4.2 Foam Removal

Under aerobic metabolism, the dissolved oxygen requirement by cells is normally high, yet the oxygen solubility is poor (ca. 7 mg/L); hence, air flow rate through the broth should be high to maintain a high driving force for mass transfer, and stirring speed should also be high to minimize concentration gradients within the liquid—both of which favor foam buildup. Foams decrease the bioreactor working volume, abstract microbial cells, and metabolites from the bulk medium, serve as vehicles of contamination, may pose toxic effects upon cell physiology, and even decrease the rate of oxygen transfer following addition of antifoaming agents.

Besides intense air bubbling, relatively high protein content of broth also serves as major driver of foaming in a bioreactor; when the rising bubbles cross the upper surface of the liquid medium, their imprisonment as foam occurs because of the lower interfacial tension brought about by proteins and other molecules present therein. Proteins appear in the medium via formulation as nutrients or as a result of cell autolysis, and are indeed prone to adsorb onto the gas/liquid interface for possessing both hydrophobic and hydrophilic side chains—especially upon medium sterilization and at pH close to their isoelectric point. Further to proteins being net contributors to foaming at levels as low as 1 mg/L, lipopolysaccharides excreted by, for example, starvation-stressed *E. coli*, or pigments excreted by, for example, *Penicillium herqueii* also serve as surfactants and thus promote foaming. However, it is important to point out the molecular differences between foam produced by low molecular weight surfactants or by proteins; the latter can change their conformation at the interface, thus providing enhanced elasticity and increased stability to the foam, with the possibility of multiple layers of proteins adsorbing at the interface—unlike observed when low molecular weight compounds are involved.

Since foam needs some energy to build up structure, reduction of the energy in transit through the bioreactor may help mitigate said phenomenon. Effective strategies include decrease of air flow rate or larger sparger nozzles, or else use of an alternative carrier for oxygen as in membrane-mediated aeration; reduction of operating temperature and decrease in rate of stirring also serve the above purpose.

Mechanical collapse of bubbles can be decomposed into two steps, which sometimes coincide in time: bubble stretching by pressure gradients, leading to instability; followed by blade cutting of disturbed membranes, leading to rupture. Most shearing apparently comes from the low-pressure zone in the liquid created by the foam breaker movements. Paradoxically, foaming is better controlled by creating more and smaller bubbles than fusing them into fewer, yet larger bubbles. The foam resulting from the former will indeed possess a finer structure—with thicker channels and higher liquid fraction; a denser foam is normally advantageous, in that reduction of foam volume may suffice to avoid difficulties arising from spontaneous formation, while aging more quickly. The larger channels in foams are characterized by higher liquid fractions, and thus offer less resistance to draining into the bulk liquid, besides facilitating delivery of antifoam chemicals; conversely, said thicker foams raise stronger problems for shear-based foam breakers. Upon mechanical treatment, a change in rheological behavior of the foam is noticed; the higher liquid fraction of the secondary foam accordingly generated makes the lamellae less responsive to yielding, since its plastic behavior has been (partially) lost. Therefore, a foam breaker should be able to push out the secondary foam for better performance, and cyclic operation may be utilized as alternative.

The aforementioned foam breakers are devices that resort to a wide variety of mechanical means to accelerate collapse of foam; instead of using chemical additives, they add stress and often require additional supply of power to the bioreactor. These devices can be designed for use in line, in the drain, or outside the bioreactor, and normally take advantage of centrifugal force, shearing, or ultrasound; examples are included in Fig. 2.30. Rotating devices mounted on the stirring shaft are a common mode of mechanical defoaming (see Fig. 2.30a), although independent devices, resorting also to rotational motion, are more efficient and produce more consistent results (see Fig. 2.30b, c). An alternative mechanical approach resorts to shock waves generated by ultrasonic devices (see Fig. 2.30e, f); use of chemical antifoamers requires, in turn, pumping as subsidiary mechanical action (see Fig. 2.30d).

Foam will eventually collapse due to gravity, because of the underlying hydrostatic force that pushes the liquid down along the lamellae; centrifugal force will have a similar, yet stronger (and faster) effect. Shearing forces can arise from direct mechanical input, bulk turbulence, or distorted pressure gradients. Several types of impellers have been put to operation as defoamers, for example, six-blade turbine (or vaned disk), two-blade paddle, and rotating disk; the first and last perform better in most scenarios of practical relevance.

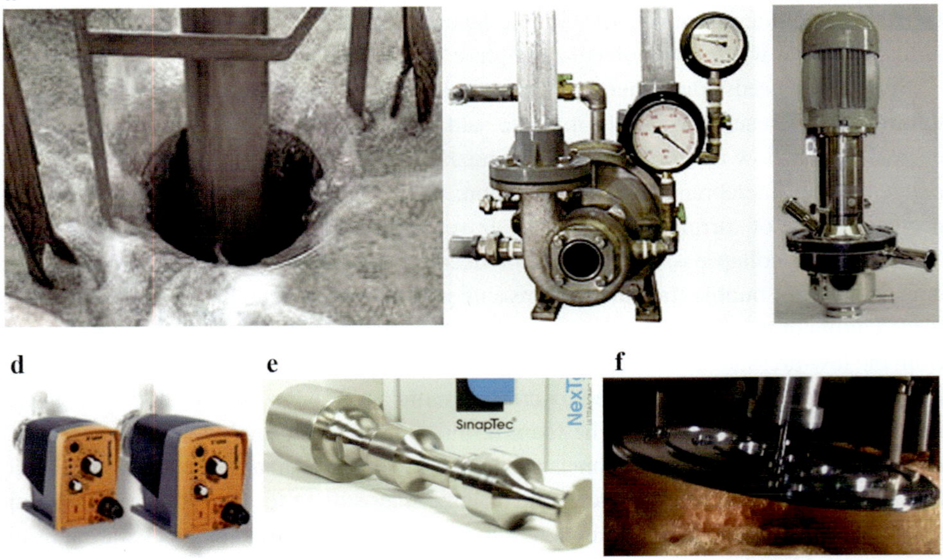

Fig. 2.30 Examples of defoaming equipment: (**a**) mechanical foam breaker (courtesy of Plantec Engineers, Shapur Nagar, India), (**b**) defoaming pump (courtesy of Yokota Manufacturing, Hiroshima-shi, Japan), (**c**) defoamer centrifuge (courtesy of Heinrich Frings GmbH, Bonn, Germany), (**d**) defoamer dosing pump (courtesy of EWAC, Ceské Budejovice, Czech Republic), (**e**) acoustic nozzle (courtesy from Sinaptec, Lezennes, France), and (**f**) ultrasonic flat cell (courtesy of Pusonics, Madrid, Spain)

An interesting improvement resorts to perforations throughout the disk, which create a pressure gradient able to stretch the bubbles beyond Gibbs' elastic capacity, thus bringing them to collapse. Pumping flow, as influenced by rotational frequency and impeller size, plays the dominant role in foam breaker efficiency, and may support performances above those associated with addition of chemical defoamers. Increased shearing via narrower passages between blades, sharper corners, or even needles make defoaming more efficient (especially under wider clearance between lamellae) than using widespread disk-based defoamers, with similar results attained at the expense of less power; addition of a stator to a rotary foam breaker further improves efficiency. An inverted cone configuration has also been proposed; as the cone spins, the foam is pumped in and travels as a thin layer along its surface. At the top edge of the cone, the foam gets projected as a film, ligaments, or drops; the cone shape also induces sufficient centrifugal force on the foam to drain water off the channel. Some foam breakers have resorted to Coanda's effect by pushing the foam through a high-velocity nozzle; a depressurized zone arises, and the foam is broken by the high-pressure gradient accordingly generated. Despite their efficiency, these apparatuses cannot be used with cell-containing medium intended for recycling back to the bioreactor, due to excessive shearing.

Mechanical vibrations carried by ultrasound can accelerate foam drainage, and ultimately rupture foam; this technique is less intrusive than classical mechanical operations, and no additional devices may be needed to come into contact with the liquid surface. Vibration can be produced at a distance via a whistle or a horn, or by causing the surface to vibrate via contact with the transducer. The latter is intrinsically more efficient but adds to operational costs, while a horn or a whistle are less expensive; nevertheless, its influence is limited to a thin layer in front of the device. Surface waves apparently push the liquid into the film; acoustic pressure deforms the lamella by creating thin and thick zones, while Gibbs' effect pushes the liquid back to the thin section thereof. If the amplitude of this phenomenon is high enough, the film will break.

When management of physical factors is not feasible or sufficient, then antifoaming chemical agents are to be added, either as part of the original broth composition or continuously (as needed) via some distribution system. Such agents are usually nonpolar and polar oils, or hydrophobic solid particles of organic or inorganic nature, or a mixture of both; they can be utilized to prevent or mitigate excessive formation of foam, and disturb or destroy the already formed foam. Nonpolar oils are based either on silicone oils (e.g., polydimethylsiloxane) or organic oils; the most important organic nonpolar oils are insoluble paraffinic mineral oils, vegetable oils such as from tall, soybean, or rapeseed, and polymeric oil such as polyisobutylene. Although a few nonpolar oils exhibit some degree of antifoaming ability on their own, mixtures of nonpolar oils with hydrophobic solid particles function much better than each component per se; for instance, neat silicone oil acts poorly as antifoaming agent, yet its performance is dramatically enhanced upon mixing with treated silica or finely divided, high-melting point waxes. The hydrophobic solid particles most widely utilized as antifoamers consist indeed of treated silica (either fumed or precipitated), organic waxes (e.g., ethylene bis-stearamide), and metallic soap (e.g., Al- or Ca-stearate). There is a whole class of soluble foam control agents, chiefly based on polypropylene (PO) or polyethylene-oxide (EO), as well as such copolymers as EO–PO–EO or ethoxylated fatty alcohols; they operate via phase separation above their cloud point to form insoluble droplets of a hydrophobic liquid.

The most widely accepted mechanism for the antifoaming action of oil takes place inside the liquid phase in a fully built foam; when present, oil disturbs the surface due to occupation thereof on a competitive basis, stops the flow, increases the disjoining forces, and eventually leads to membrane collapse. Some soluble polymeric surfactants are thought to destabilize foam, otherwise maintained by proteins adsorbed onto the interface, by entering the surface as a soluble surface-active species and concomitantly decreasing surface elasticity. Antifoam particles are known to slip inside the lamellae and dewet them, which will ultimately lead to their rupture. A hydrophobic nature is required for effective surface bending with rapid bursting should they be spherical, and exhibit a contact angle in excess of 90°; for other particle shapes, the dewetting effect will depend on particle orientation, and polyhedral shapes may even stabilize the membranes if located along the surface.

Many foam control agents are added to the foam-containing medium upon predispersion in water, either as self-dispersible neat materials or mixtures, or as oil-in-water emulsions; the rationale for such procedures comes from realization that the antifoamer is to be dispersed as small droplets in the liquid medium to efficiently play its role. In fact, it is critical that a hydrophobic droplet emerges from the aqueous phase into the air/water interface in order to destroy a foam film, along a process known as entering, and characterized by constant k_{ent}; the oil contributed by the inserted droplet will then be able to spread onto the interface, in a subsequent process known as spreading, and described by constant k_{spr}. The entering coefficient abides to

$$k_{ent} = \sigma_{A/W} + \sigma_{O/W} - \sigma_{O/A}; \tag{2.998}$$

a positive value for k_{ent} means that surface tension of the antifoamer, $\sigma_{O/A}$, is lower than the sum of surface tension of the foaming solution, $\sigma_{A/W}$, and surface tension between antifoamer and foaming solution, $\sigma_{O/W}$. When an oil drop is spread over an air/water interface, the change in surface tension is then measured by k_{spr}, according to

$$k_{spr} = \sigma_{A/W} - \sigma_{O/W} - \sigma_{O/A}; \tag{2.999}$$

during this stage, the water surface is progressively replaced by oil, so a positive value for k_{spr} implies that $\sigma_{A/W}$ exceeds the sum of $\sigma_{O/A}$ and $\sigma_{O/W}$. Entering is obviously essential to rupture the foam, so it is generally recognized that $k_{ent} > 0$ causes such rupture; in order to get positive values for both k_{ent} and k_{spr}, it is important to have a liquid possessing a low surface tension—a condition met, for instance, by silicone oils.

Silicone oils, and more specifically polydimethylsiloxane materials (PDMS: 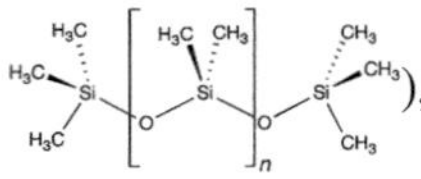

), combine several unusual features owing to their molecular characteristics, such as flexibility of Si–O–Si backbone and very low cohesive energy involving methyl groups; they exhibit low surface tension, and most of them are nonvolatile, remain liquid even at high degrees of polymerization, and are strongly insoluble in water. Despite the considerable polarity of the siloxane backbone, the primary surface-active entities in the polymer structure are its side groups, owing to the ease of adoption of alternative configurations as per their extreme flexibility. Methyl groups are the pendant groups in PDMS and experience merely London's dispersion forces; the low surface tension, a direct manifestation of intermolecular forces, corroborates that interactions between adjacent PDMS chains occur chiefly through their methyl groups. The polymer backbone controls organization of the side groups on the surface, and its flexibility plays a major role upon the ease with which those groups adopt preferred configurations. From a surface standpoint, the more flexible the backbone, the more readily the lowest surface energy configurations will be adopted; PDMS is a particularly favored case of very low intermolecular force pendant groups anchored along a very flexible backbone, thus allowing exposure of methyl groups to the surroundings.

Almost all commercial foam-control agents are liquid; their chemical nature will drive the associated physicochemical properties—the ease of dispersion in water ranking among them. This is indeed the main property that dictates whether the antifoamer needs to be further formulated, or can be employed as such. In the latter case, neat antifoamer will disperse in water and reach the particle size appropriate to control the foam; if the antifoamer shows a high interfacial tension with water, further formulation will be required, for example, via addition of surfactant to produce a self-dispersible blend, or conversion to emulsified form to facilitate handling.

Emulsions are meant to improve the effect of the foam control active substance throughout application, by resorting to a more or less finely dispersed form; the continuous phase takes the role of transporting or diluting agent, while the dispersed phase, accounted for by the antifoamer, appears as tiny droplets. Emulsions facilitate handling of foam control agents, which would otherwise appear greasy or at least highly viscous organic liquids. They owe their stability to proper use of surfactants, for reducing the interfacial tension between water and antifoamer in such a way that a high interfacial area can be attained between dispersed phase and water. Thickeners, for example, functionalized cellulose, reduce creaming and further stabilize the emulsion, but may have some detrimental impact upon downstream treatment of the fermentation broth.

Self-dispersible systems are blends of foam-control agents and surfactants, for example, silicone polyether or glycol, and are designed to promptly disperse into the foaming system, or readily emulsify and stabilize the surfactant in aqueous solutions prior to addition to the foaming liquid. Although simple at first sight, compatibility between foam-control agent and surface-active agent needs to be sufficiently good to assure stability of the self-dispersible system, and yet be effective enough to bring about prompt dispersion during application.

Still another means for addition of a foam-control agent is under a solid form; this is often consubstantiated in a powdery material that can be conveniently dosed or manipulated, and various approaches exist to convert a liquid antifoamer to a solid powder. Plain adsorption of a liquid on an adsorbing solid sometimes suffices to generate a free-flowing powder; should improvement of its physical properties be necessary, more complex formulation recipes and manufacture techniques would be in order—for instance, preliminary dispersion of antifoamer by spraying on powder acting as carrier, followed by evaporation of water.

Although chemical antifoams convey the most efficient strategy to control foaming in bioreactors, they are adjuvants; hence, their chemical nature and concentration may trigger several issues putatively critical for fermentation. Interaction can indeed occur between antifoams and (some) molecules present in the medium that make them less available, or with the microorganism itself that reduces its resilience. Even when the exact composition of antifoam additives is uncertain, most have traditionally been designed to interact with surfactants, which means they will themselves be composed of surfactants most of the times; therefore, their optimal efficiency hinges critically on the concentration range used.

By altering surface tension, antifoamers influence bubble size distribution; this normally means larger bubbles, which not only decreases their upward velocity, but also reduces their specific area available for mass transfer leading to a reduction in mass transfer rate. On the other hand, difficulties may also arise in downstream processing, in terms of reduction of filtration rate or coextraction with the intended product.

Problem 2.13

Foam forms on the top of a liquid medium as a consequence of reducing the surface tension of said interface due to preferential accumulation of surfactant molecules. Derive, from first principles, an expression that relates surface tension to interfacial concentration of surfactant.

Solution

The fundamental Gibbs' relationship states that

$$dG = -S\,dT + V\,dP + \sum_i \mu_i\,dn_i, \tag{2.1000}$$

when applied to the bulk of a liquid containing two or more components, where G denotes Gibbs' energy; S, T, V, and P denote entropy, temperature, volume, and pressure, respectively; and μ_i and n_i denote chemical potential and number of moles of i-th component, respectively. Equation (2.1000) is to be complemented by

$$dG_I = -S_I\,dT + V_I\,dP + \sum_i \mu_i\,dn_{i,I} + \sigma\,dA_I, \tag{2.1001}$$

valid for the interface (and referred to by subscript $_I$). At any given temperature and pressure, the definition of chemical potential, namely,

$$\mu_i \equiv \left(\frac{\partial G}{\partial n_i}\right)_{P,T,n_{j\neq i}}, \tag{2.1002}$$

implies that

$$G = \sum_i n_i \mu_i \tag{2.1003}$$

applicable to the bulk, since μ_i plays the role of partial Gibbs' energy. By the same token,

(continued)

Problem 2.13 (continued)

$$G_I = \sum_i n_{i,I}\mu_i + \sigma A_I \tag{2.1004}$$

pertains specifically to the interface; differentiation of both sides of Eq. (2.1004) produces

$$dG_I = \sum_i \mu_i \, dn_{i,I} + \sum_i n_{i,I} \, d\mu_i + A_I \, d\sigma + \sigma \, dA_I. \tag{2.1005}$$

Elimination of dG_I between Eqs. (2.1001) and (2.1005) leads to

$$\begin{aligned} &- S_I \, dT + V_I \, dP + \sum_i \mu_i \, dn_{i,I} + \sigma dA_I \\ &= \sum_i \mu_i \, dn_{i,I} + \sum_i n_{i,I} \, d\mu_i + A_I \, d\sigma + \sigma \, dA_I, \end{aligned} \tag{2.1006}$$

where common terms in both sides will be dropped to get

$$- S_I \, dT + V_I \, dP = \sum_i n_{i,I} \, d\mu_i + A_I \, d\sigma; \tag{2.1007}$$

under constant pressure and temperature, Eq. (2.1007) reduces further to

$$\left(A_I \, d\sigma = -\sum_i n_{i,I} \, d\mu_i \right)_{P,T} . \tag{2.1008}$$

After dividing both sides by A_I, Eq. (2.1008) becomes

$$\left(d\sigma = -\frac{1}{A_I} \sum_i n_{i,I} \, d\mu_i \right)_{P,T} , \tag{2.1009}$$

and A_I may still be factored in as

$$\left(d\sigma = -\sum_i \frac{n_{i,I}}{A_I} \, d\mu_i \right)_{P,T} ; \tag{2.1010}$$

(continued)

Problem 2.13 (continued)

$n_{i,I}/A_I$ represents an areal concentration, also known as excess concentration of i-th component, which corresponds to its concentration in the system not accounted for by bulk concentration. After setting

$$\Gamma_i \equiv \frac{n_{i,I}}{A_I}, \tag{2.1011}$$

Equation (2.1010) will be redone as

$$\left(d\sigma = -\sum_i \Gamma_i \, d\mu_i\right)_{P,T}, \tag{2.1012}$$

and has been widely termed Gibbs' isotherm for encompassing isothermal (and usually atmospheric, and thus isobaric) conditions. Consider now, for simplicity, that the liquid medium is formed only by two components, solvent (referred to by subscript $_1$) and soluble surfactant (referred to by subscript $_2$); Eq. (2.1012) will accordingly read

$$d\sigma = -(\Gamma_1 \, d\mu_1 + \Gamma_2 \, d\mu_2). \tag{2.1013}$$

Surfactant tends to accumulate at the interface, so $\Gamma_1 << \Gamma_2$; this prompts simplification of Eq. (2.1013) to

$$d\sigma \approx -\Gamma_2 \, d\mu_2 \tag{2.1014}$$

or, equivalently,

$$\frac{d\sigma}{d\mu_2} = -\Gamma_2. \tag{2.1015}$$

On the other hand, the functional dependence of chemical potential on actual concentration follows

$$\mu_2 = \mu_2^\theta + \mathscr{R}T \ln a_2, \tag{2.1016}$$

where superscript $^\theta$ refers to some reference state and a denotes activity (a function, in turn, of concentration); differentiation of both sides leaves

(continued)

Problem 2.13 (continued)

$$d\mu_2 = \mathscr{R}T\, d\ln a_2. \tag{2.1017}$$

Insertion of Eq. (2.1017) transforms Eq. (2.1015) to

$$\frac{d\sigma}{\mathscr{R}T\, d\ln a_2} = -\Gamma_2, \tag{2.1018}$$

while multiplication of both sides by $\mathscr{R}T$ produces

$$\frac{d\sigma}{d\ln a_2} = -\mathscr{R}T\Gamma_2; \tag{2.1019}$$

for an ideal solution, a_2 can be approximated by dimensionless molarity C_2, so Eq. (2.1019) degenerates to

$$\frac{d\sigma}{d\ln C_2} = -\mathscr{R}T\Gamma_2. \tag{2.1020}$$

Equation (2.1020) accordingly provides the slope of the σ-vs.-ln C_2 curve, and is valid until micelles start forming, that is, before the solution reaches the critical micellar concentration of surfactant.

Further Reading

Aguilar, D., Galceran, J., Companys, E., Puy, J., Parat, C., Authier, L., & Potin-Gautier, M. (2013). Non-purged voltammetry explored with AGNES. *Physical Chemistry and Chemical Physics, 15*, 17510–17521.

Atkinson, B., & Mavituna, F. (1991). *Biochemical engineering and biotechnology handbook*. MacMillan.

Blanchard, J. S., & Jakoby, W. B. (1984). Buffers for enzymes. In N. P. Kaplan, W. B. Jakoby, & N. P. Colowick (Eds.), *Methods in enzymology* (Vol. 104—*Enzyme purification and related techniques* (part C), pp. 404–414). Elsevier.

Bolton, F., & Weaire, D. (1990). Rigidity loss transition in a 2D froth. *Physical Review Letters, 65*, 3449–3451.

Boon, L. A., Hoeks, F. W. J. M. M., van der Lans, R., Bujalski, W., Wolff, M. O., & Nienow, A. W. (2002). Comparing a range of impellers for "stirring as foam disruption". *Biochemical Engineering Journal, 10*, 183–195.

Boon, L. A., Hoeks, F. W. J. M. M., van der Laans, R., Bujalski, W., & Nienow, A. W. (2000). Instabilities when using a standard (T/3) Rushton turbine for stirring as foam disruption. *Canadian Journal of Chemical Engineering, 78*, 884–891.

Boyles, W. A., & Lincoln, R. E. (1958). Separation and concentration of bacterial spores and vegetative cells by foam flotation. *Applied Microbiology, 6*, 327–334.

Butler, J. N. (1998). *Ionic equilibrium: Solubility and pH calculations*. Wiley.

Calik, P., Ileri, N., & Erdinc, B. (2005). Novel antifoam for fermentation process: Fluorocarbon-hydrocarbon hybrid unsymmetrical bolaform surfactant. *Langmuir, 21*, 8613–8619.

Cantat, I. (2013). Liquid meniscus friction on a wet plate: Bubbles, lamellae, and foams. *Physics of Fluids, 25*, 031303.

Carbonaro, R. F., & Stone, A. T. (2005). Speciation of chromium(III) and cobalt(III) (amino)-carboxylate complexes using capillary electrophoresis. *Analytical Chemistry, 77*, 155–164.

Cherry, R., & Hulle, C. (1992). Cell death in the thin films of bursting bubbles. *Biotechnology Progress, 8*, 11–18.

Clarke, C., Lazidis, A., Spyropoulos, F., & Norton, I. (2019). Measuring the impact of channel length on liquid flow through an ideal plateau border and node system. *Soft Matter, 15*, 1879–1889.

Cremer, M. (1906). Uber die Ursache der elektromotorischen Eigenschaften der Gewebe, zugleich ein Beitrag zur Lehre von Polyphasischen Elektrolytketten. *Zeitschrift fur Biologie, 47*, 562–608.

Cui, X., Shi, C., Zhang, S., & Xie, L. (2017). Probing the effect of salinity and pH on surface interactions between air bubbles and hydrophobic solids: Implications for colloidal assembly at air/water interfaces. *Chemistry, 12*, 201700388.

Culick, F. E. C. (1960). Comments on a ruptured soap film. *Journal of Applied Physics, 31*, 1128–1129.

Delacotte, J., Montel, L., Restagno, F., Scheid, B., Dollet, B., Stone, H. A., Langevin, D., & Rio, E. (2012). Plate coating: Influence of concentrated surfactants on the film thickness. *Langmuir, 28*, 3821–3830.

Delvigne, F., & Lecomte, J.-P. (2009). Foam formation and control in bioreactors. In M. C. Flickinger (Ed.), *Encyclopedia of industrial biotechnology*. Wiley.

Denkov, N. D. (2004). Mechanisms of foam destruction by oil-based antifoams. *Langmuir, 20*, 9463–9505.

Deshpande, N. S., & Barigou, M. (1999). Performance characteristics of novel mechanical foam breakers in a stirred tank reactor. *Journal of Chemical Technology and Biotechnology, 79*, 979–987.

Dey, A., Kapur, P. C., & Mehrotra, S. P. (1989). A search strategy for optimization of flotation circuits. *International Journal of Mineral Processing, 26*, 73–93.

Ferguson, W. J., Braunschweiger, K. I., Braunschweiger, W. R., Smith, J. R., McCormick, J. J., Wasmann, C. C., Jarvis, N. P., Bell, D. H., & Good, N. E. (1980). Hydrogen ion buffers for biological research. *Analytical Biochemistry, 104*, 300–310.

Ferreira, C. M. H., Pinto, I. S. S., Soares, E. V., & Soares, H. M. V. M. (2015). (Un)suitability of the use of pH buffers in biological, biochemical and environmental studies and their interaction with metal ions—A review. *Royal Society of Chemistry Advances, 5*, 30989–31003.

Garret, P. R. (1993). The mode of action of antifoams. In P. R. Garrett (Ed.), *Defoaming: Theory and industrial applications* (pp. 1–118). Marcel Dekker.

Ghildyal, N. P., Lonsane, B. K., & Karanth, N. G. (1988). Foam control in submerged fermentation: State of the art. *Advances in Applied Microbiology, 33*, 173–222.

Ginocchio, R., de la Fuente, L. M., Sánchez, P., Bustamante, E., Silva, Y., Urrestarazu, P., & Rodríguez, P. H. (2009). Soil acidification as a confounding factor on metal phytotoxicity in soils spiked with copper-rich mine wastes. *Environmental Toxicology and Chemistry, 28*, 2069–2081.

Good, N. E., & Isawa, S. (1972). Hydrogen ion buffers. *Methods in Enzymology, 24*, 53–68.

Good, N. E., Winget, G. D., Winter, W., Connolly, T. N., Izawa, S., & Singh, R. M. M. (1966). Hydrogen ion buffers for biological research. *Biochemistry, 5*, 467–477.

Guitian, J., & Joseph, D. (1998). How bubbly mixtures foam and foam control using a fluidized bed. *International Journal of Multiphase Flow, 24*, 1–16.

Gulden, S. J., Riedele, C., Mueller, S., Kopf, M.-H., & Nirschl, H. (2019). Combining small-scale screening method to predict microorganism floatability. *Chemical Engineering Science, 207*, 1353–1363.

Hall, D. G. (1996). Ion-selective membrane electrodes: A general limiting treatment of interference effects. *Journal of Physical Chemistry, 100*, 7230–7236.

Hoeks, F. W. J. M. M., Wees-Tangerman, V., Gasser, K., Mommers, H. M., Schmid, S., & Luyben, K. C. A. M. (1997). Stirring as foam disruption (SAFD) technique in fermentation processes. *Canadian Journal of Chemical Engineering, 75*, 1018–1029.

Jackson, J. (1962). *Classical electrodynamics* (p. 15). Wiley.

Jameson, G. J. (2010). New directions in flotation machine design. *Materials Engineering, 23*, 835–841.

Joshi, K., Jeelani, S., Bicenstorfer, C., Naegeli, C., Oliviero, C., & Windhab, E. (2006). Nonionic block copolymer antifoams. *Langmuir, 22*, 6893–6904.

Junker, B. (2007). Foam and its mitigation in fermentation systems. *Biotechnology Progress, 23*, 767–784.

Kerridge, P. M. T. (1925). The use of the glass electrode in biochemistry. *Biochemical Journal, 19*, 611–617.

Kilambi, K. P., & Gray, J. J. (2012). Rapid calculation of protein pK_a values using Rosetta. *Biophysical Journal, 103*, 587–595.

Kitchener, J. A., & Cooper, C. F. (1959). Current concepts in the theory of foaming. *Quarterly Reviews of Chemical Society, 13*, 71–97.

Koerner, M. M., Palacio, L. A., Wright, J. W., Schweitzer, K. S., Ray, B. D., & Petrache, H. I. (2011). Electrodynamics of lipid membrane interactions in the presence of zwitterionic buffers. *Biophysics Journal, 173*, 362–369.

Kulkarni, R. D., Goddadr, E. D., & Chandar, P. (1996). Science and technology of silicone antifoam. In R. K. Prud'homme & S. A. Khan (Eds.), *Foams: Theory, measurements and applications* (pp. 1–31). Marcel Dekker.

Kunas, K. T., & Papoutsakis, E. T. (1990). Damage mechanisms of suspended animal cells in agitated bioreactors with and without bubble entrainment. *Biotechnology and Bioengineering, 36*, 476–483.

Kusner, R. (1992). The number of faces in a minimal foam. *Proceedings of Royal Society (London) A, 439*, 683–686.

Lemieux, G. S.-P., Groleau, D., & Proulx, P. (2019). Introduction on foam and its impact in bioreactors. *Canadian Journal of Biotechnology, 3*, 143–157.

Leonard, R. A., & Lemlich, R. (1965). A study of interstitial liquid flow in foam. Part I. Theoretical model and application to foam fractionation. *AIChE Journal, 11*, 18–25.

Madigan, M., & Martinko, J. (Eds.). (2005). *Brock biology of microorganisms*. Prentice Hall.

Modini, R. L., Russel, L. M., Deane, G. B., & Stokes, M. D. (2013). Effect of soluble surfactant on bubble persistence and bubble-produced aerosol particles. *Journal of Geophysics Research—Atmospheric, 118*, 1388–1400.

Morey, M. D., Deshpande, N. S., & Barigou, M. (1999). Foam destabilization by mechanical and ultrasonic vibrations. *Journal of Colloid and Interface Science, 219*, 90–98.

Mysels, K. J., Shinoda, K., & Frankel, S. (1959). *Soap films: Studies of their thinning*. Pergamon Press.

Nagira, K., Hayashida, M., Shiga, M., Sasamoto, K., Kina, K., Osada, K., Sugahara, T., & Murakami, H. (1995). Effects of organic pH buffers on a cell growth and an antibody production of human-human hybridoma HB_4C_5 cells in a serum-free culture. *Cytotechnology, 17*, 117–125.

Nicklin, D. J. (1962). Two-phase bubble flow. *Chemical Engineering Science, 17*, 693–702.

Nienow, A. W. (1990). Gas dispersion performance in fermenter operation. *Chemical Engineering Progress, 86*, 61–71.

Noble, I., Collins, M., Porter, N., & Varley, J. (1994). An investigation on the physicochemical basis of foaming in fungal fermentations. *Biotechnology and Bioengineering, 44*, 801–807.

Opsomer, E., & Vandewalle, N. (2016). Novel structures for optimal space partitions. *New Journal of Physics, 18*, 103008.

Pahari, S., Sun, L., & Alexov, E. (2019). PKAD: A database of experimentally measured pK_a values of ionizable groups in proteins. *Database (Oxford), 2019*, baz024.

Pelton, R. (2002). A review of antifoam mechanisms in fermentation. *Journal of Industrial Microbiology and Biotechnology, 29*, 149–154.

Princen, H. M., & Kiss, A. D. (1987). Osmotic pressure of foams and highly concentrated emulsions. 2. Determination from the variation in volume fraction with height in an equilibrated column. *Langmuir, 3*, 36–41.

Prins, A., & R. K. (1987). Protein and surface effects in fermentation: Foam, antifoam and mass transfer. *Trends in Biotechnology, 5*, 296–301.

Rahmawati, S., & Bundjali, B. (2012). Kinetics of the oxidation of vitamin C. *Indonesian Journal of Chemistry, 12*, 535–546.

Razafindralambo, H., Paquot, M., Baniel, A., Popineau, Y., Hbid, C., Jacques, P., & Thonart, P. (1996). Foaming properties of surfactin, a lipopeptide surfactant from *Bacillus subtilis*. *Journal of American Oil Chemists' Society, 73*, 149–151.

Renganathan, M., & Bose, S. (1990). Inhibition of photosystem II activity by Cu^{++} ion. Choice of buffer and reagent is critical. *Photosynthesis Research, 23*, 95–99.

Rodríguez, G., Riera, E., Gallego-Juárez, J. A., Acosta, V. M., Pinto, A., Martínez, I., & Blanco, A. (2010). Experimental study of defoaming by air-borne power ultrasonic technology. *Physics Procedia, 3*, 135–139.

Russel, D. W., & Sambrook, J. (2001). *Molecular cloning: A laboratory manual*. Cold Spring Harbor Laboratory Press.

Schugerl, K. (1985). Foam formation, foam suppression and the effect of foam on growth. *Process Biochemistry, 20*, 122–123.

Schugerl, K. (2000). Recovery of protein and microorganisms from cultivation media by foam flotation. *Advances in Biochemical Engineering and Biotechnology, 68*, 191–233.

Scorpio, R. (2000). *Fundamentals of acids, bases, buffers and their applications to biochemical systems*. Kendall/Hunt Publishing.

Skoog, D. A., West, D. M., Holler, F. J., & Crouch, S. R. (2014) *Fundamentals of analytical chemistry*. Brooks/Cole.

Stocks, S. M., Cooke, M., & Heggs, P. J. (2005). Inverted hollow spinning cone as a device for controlling foam and hold-up in pilot scale gassed agitated fermentation vessels. *Chemical Engineering Science, 60*, 2231–2238.

Takesono, S., Onodera, M., Nagai, J., Yamagiwa, K., Mori, A., & Okhawa, A. (1993). Relation between mechanical foam-breaking difficulty and the foaming characteristics of solutions. *Journal of Fermentation and Bioengineering, 75*, 314–318.

Takesono, S., Onodera, M., Yamagiwa, K., Mori, A., & Ohkawa, A. (1994). Relation between critical disk speed for foam-breaking of rotating-disk mechanical foam-breakers and foaming characteristics of solutions. *Journal of Fermentation and Bioengineering, 77*, 221–223.

Takesono, S., Onodera, M., Yoshida, M., Yamagiwa, K., & Okahwa, A. (2002). Performance characteristics of mechanical foam-breakers fitted to a stirred-tank reactor. *Journal of Chemical Technology and Biotechnology, 78*, 48–55.

Taylor, G. I., & Michael, D. H. (1973). On making holes in a sheet of fluid. *Journal of Fluid Mechanics, 58*, 625–639.

Taylor, J. (1976). The structure and singularities in soap bubble-like and soap-film-like minimal surfaces. *Annals of Mathematics, 103*, 489–539.

Thiel, T., Liczkowski, L., & Bissen, S. T. (1998). New zwitterionic butanesulfonic acids that extend the alkaline range of four families of Good buffers: Evaluation for use in biological systems. *Journal of Biochemical and Biophysical Methods, 37*, 117–129.

Thomson, W. (1887). On the division of space with minimum partitional area. *Acta Mathematica, 11*, 121–134.

Tyagi, R. D., Tran, F. T., & Chowdhury, A. K. M. M. (1991). Performance of RBC coupled to a polyurethane foam to biodegrade petroleum refinery wastewater. *Environmental Pollution, 76*, 61–70.

Urbansky, E. T., & Schock, M. R. (2000). Understanding, deriving, and computing buffer capacity. *Journal of Chemical Education, 77*, 1640–1645.

Vardar-Sukan, F. (1998). Foaming: Consequences, prevention and destruction. *Biotechnology Advances, 16*, 913–948.

Verbist, G., Weaire, D., & Kraynik, A. M. (1996). The foam drainage equation. *Journal of Physics: Condensed Matter, 8*, 3715–3731.

Vetoshkin, A. G. (2003). Modeling of centrifugal rotary plate foam breakers. *Theoretical Foundations of Chemical Engineering, 37*, 372–377.

Vitasari, D., Cox, S., Grassia, P., & Rosario, R. (2020). Effect of surfactant redistribution on the flow and stability of foam films. *Proceedings of Royal Society A, 476*, 20190637.

Weaire, D., & Fortes, M. A. (1994). Stress and strain in liquid and solid foams. *Advances in Physics, 43*, 685–738.

Weaire, D., Pittet, N., Hutzler, S., & Pardal, D. (1993). Steady-state drainage of an aqueous foam. *Physical Review Letters, 71*, 2670–2673.

Wongsamuth, R., & Doran, P. M. (1994). Foaming and cell flotation in suspended plant cell cultures and the effect of chemical antifoams. *Biotechnology and Bioengineering, 44*, 481–488.

Zlokarnik, M., & Judat, H. (1988). Stirring. In W. Gerhartz (Ed.), *Ullmann's encyclopedia of industrial chemistry* (Vol. B2, pp. 25-1-33). VCH.

Medium Sterilization 3

Prior to inoculation, sterile medium and sterile (parts of) equipment coming into contact therewith must be guaranteed; subsequent handling will require quantitative exclusion of unwanted or contaminating microorganisms, via implementation of aseptic techniques. Sterilization implies full destruction of all microorganisms, in both vegetative and spore forms—at the expense of heat, chemicals, radiation, or filtration.

When operating batchwise, the liquid medium is typically sterilized in the vessel where it will be used following inoculation; (water) steam above atmospheric pressure is accordingly passed through the jacket of the vessel, or tubing coils placed outside or inside said vessel, or sparged directly into the medium. Fermentation processes at industrial scale normally encompass amounts of liquid medium of the order of m^3, and of air the order of m^6; since work with axenic cultures is again intended, and for better control and conformity with product specifications, the aforementioned raw materials must be supplied, on a continuous basis, free of contaminating microorganisms (and the like).

There are several possibilities to carry out sterilization of medium, ranging from chemical treatments (e.g., addition of some disinfectant or antiseptic agent), through wave treatments (e.g., exposure to *UV*, γ, or *X* rays, ultrasound), to mechanical treatments (e.g., filtration) and thermal treatments (e.g., contact, dielectric, ohmic heating).

Chemicals are the option chosen for sterilization of delicate equipment, such as optical instruments and electrical devices; this mode of operation is not so popular due to toxicity of the typical compounds employed. The most common chemicals are gaseous ethylene oxide, which alkylates amino, sulfhydryl, carboxyl, and hydroxyl groups in cell constituents, thus rendering them unviable; formaldehyde, used as fumigant, which acts by alkylating amino and sulfhydral side groups in proteins and ring nitrogen atoms of purine bases in DNA; and hydrogen peroxide, used in aseptic packaging, which acts

F. X. Malcata, *Operation Fundamentals in Bioreactor Engineering—Medium Handling*, https://doi.org/10.1007/978-3-032-03172-3_3

indirectly as generator of such oxidative species as superoxide ions and hydroxyl radicals through Fenton's reaction, implicated in oxidative damage of proteins and DNA.

Radiation is recommended for heat-sensitive materials and whenever structural changes caused by high temperature are to be avoided; the two major forms are ultraviolet (*UV*) radiation and ionizing radiation. The former initiates excitation of atoms, which leads to mutations in nucleic acids that irreversibly compromise functional protein synthesis and cell replication; since *UV* are unable to penetrate materials in depth, for being promptly absorbed through excitation of outer electrons, it is restricted to surface treatments. Due to the much higher energy content of the supporting quanta, ionizing radiation penetrates materials much deeper and brings about metabolism-incompatible ionization within the cells, even though microorganisms may exhibit enhanced resistance thereto under anoxic conditions or frozen state; it usually employs γ radiation, generated by a ^{60}Co source, due to its brief half-life, and emission of two high-energy (useful) photons.

Filtration-mediated sterilization operates by excluding rather than destroying microorganisms; besides being completely safe for the human user, it is effective to treat heat-sensitive liquids and gases. Three types of filters are currently in use—depth filters, membrane filters, and nucleating track filters. The first ones consist of columns packed with such fibrous materials as glass or cotton wool; their twisting and turning fibers entrap particles, but show little resistance to flow in the case of gases—and may also be used in a prefilter role, prior to membrane filters highly susceptible to clogging. Membrane filters act by screening out suspended particles, and are common in processing with vitamin solutions owing to their heat lability; they need, in turn, to be heat-sterilized in advance. Filtration effectiveness depends on pore size and nature/intensity of electrostatic interactions; yeasts are screened out by pore sizes within the range 0.45–1.2 μm, bacteria by 0.2 μm, and viruses and mycoplasmas by 0.01–0.1 μm. The most common are manufactured from cellulose acetate or nitrate. Nucleation track filters consist of very thin polycarbonate films, previously treated with nuclear radiation and then etched with a specific chemical to create quite uniform vertical holes; an alternative to membrane filters, they are however more easily clogged.

Thermal processing at high temperature denatures and coagulates vital proteins, and resorts to red heat, moist heat, or dry heat. The former is used at lab scale for culture manipulation, and consists in sterilization of inoculating wires or loops by holding them in a Bunsen flame until they become red-hot. Sterilization of culture media and destruction of discarded cultures resort usually to moist heat, since bacteria are more readily destroyed by steam-conveyed heat than plain dry heat. Air is first to be removed from the bioreactor or the autoclave via replacement by saturated steam, continuously supplied at 121 °C, with holding for no less than 15 min. Dry heat is employed when materials are susceptible to corrosion by steam, or must remain dry before use; this is the case of metal instruments, glass Petri dishes, flasks and pipettes, and cotton wool. In practice, this mode of sterilization requires higher temperatures and longer time intervals; for instance, dry heating to 160 °C for 120 min produces the same results as moist heating to 121 °C for 15 min. All these possibilities consubstantiate an imposed-temperature approach, because the temperature of the liquid medium cannot exceed the (given) temperature of the steam supplied; the accompanying safety of operation is then to be balanced against the slower and slower heating as temperature increases, due to reduction in driving force as per Fourier's law. In the case of direct steam injection, provision is to be made for medium dilution owing to steam condensation, which may easily reach 20%(*v/v*); the steam utilized must also be of high quality to avoid (chemical) contamination of the medium with inorganic salts or organic compounds.

Depending on the flow rate of steam supplied, heating of a whole batch of liquid medium may take a relatively long time; once the target sterilization temperature is attained, it is kept essentially constant at that value for some prespecified holding time to guarantee enough reduction in viable counts. Cooling is then in order, to room temperature or the desired processing temperature, normally at the expense of cool water flown though the jacket or the tube coils, when indirect heating was performed in the first place. In the case of direct heating, the temperature of liquid in the vessel will have eventually reached the (saturation) temperature, T_{st}, of the steam bubbled therein; hence, further bubbling of steam will be inconsequential during the holding stage (as long as the holding tube is well insulated), in view of the absence of driving force for heat transfer between heating steam and liquid medium, kept necessarily at the same pressure. If such a high pressure, $P_w^\sigma\{T_{st}\}$, is suddenly released to below atmospheric pressure, $P_w^\sigma\{T_0\}$ with T_0 denoting room temperature, then boiling of water will be enforced, and the corresponding latent heat of vaporization will be extracted from the liquid medium as sensible heat (under adiabatic operation) until its temperature approaches T_0. This process has classically been termed flash cooling and is illustrated in Fig. 3.1, overlaid on the vapor pressure curve of water versus temperature.

In the generation of the P_w^σ-vs.-T curve pertaining to water, Antoine's equation, that is,

$$P_w^\sigma = 10^{A - \frac{B}{C+T}}, \tag{3.1}$$

was used as a basis, with $A = 5.25938$, $B = 1810.94$, and $C = 244.485$ (pressure in Pa, temperature in °C). Release of pressure to just atmospheric pressure through plain venting is easy and relatively inexpensive to accomplish, yet it will bring liquid temperature rapidly to 100 °C (as normal boiling point of water); the remaining cooling of liquid medium to room temperature would take place via evaporation of water as per Fick's law, along a favorable gradient in partial pressure of water vapor (possibly combined with heat loss as per Fourier's law). This is why convective boiling, brought about by a total pressure difference in the presence of plain steam and following Darcy's law, is preferred for this second stage—even though it requires a vacuum-producing apparatus, instead of resorting to plain diffusional evaporation, spontaneous through atmospheric air acting as solvent but excessively slow to prevent unwanted thermal decay. In either case, however, atmospheric air will have to eventually be admitted into the vessel for aerobic culturing, during venting

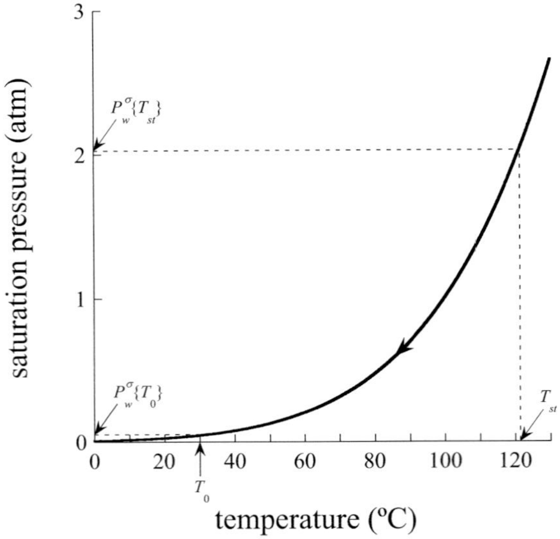

Fig. 3.1 Variation of saturation pressure, P^σ_w, versus temperature, of pure water, with indication of temperature drop from T_{st} to T_0, due to flash cooling

proper or after flash cooling is over; hence, special care is to be exercised to assure aseptic conditions for the inlet air stream. Since the latent heat of vaporization is a relatively weak function of temperature, and the final target temperature is close to room temperature, one will end up with the amount of steam condensed inside the liquid medium essentially reversed as amount of steam withdrawn by flash vaporization.

Cell death occurs at all times during batch sterilization, including heating-up and cooling-down periods. If direct injection of steam is used for heating and flash boiling for cooling, then the accompanying transients are very short, and sometimes not considered at all to simplify lethality calculations; such extra periods accordingly play the role of safety factor for the calculated holding time, particularly useful when the raw liquid medium is somehow ill-defined, in terms of qualitative and quantitative microbial contamination. Although cells die at a relatively slow pace below the normal boiling point of water, and most lethality occurs at the reference sterilization temperature of 121 °C, large bioreactor volumes frequently imply heating-up and cooling-down periods that last for hours, when the corresponding holding time may be just several minutes; under these circumstances, reduction in viability brought about during the initial and final transients should not be neglected in process simulation, otherwise gross overestimation of sterilization time might result. Therefore, a priori knowledge of the temperature profile prevailing during heating up and cooling down is relevant, and will be discussed below to some length for well-stirred liquid medium under three distinct modes of heating and two distinct modes of cooling.

Electric heating constitutes another form of thermal sterilization of a liquid medium in situ, yet it materializes an imposed-flux approach; in this case, heating will take place at constant rate if electrical power supplied is also constant, so temperature of the electrical resistor increases as much as required to guarantee that the heat flux obeying Fourier's law is equal to the electrical power dissipated due to Joule's effect. Since temperature is not upper-limited in this case, the liquid medium may boil out in full, and the resistor material may even melt down with dangerous and irreversible outcomes; hence, this modus operandi is restricted to small-scale equipment, as damages would be limited anyway. Furthermore, the cost of burning fuel to produce steam still lies well below the cost of electrical power for the same amount of heat supplied, thus also helping justify choice of steam to convey heating in industrial settings.

The design procedures for sterilization usually assume a homogeneous liquid medium, exhibiting a uniform temperature as an outcome of good stirring; however, temperature gradients may develop locally due to stagnant spots, or when the liquid contains particles in suspension of either inorganic nature (e.g., salt clumps) or organic nature (e.g., cell flocs or pellets). These solid materials frequently possess lower thermal conductivities than the liquid itself, meaning that temperature in the innermost portion of those particles will be lower than that prevailing in the liquid surrounding them; hence, that portion will remain below sterilization temperature for a significant portion of the processing period. Consequently, the extent of lethality of contaminating cells and spores inside the particles is expected to be lower, and longer holding times will be necessary for efficient thermal treatment whenever particulate media (or even submerged solid substrates) are at stake.

On the other hand, when sterilization of larger volumes of liquid is envisaged, longer thermal treatments are needed because temperature of the sterilizing steam cannot increase in the same proportion, and the ratio of outer area to inner volume decreases in the case of jacketed bioreactors. Sustained elevated temperatures during heating (and cooling, to a lesser degree) can damage heat-labile components of the medium (especially proteins, vitamins, and sugars), so the nutritional quality of the medium may be compromised when scaling up.

Despite all the above possibilities, only filter- and heat-mediated sterilization have met with success in common practice, with the latter materialized in direct or indirect heat transfer from an external source; just those possibilities will accordingly be explored to some length in the following subsections.

Problem 3.1

Estimate the flow rates of steam extracted from a batch of hot liquid medium at 121 °C, using either Darcyan boiling to a vacuum of 3 kPa or Fickian evaporation below 100 °C at 1 atm, in a vessel connected to the outside environment via a cylindrical pipe 1 cm-long and 1 cm-wide, and compare their orders of magnitude using water as reference.

Solution

According to Hagen and Poiseuille's law, the (viscous) pressure drop, ΔP, of a fluid of viscosity μ, along a cylindrical tube of length L and radius R, is given by

(continued)

Problem 3.1 (continued)

$$\Delta P = \frac{8\mu L}{R^2} v, \tag{3.2}$$

where μ and v denote viscosity and (average) velocity of fluid; laminar flow conditions are assumed for simplicity, knowing that turbulent regime would lead to, at most, one order of magnitude deviation in the estimation process. Since volumetric flow rate, Q_d, and velocity are hereby related by

$$Q_d = \pi R^2 v \tag{3.3}$$

due to the cylindrical shape at stake, one ends up with

$$Q_d = \pi R^2 \frac{R^2 \Delta P}{8\mu L} = \frac{\pi R^4 (P_{in} - P_{out})}{8\mu L} \tag{3.4}$$

upon insertion of Eq. (3.2) and explicitation of ΔP as difference between inside and outside pressures, P_{in} and P_{out}, respectively. For $\mu = 20$ μPa.s in the case of water vapor at 121 °C, $L = 1$ cm and $R = 5$ mm by hypothesis, besides $P_{in} = 0.2$ MPa as saturation pressure at 121 °C, and $P_{out} = 3$ kPa as given, Eq. (3.4) will look like

$$Q_d = \frac{\pi \left(5 \times 10^{-3}\right)^4 \left(0.2 \times 10^6 - 3 \times 10^3\right)}{8 \cdot 20 \times 10^{-6} \cdot 1 \times 10^{-2}} = 242 \text{ m}^3\text{s}^{-1} \tag{3.5}$$

as descriptor of flow rate of steam extracted via flash cooling. Assume now that plain diffusional flow takes place according to Fick's law, namely,

$$N = \mathcal{D}_{W,air} \frac{\Delta C}{L}, \tag{3.6}$$

where N denotes molar flux of water vapor, $\mathcal{D}_{W,air}$ denotes diffusivity of water vapor in air, and ΔC denotes difference in molar concentration of water vapor between tube ends; after multiplying both sides by the cross-sectional area, Eq. (3.6) gives rise to

$$M_f = \mathcal{D}_{W,air} \frac{\Delta C}{L} \pi R^2 = \frac{\pi R^2 \mathcal{D}_{W,air} (C_{in} - C_{out})}{L}, \tag{3.7}$$

where the molar flow rate is denoted by M_f, and ΔC is given by inside concentration, C_{in}, abstracted of outside concentration, C_{out}, of water vapor. Recall the ideal gas equation, that is,

(continued)

Problem 3.1 (continued)

$$PV = n\mathcal{R}T, \tag{3.8}$$

where P denotes pressure, V denotes volume, n denotes number of moles, T denotes absolute temperature, and $\mathcal{R}$ denotes ideal gas constant; upon division of both sides by $V\mathcal{R}T$, Eq. (3.8) yields

$$C = \frac{P}{\mathcal{R}T}, \tag{3.9}$$

since molar concentration, C, abides to

$$C \equiv \frac{n}{V}. \tag{3.10}$$

The vapor is at atmospheric pressure, P_0, on the inlet side, that is,

$$C_{in} \equiv \frac{P_0}{\mathcal{R}T_{in}}; \tag{3.11}$$

whereas its partial pressure on the outside would be, at most, the saturation pressure of water at room temperature, $P_w^\sigma\{T_0\}$ (corresponding to a relative humidity of 100%), that is,

$$C_{out} \equiv \frac{P_w^\sigma\{T_0\}}{\mathcal{R}T_0}; \tag{3.12}$$

insertion of Eqs. (3.11) and (3.12) converts Eq. (3.7) to

$$M_f = \frac{\pi R^2 \mathcal{D}_{W,air}\left(\frac{P_0}{\mathcal{R}T_{in}} - \frac{P_w^\sigma\{T_0\}}{\mathcal{R}T_0}\right)}{L} = \frac{\pi R^2 \mathcal{D}_{W,air}}{\mathcal{R}L}\left(\frac{P_0}{T_{in}} - \frac{P_w^\sigma\{T_0\}}{T_0}\right), \tag{3.13}$$

along with factoring out of $\mathcal{R}$. Equation (3.8) may instead be taken on a per unit time basis, according to

$$P_0 Q_f = M_f \mathcal{R} T_0, \tag{3.14}$$

(continued)

Problem 3.1 (continued)

where Q_f denotes volumetric flow rate, evaluated at outlet conditions, namely, pressure P_0 and temperature T_0; elimination of M_f between Eqs. (3.13) and (3.14) gives rise to

$$\frac{P_0 Q_f}{\mathscr{R} T_0} = \frac{\pi R^2 \mathcal{D}_{W,air}}{\mathscr{R} L} \left(\frac{P_0}{T_{in}} - \frac{P_w^\sigma\{T_0\}}{T_0} \right), \tag{3.15}$$

or else

$$Q_f = \frac{\mathscr{R} T_0}{P_0} \frac{\pi R^2 \mathcal{D}_{W,air}}{\mathscr{R} L} \left(\frac{P_0}{T_{in}} - \frac{P_w^\sigma\{T_0\}}{T_0} \right) = \frac{\pi R^2 \mathcal{D}_{W,air}}{L} \left(\frac{T_0}{T_{in}} - \frac{P_w^\sigma\{T_0\}}{P_0} \right) \tag{3.16}$$

upon isolation of Q_f, followed by factoring in of T_0/P_0 and dropping of common factors between numerator and denominator afterward. For $\mathcal{D}_{W,air} = 0.45$ cm^2 s^{-1} at 100 °C, $L = 1$ cm, $R = 5$ mm, $P_0 = 0.1$ MPa, $T_0 = 20$ °C, and $T_{in} = 100$ °C, besides $P_w^\sigma\{T_0\} = 3$ kPa, Eq. (3.16) leads to

$$Q_f = \frac{\pi (5 \times 10^{-3})^2 0.45 \times 10^{-4}}{1 \times 10^{-2}} \left(\frac{20 + 273}{100 + 273} - \frac{3 \times 10^3}{0.1 \times 10^6} \right) = 2.67 \times 10^{-7} \ \mathrm{m^3 s^{-1}} \tag{3.17}$$

for an assumed environmental temperature of 20 °C. Comparison of the results conveyed by Eqs. (3.5) and (3.17) unfolds diffusive evaporation as being nine orders of magnitude slower than Darcyan boiling, thus fully justifying use of flash cooling in industrial processing. Although completely dry air could have been used (with nil relative humidity), thus turning the second term in parenthesis in Eq. (3.17) to zero, the resulting Q_f would increase to just 2.78×10^{-7} m^3 s^{-1}; this slight increase does not materially change the previous conclusion. Such correction would also be essentially unaffected by the temperature of water inside the vessel gradually dropping from 100 °C to eventually 20 °C, because the increase of the first term in parenthesis of Eq. (3.17) from 0.786 to 1 (corresponding to $T_{in} = 20$ °C) would cause a balancing variation in factor $\mathcal{D}_{W,air}$, from 0.45 cm^2 s^{-1} down within the very same temperature interval.

3.1 Cell Death

The rate of dying of contaminant cells, at a given temperature, has classically been modelled as

$$\frac{dC_Y}{dt} = -k_{d,Y}C_Y, \tag{3.18}$$

where C_Y denotes concentration of viable cells, t denotes time, and $k_{d,Y}$ denotes a first-order rate constant; this differential equation is subjected to

$$C_Y|_{t=0} = C_{Y,0} \tag{3.19}$$

for initial condition. Since $k_{d,Y} \equiv k_{d,Y}\{T\}$, it is convenient to make explicit such a functional dependence, namely, via Arrhenius' law, according to

$$k_{d,Y} = k_{0,Y}\mathrm{e}^{-\frac{E_{act,Y}}{\mathscr{R}T}} \tag{3.20}$$

with $k_{0,Y}$ denoting pre-exponential factor and $E_{act,Y}$ denoting activation energy associated with the death process. It should be emphasized that the concept of activation energy applies strictly to a chemical reaction, in which a ubiquitous hill in the energy domain, associated to an activated complex, appears along the reaction coordinate; it is located between reactant and product, and exhibits height equal to said activation energy. If the (absolute) temperature increased without bound, then the average kinetic energy of the molecules in any system would exceed by far said energy barrier, irrespective of its height; this asymptotic situation would then be driven by

$$\lim_{T\to\infty} k_{d,Y} = \lim_{T\to\infty} k_{0,Y}\mathrm{e}^{-\frac{E_{act,Y}}{\mathscr{R}T}} = k_{0,Y}\mathrm{e}^{-\frac{E_{act,Y}}{\mathscr{R}\infty}} = k_{0,Y}\mathrm{e}^{0} = k_{0,Y}, \tag{3.21}$$

stemming from Eq. (3.20). Therefore, $k_{0,Y}$ may be viewed as the limiting value of $k_{d,Y}$ if the energy barrier (measured by $E_{act,Y}$) were negligible (or nonexistent) compared to the average molecular energy (measured by $\mathscr{R}T$). In other words, product molecule(s) would automatically (and necessarily) form upon collision between or rearrangement of reactant molecule(s), with an accompanying full efficiency of chemical transformation; this is why $k_{0,Y}$ is sometimes termed frequency factor, since the reaction would proceed at the very same rate or frequency of molecular collisions. Cell metabolism consists of a coordinated array of (bio)chemical reactions, rather than a single chemical reaction; however, if cell survival depends on a critical enzyme that undergoes thermal deactivation (a known first-order decay process), then the rate of destruction of viable cells will indeed be proportional to the rate of decay of such enzyme, and this argument suffices to justify the functional form of Eqs. (3.18) and (3.20) describing cell death.

Insertion of Eq. (3.20) converts Eq. (3.18) to

$$\frac{dC_Y}{dt} = -k_{0,Y} e^{-\frac{E_{act,Y}}{\mathcal{R}T}} C_Y, \tag{3.22}$$

where integration via separation of variables ensues as

$$\int_{C_{Y,0}}^{C_Y} \frac{d\widetilde{C}_Y}{\widetilde{C}_Y} = -k_{0,Y} \int_0^t e^{-\frac{E_{act,Y}}{\mathcal{R}T\{\tilde{t}\}}} d\tilde{t} \tag{3.23}$$

with the aid of Eq. (3.19); the left-hand side may be further transformed to

$$\ln \widetilde{C}_Y \Big|_{C_{Y,0}}^{C_Y} = -k_{0,Y} \int_0^t e^{-\frac{E_{act,Y}}{\mathcal{R}T\{\tilde{t}\}}} d\tilde{t}, \tag{3.24}$$

which degenerates to

$$\ln \frac{C_Y}{C_{Y,0}} = -k_{0,Y} \int_0^t e^{-\frac{E_{act,Y}}{\mathcal{R}T\{\tilde{t}\}}} d\tilde{t}. \tag{3.25}$$

Upon application of exponentials to both sides, Eq. (3.25) becomes

$$\frac{C_Y}{C_{Y,0}} = \exp\left\{ -k_{0,Y} \int_0^t e^{-\frac{E_{act,Y}}{\mathcal{R}T\{\tilde{t}\}}} d\tilde{t} \right\}; \tag{3.26}$$

calculation of $C_Y/C_{Y,0}$ cannot proceed any further, unless $T \equiv T\{t\}$ is made available, which depends on process envisaged and equipment employed. In order to deal with only dimensionless variables, with similar orders of magnitude, it is convenient to make $e^{E_{act,Y}/\mathscr{R}T_0}$ appear explicitly in Eq. (3.26), according to

$$\frac{C_Y}{C_{Y,0}} = \exp\left\{ - \int_0^t k_{0,Y} e^{\frac{E_{act,Y}}{\mathscr{R}T_0}\left(-1+1-\frac{T_0}{T\{t\}}\right)} d\tilde{t} \right\} = \exp\left\{ - \int_0^t k_{0,Y} e^{-\frac{E_{act,Y}}{\mathscr{R}T_0}} e^{\frac{E_{act,Y}}{\mathscr{R}T_0}\left(1-\frac{T_0}{T\{t\}}\right)} d\tilde{t} \right\}, \tag{3.27}$$

where $k_{0,Y}$ was meanwhile taken back into the kernel, and unity added to and subtracted from the argument of the exponential function; T_0 hereby denotes some reference (usually initial) temperature. After setting

$$k_{d,Y,0} = k_{0,Y} e^{-\frac{E_{act,Y}}{\mathscr{R}T_0}} \tag{3.28}$$

using Eq. (3.20) for template and resorting to $T = T_0$, one may redo Eq. (3.27) as

$$\frac{C_Y}{C_{Y,0}} = \exp\left\{ - \int_0^t k_{d,Y,0} e^{\frac{E_{act,Y}}{\mathscr{R}T_0}\left(1-\frac{T_0}{T\{t\}}\right)} d\tilde{t} \right\}, \tag{3.29}$$

or else

$$\frac{C_Y}{C_{Y,0}} = \exp\left\{ - \int_0^{k_{d,Y,0}t} e^{\frac{E_{act,Y}}{\mathscr{R}T_0}\left(1-\frac{T_0}{T\{t\}}\right)} d\left(k_{d,Y,0}\tilde{t}\right) \right\} \tag{3.30}$$

since $k_{d,Y,0}$ is a constant.

In the case of constant temperature T, Eq. (3.30) will read

$$\frac{C_Y}{C_{Y,0}} = \exp\left\{ - e^{\frac{E_{act,Y}}{\mathscr{R}T_0}\left(1-\frac{T_0}{T}\right)} \int_0^{k_{d,Y,0}t} d\left(k_{d,Y,0}\tilde{t}\right) \right\}, \tag{3.31}$$

which still reduces to

$$\frac{C_Y}{C_{Y,0}} = \exp\left\{ -e^{\frac{E_{act,Y}}{\mathcal{R}T_0}\left(1-\frac{T_0}{T}\right)} k_{d,Y,0}\tilde{t}\,\Big|_0^{k_{d,Y,0}t} \right\} \tag{3.32}$$

upon trivial calculation of the outstanding integral; Eq. (3.32) may be replaced by just

$$\frac{C_Y}{C_{Y,0}} = \exp\left\{ -e^{\frac{E_{act,Y}}{\mathcal{R}T_0}\left(1-\frac{T_0}{T}\right)} k_{d,Y,0}t \right\}, \tag{3.33}$$

as illustrated in Fig. 3.2. Note the linear pattern of the $C_Y/C_{Y,0}$-vs.-$k_{d,Y,0}t$ curves, arising from the functional form of Eq. (3.33) and the log/lin scale elected, with slope given by $-\exp\{E_{act,Y}/\mathcal{R}T_0(1 - T_0/T)\}/\ln 10$, apparent after taking natural logarithms of both sides of Eq. (3.33) and then converting to decimal logarithms; hence, increases in $E_{act,Y}/\mathcal{R}T_0$ or in T/T_0 imply faster rates of decay, the former owing to a magnified sensitiveness to temperature and the latter to a higher temperature employed. The softness of treatment at room temperature, $T = T_0$, is also worth noticing, as well as the great lethality associated to $T/T_0 = 1.35$, thus justifying choice of 121 °C for sterilization temperature of a system at room temperature (20 °C).

Although many contaminating microorganisms (and their spores), meant to be eliminated from liquid media through sterilization, are particularly sensitive to temperature with values of $E_{act,Y}/\mathcal{R}T_0$ up to 100, the curves in Fig. 3.2 are still useful; in fact, they cover less heat-susceptible microorganisms, and thus microorganisms harder to eliminate.

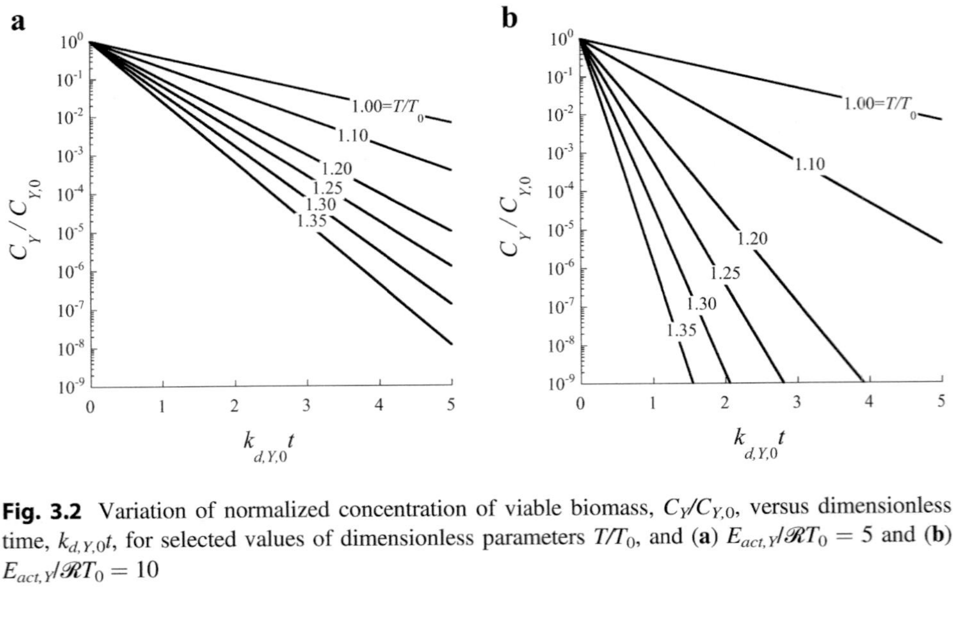

Fig. 3.2 Variation of normalized concentration of viable biomass, $C_Y/C_{Y,0}$, versus dimensionless time, $k_{d,Y,0}t$, for selected values of dimensionless parameters T/T_0, and (**a**) $E_{act,Y}/\mathcal{R}T_0 = 5$ and (**b**) $E_{act,Y}/\mathcal{R}T_0 = 10$

Problem 3.2

The classical model of first-order kinetics considers a constant probability of death of a microorganism cell, irrespective of the time when it happens and the size of the current biomass inventory. The (less restrictive) possibility of time-varying heat lability has been addressed by Weibull's model, namely,

$$\frac{dC_Y}{dt} = -k_{d,Y}p\left(\frac{t}{\delta}\right)^{p-1} C_Y, \tag{3.34}$$

where $p(t/\delta)^{p-1}$ appears as corrective factor in the original Eq. (3.18) and represents a pseudo probability density function; such correction is characterized by parameters $p > 0$ and $\delta > 0$, which relate to population behavior and destruction behavior, respectively. Discuss this model vis-à-vis with the classical model of cell death and, in particular, ascribe a significance to $p < 1$, $p = 1$, and $p > 1$.

Solution

To make Weibull's model useful, Eq. (3.34) should first be integrated (via separation of variables), that is,

$$\int_{C_{Y,0}}^{C_Y} \frac{d\tilde{C}_Y}{\tilde{C}_Y} = -\frac{k_{d,Y}}{\delta^{p-1}} \int_0^t p\tilde{t}^{p-1} d\tilde{t}, \tag{3.35}$$

where $k_{d,Y}$ and δ^{p-1} being constants was taken advantage of, and Eq. (3.19) was kept as still valid; the fundamental theorem of integral calculus may then be invoked to obtain

$$\ln \tilde{C}_Y \Bigg|_{C_{Y,0}}^{C_Y} = -\frac{k_{d,Y}}{\delta^{p-1}} p \left. \frac{\tilde{t}^p}{p} \right|_0^t = -\frac{k_{d,Y}}{\delta^{p-1}} \tilde{t}^p \Bigg|_0^t, \tag{3.36}$$

which promptly becomes

$$\ln \frac{C_Y}{C_{Y,0}} = -k_{d,Y} \frac{t^p}{\delta^{p-1}}. \tag{3.37}$$

For unit p, Eq. (3.37) degenerates to

(continued)

Problem 3.2 (continued)

$$\ln \left.\frac{C_Y}{C_{Y,0}}\right|_{p=1} = -k_{d,Y}\frac{t^1}{\delta^{1-1}} = -k_{d,Y}t, \tag{3.38}$$

and exponentials taken of both sides unfold

$$\left.\frac{C_Y}{C_{Y,0}}\right|_{p=1} = \mathrm{e}^{-k_{d,Y}t}; \tag{3.39}$$

if $T = T_0$, then $k_{d,Y} = k_{d,Y,0}$, so Eq. (3.33) will end up coincident with Eq. (3.39), meaning that first-order death kinetics is a particular case of Weibull's model, described by $p = 1$. If $p > 1$, then t^p represents a convex function of time, and so does t^p/δ^{p-1} (since the denominator is a constant); therefore, surviving cells become more susceptible to thermal treatment as time elapses, at any given (constant) temperature, probably a consequence of cumulative damage in their metabolic machinery. Conversely, $p < 1$ entails a concave shape for the t^p/δ^{p-1} curve, that is, bearing a concavity that faces down; destruction of cells becomes, in this case, slower as time elapses, that is, a more resistant fraction of cells emerges for being able to survive longer.

3.2 Heating

Heating consists of transfer of energy in the form of heat, so it represents a degraded form of energy in transit, consubstantiated in random translations, rotations, and vibrations of molecules and the like. The aforementioned transfer can take place through contact among molecules at the expense of collisions, or via photons that propagate in space at the speed of light and do not require a material medium.

Heat transfer by conduction follows Fourier's law, which in general takes the form

$$Q = -kA\frac{dT}{dx}, \tag{3.40}$$

where Q denotes heat flow rate (or energy transferred per unit time), k denotes thermal conductivity, A denotes heat transfer area, T denotes temperature, and x denotes space coordinate associated to direction of flow. When heat transfer occurs within a flowing fluid phase, as happens in medium heated via condensing steam, Eq. (3.40) still applies, yet portions of fluid move with different velocities and according to patterns that are often difficult to model. Furthermore, it has been observed that most of the underlying temperature gradient occurs within a thin layer, of thickness δ, where the fluid contacts another fluid or a solid surface. In attempts to circumvent the above hydrodynamic issues, a black-box approach has been classically followed, derived by first integrating Eq. (3.40) via separation of variables, that is,

$$Q\int_{x}^{x+\delta} dx = -kA\int_{T|_x}^{T|_{x+\delta}} dT, \tag{3.41}$$

where $T|_x$ denotes temperature at position x and $T|_{x+\delta}$ denotes temperature at position $x+\delta$; along with (hypothesized) constancy of Q, k, and A. Equation (3.41) is equivalent to

$$Q(x+\delta-x) = -kA(T|_{x+\delta} - T|_x) \tag{3.42}$$

after invoking the fundamental theorem of integral calculus, or merely

$$Q\delta = kA(T|_x - T|_{x+\delta}) \tag{3.43}$$

upon cancelling x with its negative and factoring in -1. Once both sides are divided by δ, Eq. (3.43) gives rise to

$$Q = hA\Delta T \tag{3.44}$$

as long as a heat transfer coefficient, h, is defined as

$$h \equiv \frac{k}{\delta}, \tag{3.45}$$

and the temperature drop abides to

$$\Delta T \equiv T|_x - T|_{x+\delta}; \tag{3.46}$$

note that the difficulty in estimating δ has been translated to the need of some (experimentally) validated correlation to estimate h.

Heat transfer by radiation satisfies instead Stefan and Boltzmann's law, namely,

$$Q_{12} = \sigma\left(\alpha_1\varepsilon_2 A_2 F_{21} T_2{}^4 - \alpha_2\varepsilon_1 A_1 F_{12} T_1{}^4\right), \tag{3.47}$$

where Q_{12} is the net heat flow exchange between bodies 1 and 2, with σ denoting Stefan and Boltzmann's constant; α_1 and α_2 denoting absorptivity of bodies 1 and 2, respectively; ε_1 and ε_2 denoting emissivity of bodies 1 and 2, respectively; A_1 and A_2 denoting outer area of bodies 1 and 2, respectively; viewfactor F_{21} denoting fraction of radiant energy emitted by body 2 that reaches body 1; viewfactor F_{12} denoting fraction of radiant energy emitted by body 1 that reaches body 2; and T_1 and T_2 denoting absolute temperature of bodies 1 and 2, respectively. On the other hand, Kirchhoff's law states that the absorptivity of any body i, at a given temperature, equals its emissivity at the same temperature, that is,

$$(\alpha_i = \varepsilon_i)_T; \tag{3.48}$$

in practice, however, the dependence of said equality on temperature is weak, so one takes

$$\alpha_i = \varepsilon_i \tag{3.49}$$

to essentially hold within the relatively narrow range $[T_1, T_2]$ of interest in bioreactor operation.

Combination with Eq. (3.49) converts Eq. (3.47) to

$$Q_{12} = \sigma\left(\varepsilon_1 \varepsilon_2 A_2 F_{21} T_2{}^4 - \varepsilon_2 \varepsilon_1 A_1 F_{12} T_1{}^4\right), \tag{3.50}$$

in view of $\alpha_1\{T_1\} \approx \alpha_1\{T_2\} = \varepsilon_1\{T_2\}$ and $\alpha_2\{T_2\} \approx \alpha_2\{T_1\} = \varepsilon_2\{T_1\}$, or else

$$Q_{12} = \sigma \varepsilon_1 \varepsilon_2 \left(A_2 F_{21} T_2{}^4 - A_1 F_{12} T_1{}^4\right) \tag{3.51}$$

after factoring out ε_1 and ε_2, and realizing that $\varepsilon_1\{T_2\} \approx \varepsilon_1\{T_1\} \equiv \varepsilon_1$ and $\varepsilon_2\{T_1\} \approx \varepsilon_2\{T_2\} \equiv \varepsilon_2$. Finally, the law of reciprocity of viewfactors has it that

$$A_2 F_{21} = A_1 F_{12}, \tag{3.52}$$

which prompts transformation of Eq. (3.51) to

$$Q_{12} = \sigma \varepsilon_1 \varepsilon_2 \left(A_1 F_{12} T_2{}^4 - A_1 F_{12} T_1{}^4\right); \tag{3.53}$$

Equation (3.53) breaks down to

$$Q_{12} = \sigma \varepsilon_1 \varepsilon_2 A_1 F_{12} \left(T_2{}^4 - T_1{}^4\right), \tag{3.54}$$

after having factored out A_1F_{12}.

In a sterilization process, heat transfer takes place by both conduction and radiation; however, combination of $\sigma = 5.6704 \times 10^{-8}$ W m^{-2} K^{-4} with $0 < \varepsilon_1 < 1, 0 < \varepsilon_2 < 1$, and $0 < F_{12} < 1$ leads to

$$\frac{Q_{12}}{A_1} << 5.6704 \times 10^{-8} \cdot 1 \cdot 1 \cdot 1 \left((121 + 273)^4 - (20 + 273)^4 \right) = 9.49 \times 10^2 \text{ Wm}^{-2} \tag{3.55}$$

based on Eq. (3.54), for a typical initial temperature of 20 °C for the liquid medium and 121 °C for the sterilizing steam. Conversely, the heat transfer coefficient of an aqueous medium, h_l, is of the order of 5×10^3 W m^{-2} K^{-1}, and that of condensing steam, h_s, is of the order of 10×10^3 W m^{-2} K^{-1}, while the metal wall in between often offers negligible resistance to heat transfer; hence, a typical overall heat transfer coefficient h, given by

$$h = \frac{1}{\frac{1}{h_l} + \frac{1}{h_s}} = \frac{1}{\frac{1}{5 \times 10^3} + \frac{1}{10 \times 10^3}} = 3333 \text{ Wm}^{-2}\text{K}^{-1}, \tag{3.56}$$

leads to

$$\frac{Q}{A} = 3333 \left((121 + 273) - (20 + 273) \right) = 3.37 \times 10^5 \text{ Wm}^{-2} \tag{3.57}$$

stemming from Eq. (3.44). Comparative inspection of Eqs. (3.55) and (3.57) indicates that heat transfer by radiation is more than two orders of magnitude lower than heat transfer by conduction, within the temperature range germane in common sterilization, thus justifying why only the latter is normally taken into account. In addition, the amount of heat transferred is readily accessible from the mass of steam condensed multiplied by its latent heat of condensation.

Problem 3.3

A simple theory of energy transport in (pure) liquids has been proposed by Bridgman. Molecules were thereby assumed to arrange in a cubic lattice, with center-to-center spacing ϖ_l/k_a, where ϖ_l denotes molar volume and k_a denotes Avogadro's number; energy was further assumed to transfer from one to the next lattice plane at sound velocity v_{sd} in said liquid (since sound is a purely molecular-

(continued)

Problem 3.3 (continued)

based mechanical event). Based on a reinterpretation of the rigid-sphere gas theory, which has it that

$$k_g = \frac{1}{3}\rho \widehat{c}_V \overline{v} \lambda \tag{3.58}$$

pertaining to thermal conductivity of gas, k_g, where ρ and $\widehat{c}_V$ denote mass density and specific heat capacity at constant volume, respectively, $\overline{v}$ denotes mean molecular speed, and λ denotes mean free path, prove that thermal conductivity of liquid, k_l, will be given by

$$k_l = 3k_b \left(\frac{k_a}{\varpi_l}\right)^{\frac{2}{3}} \left(\frac{\partial P}{\partial \rho_l}\right)_T^{\frac{1}{2}} \tag{3.59}$$

where $(\partial P/\partial \rho_l)_T$ is readily obtainable from isothermal compressibility measurements, or from an equation of state or compressibility correlation. *Hint*: The heat capacity per unit volume of a liquid is about the same as that of a solid.

Solution

According to Dulong and Petit's rule, the molar heat capacity of a solid, c_s, is essentially constant over temperature and approximately given by

$$M_s \widehat{c}_s \equiv c_s \approx 3\mathscr{R}, \tag{3.60}$$

where M_s denotes molar mass and $\widehat{c}_s$ denotes specific heat capacity of solid, respectively; and $\mathscr{R}$ denotes ideal gas constant, defined, in turn, by

$$\mathscr{R} \equiv k_a k_b \tag{3.61}$$

provided that k_b denotes Boltzmann's constant, whereas mass density abides to

$$\rho_s = \frac{M_s}{\varpi_s}. \tag{3.62}$$

Equations (3.60)–(3.62) can be combined to yield

$$\rho_s \varpi_s \widehat{c}_s = 3k_a k_b, \tag{3.63}$$

or else

(continued)

Problem 3.3 (continued)

$$\rho_s \widehat{c}_s = 3\frac{k_a}{\varpi_s} k_b \tag{3.64}$$

upon isolation of $\rho_s \widehat{c}_s$, whereas

$$\rho_l \widehat{c}_l \approx \rho_s \widehat{c}_s \tag{3.65}$$

as per hint, and thus

$$\rho_l \widehat{c}_l = 3\frac{k_a}{\varpi_l} k_b, \tag{3.66}$$

using Eq. (3.64) for template. On the other hand,

$$v_y = \frac{1}{3}\overline{v} \tag{3.67}$$

with v_y denoting velocity in the direction of heat flow, because of the tridimensional nature of molecular motion; hence, Eq. (3.58) may be replaced by

$$k_g = \rho \widehat{c}_V v_y \lambda. \tag{3.68}$$

By analogy to Eq. (3.68), one will resort to

$$k_l = \rho_l \widehat{c}_l v_{sd} \left(\frac{\varpi_l}{k_a} \right)^{\frac{1}{3}} \tag{3.69}$$

with k_l replaced k_g, $\rho_l \widehat{c}_l$ replaced $\rho \widehat{c}_V$, v_{sd} replaced v_y, and lattice spacing $(\varpi_l/k_a)^{1/3}$ replaced by λ; the latter is valid as long as each molecule of the liquid oscillates within a cubic cage formed by its nearest neighbors. Combination of Eq. (3.69) with Eq. (3.65) gives rise to

$$k_l = \rho_s \widehat{c}_s v_{sd} \left(\frac{\varpi_l}{k_a} \right)^{\frac{1}{3}}; \tag{3.70}$$

whereas Eq. (3.64) justifies

(continued)

Problem 3.3 (continued)

$$k_l = 3\frac{k_a}{\varpi_l}k_b v_{sd}\left(\frac{\varpi_l}{k_a}\right)^{\frac{1}{3}} \tag{3.71}$$

or, after lumping powers alike,

$$k_l = 3k_b\left(\frac{k_a}{\varpi_l}\right)^{\frac{2}{3}} v_{sd}. \tag{3.72}$$

In attempts to ascertain the velocity of propagation of sound along the, say, x-direction in a liquid of mass density ρ_l, one should to advantage consider flow through a parallelopiped of sides dx, dy, and dz in length; the mass of liquid dm_l occupying volume $dx\ dy\ dz$ is then given by

$$dm_l = \rho_l\, dx\, dy\, dz. \tag{3.73}$$

The corresponding mass flow rate in the x-direction reads

$$\frac{dm_l}{dt} = \frac{\rho_l\, dx\, dy\, dz}{dt} = \rho_l \frac{dx}{dt}\, dy\, dz = \rho_l v_x\, dy\, dz \tag{3.74}$$

upon division of both sides of Eq. (3.73) by dt, where $v_x \equiv dx/dt$ as per definition of velocity in the x-direction, and v_y and v_z have been taken as nil (due to the solid nature of the containing conduit). If a sound wave moves through an elementary length dx within said liquid, then density, velocity, pressure, and temperature will change from ρ_l, $v_x \equiv v_{sd}$, P, and T to $\rho_l + d\rho_l$, $v_{sd} + dv_{sd}$, $P + dP$, and $T + dT$, respectively, so the continuity equation (along the x-direction) enforces

$$\rho_l v_{sd}\, dy\, dz = (\rho_l + d\rho_l)(v_{sd} + dv_{sd})dy\, dz; \tag{3.75}$$

after dropping $dy\ dz$ from both sides, Eq. (3.75) reduces to

$$\rho_l v_{sd} = (\rho_l + d\rho_l)(v_{sd} + dv_{sd}), \tag{3.76}$$

where $d\rho_l$ is solely due to change in pressure should temperature be assumed constant. Elimination of parentheses converts Eq. (3.76) to

(continued)

Problem 3.3 (continued)

$$\rho_l v_{sd} = \rho_l v_{sd} + \rho_l\, dv_{sd} + v_{sd}\, d\rho_l + d\rho_l\, dv_{sd}; \tag{3.77}$$

second-order differentials can be discarded for being much smaller in magnitude than any first-order differential, and common terms can be dropped from both sides as well to leave

$$0 = \rho_l\, dv_{sd} + v_{sd}\, d\rho_l, \tag{3.78}$$

thus leading to

$$\rho_l\, dv_{sd} = -v_{sd}\, d\rho_l, \tag{3.79}$$

upon isolation of $\rho_l\, dv_{sd}$. On the other hand, the net force dF_{net} on the volume of fluid equals the sum of forces acting upstream and downstream along the direction of flow, according to

$$dF_{net} = P\,dy\,dz - (P + dP)dy\,dz, \tag{3.80}$$

where elimination of parenthesis unfolds

$$dF_{net} = P\,dy\,dz - P\,dy\,dz - dP\,dy\,dz; \tag{3.81}$$

cancellation of $P\,dy\,dz$ with its negative permits simplification to

$$dF_{net} = -dP\,dy\,dz. \tag{3.82}$$

Meanwhile, Newton's law states that

$$dF_{net} = a\,dm_l \tag{3.83}$$

with a denoting acceleration, so insertion of Eq. (3.82) yields

(continued)

Problem 3.3 (continued)

$$-dP\,dy\,dz = a\,dm_l. \tag{3.84}$$

Isolation of a in Eq. (3.84) leads to

$$a = -\frac{dP\,dy\,dz}{dm_l}; \tag{3.85}$$

hence, combination with Eq. (3.73) produces

$$a = -\frac{dP\,dy\,dz}{\rho_l\,dx\,dy\,dz}, \tag{3.86}$$

where cancellation of $dy\,dz$ between numerator and denominator leaves

$$a = -\frac{1}{\rho_l}\frac{dP}{dx}. \tag{3.87}$$

On the other hand, the definition of acceleration as $a \equiv dv/dt$ justifies

$$\frac{dv_{sd}}{dt} = -\frac{1}{\rho_l}\frac{dP}{dx} \tag{3.88}$$

as alias to Eq. (3.87), thus prompting multiplication of both sides by dt to get

$$dv_{sd} = -\frac{1}{\rho_l}\frac{dP}{dx}dt; \tag{3.89}$$

rearrangement as per Leibnitz's algebra leads to

$$dv_{sd} = -\frac{1}{\rho_l}\frac{dP}{\frac{dx}{dt}}, \tag{3.90}$$

which is equivalent to writing

$$dv_{sd} = -\frac{1}{\rho_l}\frac{dP}{v_{sd}} \tag{3.91}$$

or, following removal of denominators,

(continued)

Problem 3.3 (continued)

$$\rho_l v_{sd}\, dv_{sd} = v_{sd}\rho_l\, dv_{sd} = -dP. \tag{3.92}$$

Elimination of $\rho_l\, dv_{sd}$ between Eqs. (3.79) and (3.92) produces

$$v_{sd}(-v_{sd}\, d\rho_l) = -dP \tag{3.93}$$

or, equivalently,

$$v_{sd}^2\, d\rho_l = dP; \tag{3.94}$$

isolation of v_{sd} finally gives

$$v_{sd} = \sqrt{\left(\frac{dP}{d\rho_l}\right)_T}, \tag{3.95}$$

where subscript $_T$ was inserted to emphasize that $\rho_l \equiv \rho_l\{P\}$. Insertion of Eq. (3.95) in Eq. (3.72) finally retrieves Eq. (3.59), as originally intended.

3.2.1 Indirect Heating

Under perfect mixing, a liquid medium being heated should obey

$$UA(T_{st} - T) = V\rho c_P \frac{dT}{dt} \tag{3.96}$$

for enthalpy balance, subjected to

$$T|_{t=0} = T_0 \tag{3.97}$$

for initial condition, when the required sensible heat is obtained at the expense of latent heat released by condensing steam. Here, V, ρ, and c_P denote volume, mass density, and (isobaric) specific heat capacity, respectively, of liquid medium; U denotes overall heat transfer coefficient between steam and liquid medium; T_0 and T denote initial and current temperature, respectively, of liquid medium; and T_{st} denotes temperature of steam. Upon division of both sides by $V\rho c_P$, Eq. (3.96) becomes

$$\frac{dT}{dt} = \frac{UA}{V\rho c_P}(T_{st} - T), \tag{3.98}$$

which may resort to a simpler notation as

$$\frac{d\widehat{T}}{dt} = -\beta\widehat{T} \tag{3.99}$$

after realizing that

$$-d(T_{st} - T) = d(T - T_{st}) = d\widehat{T} \tag{3.100}$$

due to constancy of T_{st}, while defining

$$\beta \equiv \frac{UA}{V\rho c_P} \tag{3.101}$$

as auxiliary parameter and

$$\widehat{T} \equiv T - T_{st} \tag{3.102}$$

as deviation (dependent) variable; Eq. (3.97) will accordingly look like

$$\widehat{T}|_{t=0} = T_0 - T_{st}, \tag{3.103}$$

with the aid of Eq. (3.102). Integration of Eq. (3.99) ensues via separation of variables, namely,

$$\int_{T_0 - T_{st}}^{\widehat{T}} \frac{d\tilde{\widehat{T}}}{\tilde{\widehat{T}}} = -\beta \int_0^t d\tilde{t} \tag{3.104}$$

following combination with Eq. (3.103), where the fundamental theorem of integral calculus has it that

$$\ln \tilde{\widehat{T}}\Big|_{T_0 - T_{st}}^{\widehat{T}} = -\beta\, \tilde{t}\Big|_0^t; \tag{3.105}$$

Equation (3.105) is equivalent to

$$\ln \frac{\widehat{T}}{T_0 - T_{st}} = -\beta t, \tag{3.106}$$

where application of exponentials to both sides unfolds

$$\frac{\widehat{T}}{T_0 - T_{st}} = \mathrm{e}^{-\beta t}. \tag{3.107}$$

The initial notation may now be retrieved in Eq. (3.107), at the expense of Eqs. (3.101) and (3.102), namely,

$$\frac{T - T_{st}}{T_0 - T_{st}} = \mathrm{e}^{-\frac{UA}{V\rho c_P}t}, \tag{3.108}$$

or else

$$\frac{T_{st} - T}{T_{st} - T_0} = \mathrm{e}^{-\frac{UA}{V\rho c_P}t} \tag{3.109}$$

after taking negatives of both numerator and denominator (since $T_{st} > T > T_0$); solution for T finally yields

$$T = T_{st} + (T_0 - T_{st})\mathrm{e}^{-\frac{UA}{V\rho c_P}t}, \tag{3.110}$$

to serve as descriptor of the evolution in liquid medium temperature. If both sides are divided by T_0, and the argument of the exponential function is multiplied and divided by $k_{d,Y,0}$, then Eq. (3.110) becomes

$$\frac{T}{T_0} = \frac{T_{st}}{T_0} + \left(1 - \frac{T_{st}}{T_0}\right)\mathrm{e}^{-\frac{UA}{V\rho c_P k_{d,Y,0}}k_{d,Y,0}t} \tag{3.111}$$

as depicted in Fig. 3.3, with dimensionless-dependent temperature, T/T_0, plotted as a function of dimensionless time, $k_{d,Y,0}t$, as well as $UA/V\rho c_P k_{d,Y,0}$ and T_{st}/T_0 = (121 + 273.2)/(20 + 273.2) = 1.35 as germane dimensionless parameters. A reciprocal exponential trend is apparent, starting from unity, in agreement with Eq. (3.97), and gradually increasing until reaching a plateau, described by a horizontal line with T_{st}/T_0 for vertical intercept in view of $\lim_{t \to \infty} T/T_0 = T_{st}/T_0$ as per Eq. (3.111). As expected, the target sterilization temperature is reached faster when heat transfer is faster, as measured

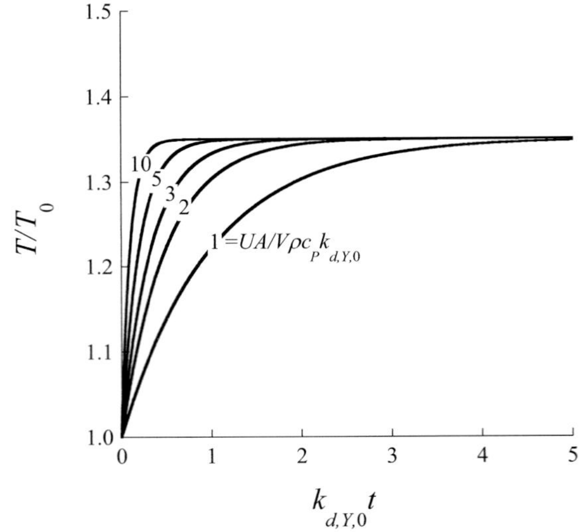

Fig. 3.3 Variation of dimensionless temperature, T/T_0, versus dimensionless time, $k_{d,Y,0}t$, for selected values of dimensionless parameters $UA/V\rho c_P k_{d,Y,0}$ and $T_{st}/T_0 = 1.35$

by higher U or A, or when thermal inertia is lower, as measured by smaller V, ρ, or c_P; either case is associated with a larger value for lumped parameter $UA/V\rho c_P k_{d,Y,0}$.

Once in possession of Eq. (3.111), one may retrieve Eq. (3.30) as

$$\frac{C_Y}{C_{Y,0}} = \exp\left\{ -\int_0^{k_{d,Y,0}t} \exp\left\{ \frac{E_{act,Y}}{\mathscr{R}T_0}\left(1 - \frac{1}{\frac{T_{st}}{T_0} + \left(1 - \frac{T_{st}}{T_0}\right) e^{-\frac{UA}{V\rho c_P k_{d,Y,0}}\left(k_{d,Y,0}\tilde{t}\right)}} \right) \right\} d\left(k_{d,Y,0}\tilde{t}\right) \right\}; \tag{3.112}$$

a graphical representation of Eq. (3.112) is provided as Fig. 3.4, for selected values of dimensionless parameters $E_{act,Y}/\mathscr{R}T_0$, $UA/V\rho c_P k_{d,Y,0}$, and T_{st}/T_0. The curves in this figure represent similar decay patterns, corresponding to an initial, brief plateau when temperature is still not much above room temperature; followed by a sudden and faster and faster decline, as temperature approaches the target sterilization temperature (see Fig. 3.3), and until an almost log-linear behavior is reached in the vicinity of the (sustained) sterilization temperature. Once again, this drop starts earlier for higher values of dimensionless parameter $UA/V\rho c_P k_{d,Y,0}$; and a higher activation energy $E_{act,Y}$, for a given pre-exponential factor $k_{0,Y}$, accelerates this process, visible as lines with steeper slopes, because the specific death rate is described by $k_{0,Y}e^{-E_{act,Y}/\mathscr{R}T}$ (see Eq. (3.20)).

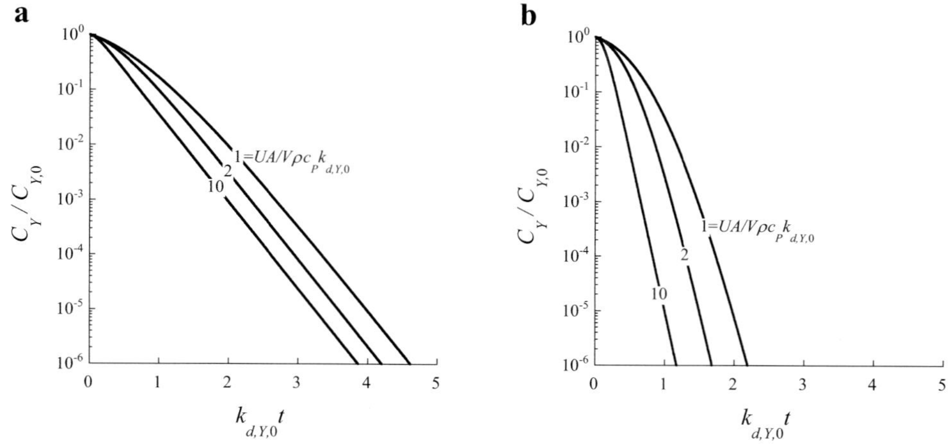

Fig. 3.4 Variation of normalized concentration of viable biomass, $C_Y/C_{Y,0}$, versus dimensionless time, $k_{d,Y,0}t$, for selected values of dimensionless parameters $UA/V\rho c_P k_{d,Y,0}$, $T_{st}/T_0 = 1.35$, and (**a**) $E_{act,Y}/\mathcal{R}T_0 = 5$ and (**b**) $E_{act,Y}/\mathcal{R}T_0 = 10$

Problem 3.4

Prove that the tangent to the trajectories in Fig. 3.4 shares the slope of the lines in Fig. 3.2, described by the same instantaneous T/T_0.

Solution

The slope of the straight-line tangent to any of the curves in Fig. 3.4 is given by the derivative of the decimal logarithm of $C_Y/C_{Y,0}$ as per Eq. (3.112), that is,

$$\frac{d\left(\log_{10}\frac{C_Y}{C_{Y,0}}\right)}{d(k_{d,Y,0}t)}=\frac{d\left(\ln\frac{C_Y}{C_{Y,0}}\log_{10}\mathrm{e}\right)}{d(k_{d,Y,0}t)}=\frac{d\left(\ln\frac{C_Y}{C_{Y,0}}\right)}{d(k_{d,Y,0}t)}\log_{10}\mathrm{e}=$$

$$=\frac{d}{d(k_{d,Y,0}t)}\left(-\int\limits_0^{k_{d,Y,0}t}\exp\left\{\frac{E_{act,Y}}{\mathcal{R}T_0}\left(1-\frac{1}{\frac{T_{st}}{T_0}+\left(1-\frac{T_{st}}{T_0}\right)\mathrm{e}^{-\frac{UA}{V\rho c_P k_{d,Y,0}}\left(k_{d,Y,0}\tilde{t}\right)}}\right)\right\}d(k_{d,Y,0}\tilde{t})\right)\log_{10}\mathrm{e}, \tag{3.113}$$

where the rule of base change of a logarithmic function was taken on board, besides realization that $\log_{10}\mathrm{e}$ is a constant; application of Leibnitz's rule for differentiation of an integral supports

(continued)

Problem 3.4 (continued)

$$\frac{d\left(\log_{10}\frac{C_Y}{C_{Y,0}}\right)}{d(k_{d,Y,0}t)}=-\exp\left\{\frac{E_{act,Y}}{\mathscr{R}T_0}\left(1-\frac{1}{\frac{T_{st}}{T_0}+\left(1-\frac{T_{st}}{T_0}\right)\mathrm{e}^{-\frac{UA}{V\rho c_p k_{d,Y,0}}\left(k_{d,Y,0}\tilde{t}\right)}}\right)\right\}\Bigg|_{k_{d,Y,0}\tilde{t}=k_{d,Y,0}t}\frac{d(k_{d,Y,0}t)}{d(k_{d,Y,0}t)}\log_{10}\mathrm{e}, \tag{3.114}$$

which is equivalent to writing

$$\frac{d\left(\log_{10}\frac{C_Y}{C_{Y,0}}\right)}{d(k_{d,Y,0}t)}=-\exp\left\{\frac{E_{act,Y}}{\mathscr{R}T_0}\left(1-\frac{1}{\frac{T_{st}}{T_0}+\left(1-\frac{T_{st}}{T_0}\right)\mathrm{e}^{-\frac{UA}{V\rho c_p k_{d,Y,0}}(k_{d,Y,0}t)}}\right)\right\}\log_{10}\mathrm{e}. \tag{3.115}$$

In view of Eq. (3.111), it is possible to redo Eq. (3.115) as

$$\frac{d\left(\log_{10}\frac{C_Y}{C_{Y,0}}\right)}{d(k_{d,Y,0}t)} = -\exp\left\{\frac{E_{act,Y}}{\mathscr{R}T_0}\left(1-\frac{1}{\frac{T\{t\}}{T_0}}\right)\right\}\log_{10}\mathrm{e}, \tag{3.116}$$

or else

$$\frac{d\left(\log_{10}\frac{C_Y}{C_{Y,0}}\right)}{d(k_{d,Y,0}t)} = -\mathrm{e}^{\frac{E_{act,Y}}{\mathscr{R}T_0}\left(1-\frac{T_0}{T\{t\}}\right)}\log_{10}\mathrm{e}. \tag{3.117}$$

If T were constant, then Eq. (3.117) would be equivalent to

$$\frac{d\left(\log_{10}\frac{C_Y}{C_{Y,0}}\right)}{d(k_{d,Y,0}t)} = \frac{d}{d(k_{d,Y,0}t)}\left(-\mathrm{e}^{\frac{E_{act,Y}}{\mathscr{R}T_0}\left(1-\frac{T_0}{T}\right)}k_{d,Y,0}t\right)\log_{10}\mathrm{e}; \tag{3.118}$$

Equation (3.118) may, in turn, be rephrased as

(continued)

Problem 3.4 (continued)

$$\frac{d\left(\log_{10}\frac{C_Y}{C_{Y,0}}\right)}{d(k_{d,Y,0}t)}=\frac{d}{d(k_{d,Y,0}t)}\left(\ln\exp\left\{-e^{\frac{E_{act,Y}}{\mathcal{R}T_0}\left(1-\frac{T_0}{T}\right)}k_{d,Y,0}t\right\}\right)\log_{10}e, \tag{3.119}$$

for exponential being the inverse function of logarithm. After taking $\log_{10}e$ into the differential operator for its constancy, Eq. (3.119) becomes

$$\frac{d\left(\log_{10}\frac{C_Y}{C_{Y,0}}\right)}{d(k_{d,Y,0}t)}=\frac{d}{d(k_{d,Y,0}t)}\left(\ln\exp\left\{-e^{\frac{E_{act,Y}}{\mathcal{R}T_0}\left(1-\frac{T_0}{T}\right)}k_{d,Y,0}t\right\}\log_{10}e\right), \tag{3.120}$$

which can be condensed as

$$\frac{d\left(\log_{10}\frac{C_Y}{C_{Y,0}}\right)}{d(k_{d,Y,0}t)}=\frac{d}{d(k_{d,Y,0}t)}\left(\log_{10}\exp\left\{-e^{\frac{E_{act,Y}}{\mathcal{R}T_0}\left(1-\frac{T_0}{T}\right)}k_{d,Y,0}t\right\}\right) \tag{3.121}$$

in view again of the rule of exchange of logarithmic base; Eq. (3.121) represents the slope of the straight-line tangent to the logarithmic plot of $C_Y/C_{Y,0}$, as given by Eq. (3.33), versus $k_{d,Y,0}t$ as independent variable, thus confirming the graphical relationship between Figs. 3.2 and 3.4.

3.2.2 Direct Heating

Heat transfer to a liquid medium, aimed at sterilizing it, is sometimes relatively slow especially in larger vessels; in fact, the underlying kinetics abides to Fourier's law where interfacial area, A, appears as multiplying factor, whereas the underlying thermodynamics indicates that the sensible effects upon the batch of medium are inversely proportional to its volume, V. In other words, the specific area, a, defined as A/V, proves the limiting factor. For a commercial cylindrical vessel, with height H typically equal to twice its radius R, one finds $A = 2\pi RH = 2\pi R2R = 4\pi R^2$ for its lateral area, and $V = \pi R^2H = \pi R^2 2R = 2\pi R^3$, thus implying $a = 4\pi R^2/2\pi R^3 = 2/R$; hence, a larger R causes a hyperbolic reduction of a.

Inspection of the functional form of Fourier's law unfolds three alternative modes to circumvent the aforementioned loss of efficiency in heat transfer: either increase overall heat transfer coefficient, or increase a, or increase the temperature difference that serves as driving force. Since the composition of the liquid medium is fixed a priori by the microorganism nutritional requirements, and the largest rate of stirring is constrained by the power of the (electrical) engine normally employed for mixing during regular operation, there is little room to improve the heat transfer coefficient of the liquid phase. The easiest way of increasing a is to decrease the radius of the bioreactor, as seen above; however, the cost of the bioreactor varies approximately with the 0.6-th power of its capacity, so this option hampers economic feasibility. Finally, the temperature of the common heating utility, that is, saturated steam, cannot be raised beyond the pressure safety threshold, owing to the exponential increase in vapor pressure with temperature (see Fig. 3.1), while the lowest (reasonable) temperature of supply of liquid medium is room temperature; hence, the driving force for heat transfer appears rather constrained.

In view of the above shortcomings, one way out is removal of the metal wall between liquid medium and condensing steam; direct bubbling of steam increases dramatically the degree of turbulence in the liquid, prone to magnify the heat transfer coefficient, due to thinning of the boundary layer. At the same time, bubbles of steam permeate the liquid medium upward and remain essentially independent of one another; if the nozzle radius of the steam diffuser placed at the bottom of the vessel is small enough, the radius R_b of the bubbles will also be very small, so their specific area a, given by the ratio of $A = 4\pi R_b^{\,2}$ to $V = 4\pi R_b^{\,3}/3$, that is, $a = 4\pi R_b^{\,2}/(4\pi R_b^{\,3}/3) = 3/R_b$, will be quite enhanced. A much higher heat transfer coefficient and a much higher specific surface area accordingly combine to substantially enhance the rate of heat transfer.

Another form of direct heating is to resort to a flux-controlled process, instead of the temperature-controlled process present when (condensing) steam is used—since the temperature ever attainable by the liquid medium is upper bounded by the steam temperature itself. When an electric resistor is at stake, the heat generated due to Joule's effect is determined only by the power, $\mathcal{P}$, of the electrical supply, that is, $I\Delta V$ as per Eq. (3.178), with I denoting intensity of current and ΔV denoting difference in electrical potential; this is equivalent to $\mathcal{R} I^2$ when Ohm's law, $\mathcal{R} = \Delta V/I$ as per Eq. (3.179), is brought on board, with $\mathcal{R}$ denoting electrical resistance. The temperature of said resistor will accordingly adapt, so that the driving force in Fourier's law becomes compatible with the heat flow generated by the electrical current; therefore, temperature will eventually attain much higher values, and an upper limit (below meltdown) will be reached only if there are heat losses to the surroundings of the vessel. This approach conveys a simple and effective control of heat supplied via acting upstream on the difference of potential, or resorting to a rheostat or a triac device to control current intensity; however, it also poses safety risks in case the medium temperature goes beyond the maximum acceptable threshold.

Problem 3.5
Consider a cylindrical vessel, of inner radius R and height H, with an electrical resistance homogeneously built in its wall of thickness δ; said vessel is designed to batch sterilize a liquid medium inside, kept well stirred.

(a) Write an enthalpic balance to the vessel, under steady-state conditions, and propose suitable boundary conditions.
(b) Prove that a (local) maximum of temperature will be attained somewhere inside the vessel wall.
(c) Estimate the evolution of actual temperature prevailing in the bulk liquid.

Solution

(a) A differential enthalpic balance to the wall matrix takes the form

$$-\left(k\frac{dT}{dr}2\pi rH\right)\bigg|_{r} + \mathcal{P}2\pi rH\,dr = -\left(k\frac{dT}{dr}2\pi rH\right)\bigg|_{r+dr}, \tag{3.122}$$

where k denotes thermal conductivity of wall material, T denotes temperature, r denotes generic radius, and $\mathcal{P}$ denotes rate of production of heat per unit volume, taken as independent of temperature and position. Assuming that k is also the same irrespective of temperature and position, Eq. (3.122) may be rewritten as

$$2\pi kH\frac{\left(r\frac{dT}{dr}\right)\Big|_{r+dr} - \left(r\frac{dT}{dr}\right)\Big|_{r}}{dr} = -2\pi\mathcal{P}Hr, \tag{3.123}$$

following condensation of terms alike, and division of both sides by dr afterward; upon cancelling $2\pi H$ between sides and recalling the definition of derivative, one obtains

$$k\frac{d}{dr}\left(r\frac{dT}{dr}\right) = -\mathcal{P}r. \tag{3.124}$$

Integration of Eq. (3.124) ensues via separation of variables, namely,

(continued)

Problem 3.5 (continued)

$$\int d\left(r\frac{dT}{dr}\right) = -\frac{\mathcal{P}}{k}\int r\,dr; \tag{3.125}$$

Equation (3.125) breaks down to

$$r\frac{dT}{dr} = -\frac{\mathcal{P}}{k}\frac{r^2}{2} + \kappa_1 \tag{3.126}$$

with κ_1 denoting an arbitrary constant, or else

$$\frac{dT}{dr} = -\frac{\mathcal{P}}{2k}r + \frac{\kappa_1}{r} \tag{3.127}$$

after division of both sides by r. A second integration by separation of variables is in order, which transforms Eq. (3.127) to

$$\int dT = \int\left(-\frac{\mathcal{P}}{2k}r + \frac{\kappa_1}{r}\right)dr \tag{3.128}$$

or, equivalently,

$$T = -\frac{\mathcal{P}}{2k}\frac{r^2}{2} + \kappa_1 \ln r + \kappa_2; \tag{3.129}$$

Equation (3.129) may take the alternative form

$$T = -\frac{\mathcal{P}}{4k}r^2 + \kappa_1 \ln r + \kappa_2, \tag{3.130}$$

where another arbitrary constant, κ_2, emerged—knowing that both κ_1 and κ_2 are to be determined at the expense of suitable boundary conditions. One such condition will be setup as

$$T|_{r=R+\delta} = T_0, \tag{3.131}$$

with T_0 denoting outer (constant) room temperature, and

(continued)

Problem 3.5 (continued)

$$-k\left.\frac{dT}{dr}\right|_{r=R}=h\left(T_l-T|_{r=R}\right) \tag{3.132}$$

will serve as another boundary condition, provided that T_l denotes bulk temperature in liquid and h denotes heat transfer coefficient therein. Compatibility between Eqs. (3.130) and (3.131) requires

$$T_0=-\frac{\mathcal{P}}{4k}(R+\delta)^2+\kappa_1\ln\{R+\delta\}+\kappa_2, \tag{3.133}$$

with addition of $\mathcal{P}(R+\delta)^2/4k$ to both sides leaving

$$\kappa_1\ln\{R+\delta\}+\kappa_2=T_0+\frac{\mathcal{P}(R+\delta)^2}{4k}; \tag{3.134}$$

by the same token, compatibility between Eqs. (3.127), (3.130), and (3.132) enforces

$$-k\left(-\frac{\mathcal{P}}{2k}R+\frac{\kappa_1}{R}\right)=h\left(T_l-\left(-\frac{\mathcal{P}}{4k}R^2+\kappa_1\ln R+\kappa_2\right)\right) \tag{3.135}$$

or, after factoring in k and h,

$$\frac{\mathcal{P}R}{2}-\frac{k}{R}\kappa_1=hT_l+\frac{h\mathcal{P}R^2}{4k}-h\kappa_1\ln R-h\kappa_2, \tag{3.136}$$

where algebraic reorganization unfolds

$$\kappa_1\left(h\ln R-\frac{k}{R}\right)+\kappa_2 h=hT_l+\frac{h\mathcal{P}R^2}{4k}-\frac{\mathcal{P}R}{2}. \tag{3.137}$$

Application of Cramer's rule to the set of Eqs. (3.134) and (3.137) is in order, and leads to

(continued)

Problem 3.5 (continued)

$$\kappa_1 = \frac{\begin{vmatrix} T_0 + \dfrac{\mathcal{P}(R+\delta)^2}{4k} & 1 \\ hT_l + \dfrac{h\mathcal{P}R^2}{4k} - \dfrac{\mathcal{P}R}{2} & h \end{vmatrix}}{\begin{vmatrix} \ln\{R+\delta\} & 1 \\ h\ln R - \dfrac{k}{R} & h \end{vmatrix}} \tag{3.138}$$

and

$$\kappa_2 = \frac{\begin{vmatrix} \ln\{R+\delta\} & T_0 + \dfrac{\mathcal{P}(R+\delta)^2}{4k} \\ h\ln R - \dfrac{k}{R} & hT_l + \dfrac{h\mathcal{P}R^2}{4k} - \dfrac{\mathcal{P}R}{2} \end{vmatrix}}{\begin{vmatrix} \ln\{R+\delta\} & 1 \\ h\ln R - \dfrac{k}{R} & h \end{vmatrix}}, \tag{3.139}$$

pertaining to κ_1 and κ_2, respectively. The definition of second-order determinant justifies conversion of Eq. (3.138) to

$$\kappa_1 = \frac{h\left(T_0 + \frac{\mathcal{P}(R+\delta)^2}{4k}\right) - 1\left(hT_l + \frac{h\mathcal{P}R^2}{4k} - \frac{\mathcal{P}R}{2}\right)}{h\ln\{R+\delta\} - 1\left(h\ln R - \frac{k}{R}\right)} = \frac{hT_0 + \frac{h\mathcal{P}(R+\delta)^2}{4k} - hT_l - \frac{h\mathcal{P}R^2}{4k} + \frac{\mathcal{P}R}{2}}{h\ln\{R+\delta\} - h\ln R + \frac{k}{R}} \tag{3.140}$$

along with elimination of parentheses; and still to

$$\kappa_1\{T_l\} = \frac{h(T_0 - T_l) + \frac{h\mathcal{P}}{4k}\left(R^2 + 2R\delta + \delta^2 - R^2\right) + \frac{\mathcal{P}R}{2}}{h(\ln\{R+\delta\} - \ln R) + \frac{k}{R}} = \frac{h(T_0 - T_l) + \frac{h\mathcal{P}\delta(2R+\delta)}{4k} + \frac{\mathcal{P}R}{2}}{h\ln\frac{R+\delta}{R} + \frac{k}{R}} \tag{3.141}$$

once h, $h\mathcal{P}/4k$, or δ (as appropriate) have been factored out, Newton's binomial theorem recalled, symmetrical terms cancelled, and logarithmic functions collapsed. Equation (3.139) similarly leads to

(continued)

Problem 3.5 (continued)

$$\kappa_2 = \frac{\left(hT_l + \frac{h\mathcal{P}R^2}{4k} - \frac{\mathcal{P}R}{2}\right)\ln\{R+\delta\} - \left(h\ln R - \frac{k}{R}\right)\left(T_0 + \frac{\mathcal{P}(R+\delta)^2}{4k}\right)}{h\ln\{R+\delta\} - 1\left(h\ln R - \frac{k}{R}\right)}$$
$$= \frac{\left(hT_l + \frac{h\mathcal{P}R^2}{4k} - \frac{\mathcal{P}R}{2}\right)\ln\{R+\delta\} - \left(h\ln R - \frac{k}{R}\right)\left(T_0 + \frac{\mathcal{P}(R+\delta)^2}{4k}\right)}{h\ln\{R+\delta\} - h\ln R + \frac{k}{R}} \tag{3.142}$$

owing to the definition of second-order determinant, with parenthesis in denominator meanwhile removed; Eq. (3.142) is equivalent to

$$\kappa_2\{T_l\} = \frac{\left(hT_l + \frac{h\mathcal{P}R^2}{4k} - \frac{\mathcal{P}R}{2}\right)\ln\{R+\delta\} - \left(h\ln R - \frac{k}{R}\right)\left(T_0 + \frac{\mathcal{P}(R+\delta)^2}{4k}\right)}{h\ln\frac{R+\delta}{R} + \frac{k}{R}}, \tag{3.143}$$

via replacement of difference of logarithms by logarithm of their quotient. Equation (3.130) will accordingly serve as descriptor of $T \equiv T\{r\}$, as long as Eqs. (3.141) and (3.143) are taken on board—both dependent on the steady-state value of T_l.

(b) The maximum temperature attained inside the vessel wall should satisfy

$$\frac{dT}{dr} = 0 \tag{3.144}$$

as necessary condition, which leads to

$$-\frac{\mathcal{P}}{2k}r + \frac{\kappa_1}{r} = 0 \tag{3.145}$$

upon combination with Eq. (3.127); following multiplication of both sides by r, Eq. (3.145) becomes

$$-\frac{\mathcal{P}}{2k}r^2 + \kappa_1 = 0, \tag{3.146}$$

so r^2 will be given by

(continued)

Problem 3.5 (continued)

$$r^2 = \frac{\kappa_1}{\frac{\mathcal{P}}{2k}} \tag{3.147}$$

or, once square roots are taken of both sides,

$$r_{opt} = \sqrt{\frac{2k\kappa_1}{\mathcal{P}}}. \tag{3.148}$$

The above extremum, r_{opt}, corresponds indeed to a maximum because the second-order derivative of T with regard to r, that is,

$$\left.\frac{d^2T}{dr^2}\right|_{r=r_{opt}} \equiv \left.\frac{d}{dr}\left(\frac{dT}{dr}\right)\right|_{r=r_{opt}} = \left.\frac{d}{dr}\left(-\frac{\mathcal{P}}{2k}r + \frac{\kappa_1}{r}\right)\right|_{r=r_{opt}} = \left.\left(-\frac{\mathcal{P}}{2k} - \frac{\kappa_1}{r^2}\right)\right|_{r=r_{opt}} \tag{3.149}$$

in agreement with Eq. (3.127), is negative irrespective of r; in fact, Eq. (3.149) proves equivalent to

$$\left.\frac{d^2T}{dr^2}\right|_{r=r_{opt}} = -\frac{\mathcal{P}}{2k} - \frac{\kappa_1}{\left(\sqrt{\frac{2k\kappa_1}{\mathcal{P}}}\right)^2} = -\frac{\mathcal{P}}{2k} - \frac{\kappa_1\mathcal{P}}{2k\kappa_1} = -\frac{\mathcal{P}}{2k} - \frac{\mathcal{P}}{2k} = -\frac{\mathcal{P}}{k} < 0, \tag{3.150}$$

after retrieving Eq. (3.148) and performing straightforward algebraic manipulation. The corresponding maximum temperature attained in the wall, T_{max}, is given by

$$T_{max} \equiv T\big|_{r=r_{opt}} = -\frac{\mathcal{P}}{4k}{r_{opt}}^2 + \kappa_1 \ln r_{opt} + \kappa_2 \tag{3.151}$$

stemming from Eq. (3.130); insertion of Eq. (3.148) gives rise to

$$T_{max} = -\frac{\mathcal{P}}{4k}\left(\sqrt{\frac{2k\kappa_1}{\mathcal{P}}}\right)^2 + \kappa_1 \ln \sqrt{\frac{2k\kappa_1}{\mathcal{P}}} + \kappa_2, \tag{3.152}$$

which reduces to

$$T_{max} = -\frac{\mathcal{P}}{4k}\frac{2k\kappa_1}{\mathcal{P}} + \frac{\kappa_1}{2}\ln\frac{2k\kappa_1}{\mathcal{P}} + \kappa_2 \tag{3.153}$$

(continued)

Problem 3.5 (continued)

due to composition of square power with square root, complemented by the operational features of logarithms.

(c) It should first be emphasized that Eq. (3.130) is strictly valid under steady state, which accordingly requires a constant T_l in Eq. (3.132); however, batchwise operation with regard to liquid and externally imposed electrical power with regard to heat source imply $T_l \equiv T_l\{t\}$. On the other hand, reorganization of the temperature profile in the vessel wall occurs intrinsically much faster than change in temperature of the liquid medium—due not only to the high thermal conductivity of the metallic material of the vessel compared to the conductivity of the aqueous medium, but also to the high specific heat capacity, c_P, of the aqueous medium of density ρ compared to the specific heat capacity of the metallic thin wall (with $\delta << R$); under these circumstances, a pseudo steady state may be claimed. An enthalpic balance to the liquid will accordingly take the form

$$\pi R^2 H\rho c_P \frac{dT_l}{dt} = h\left(T|_{r=R} - T_l\right), \tag{3.154}$$

after deliberately discarding the accumulation term of enthalpy in the vessel wall, so Eq. (3.130) will be called upon to write

$$\pi R^2 H\rho c_P \frac{dT_l}{dt} = h\left(-\frac{\mathcal{P}R^2}{4k} + \kappa_1\{T_l\}\ln R + \kappa_2\{T_l\} - T_l\right); \tag{3.155}$$

integration is thus in order, via separation of variables, as

$$\int\limits_{T_{l,0}}^{T_l} \frac{d\widetilde{T}_l}{-\frac{\mathcal{P}R^2}{4k} + \kappa_1\{\widetilde{T}_l\}\ln R + \kappa_2\{\widetilde{T}_l\} - \widetilde{T}_l} = \frac{h}{\pi R^2 H\rho c_P}\int\limits_0^t d\widetilde{t}, \tag{3.156}$$

again assuming that h, ρ, and c_P do not significantly change with time or temperature, with

$$T_l|_{t=0} = T_{l,0} \tag{3.157}$$

(continued)

Problem 3.5 (continued)

serving as suitable initial condition (provided that $T_{l,0}$ is given). The integral in its right-hand side is trivial, thus meaning that Eq. (3.156) can be replaced by

$$\int_{T_{l,0}}^{T_l} \frac{d\widetilde{T}_l}{-\frac{\wp R^2}{4k}+\kappa_1\{\widetilde{T}_l\}\ln R+\kappa_2\{\widetilde{T}_l\}-\widetilde{T}_l} = \frac{h}{\pi R^2 H\rho c_P}t. \tag{3.158}$$

Note that $k_1 \equiv k_1\{T_l\}$ and $k_2 \equiv k_2\{T_l\}$ are linear functions of T_l, according to

$$\kappa_1\{T_l\} = \frac{hT_0+\frac{h\wp\delta(2R+\delta)}{4k}+\frac{\wp R}{2}}{h\ln\frac{R+\delta}{R}+\frac{k}{R}} - \frac{h}{h\ln\frac{R+\delta}{R}+\frac{k}{R}}T_l \tag{3.159}$$

as per Eq. (3.141), and

$$\kappa_2\{T_l\} = \frac{\left(\frac{h\wp R^2}{4k}-\frac{\wp R}{2}\right)\ln\{R+\delta\}-\left(h\ln R-\frac{k}{R}\right)\left(T_0+\frac{\wp(R+\delta)^2}{4k}\right)}{h\ln\frac{R+\delta}{R}+\frac{k}{R}} + \frac{h\ln\{R+\delta\}}{h\ln\frac{R+\delta}{R}+\frac{k}{R}}T_l \tag{3.160}$$

as per Eq. (3.143). Notation may be conveniently simplified as

$$\kappa_1\{T_l\} = a - bT_l \tag{3.161}$$

and

$$\kappa_2\{T_l\} = \alpha + \beta T_l, \tag{3.162}$$

provided that the germane (true) constants abide to

$$a \equiv \frac{hT_0+\frac{h\wp\delta(2R+\delta)}{4k}+\frac{\wp R}{2}}{h\ln\frac{R+\delta}{R}+\frac{k}{R}}, \tag{3.163}$$

(continued)

Problem 3.5 (continued)

$$b \equiv \frac{h}{h \ln \frac{R+\delta}{R} + \frac{k}{R}}, \tag{3.164}$$

$$\alpha \equiv \frac{\left(\frac{h\mathcal{P}R^2}{4k} - \frac{\mathcal{P}R}{2}\right) \ln\{R+\delta\} - \left(h \ln R - \frac{k}{R}\right)\left(T_0 + \frac{\mathcal{P}(R+\delta)^2}{4k}\right)}{h \ln \frac{R+\delta}{R} + \frac{k}{R}}, \tag{3.165}$$

and

$$\beta \equiv \frac{h \ln\{R+\delta\}}{h \ln \frac{R+\delta}{R} + \frac{k}{R}}; \tag{3.166}$$

in this case, Eq. (3.158) will take the form

$$\int_{T_{l,0}}^{T_l} \frac{d\widetilde{T}_l}{-\frac{\mathcal{P}R^2}{4k} + \left(a - b\widetilde{T}_l\right) \ln R + \alpha + \beta\widetilde{T}_l - \widetilde{T}_l} = \frac{h}{\pi R^2 H \rho c_P} t \tag{3.167}$$

with the aid of Eq. (3.161), which degenerates to

$$\int_{T_{l,0}}^{T_l} \frac{d\widetilde{T}_l}{a \ln R + \alpha - \frac{\mathcal{P}R^2}{4k} - (1 + b \ln R - \beta)\widetilde{T}_l} = \frac{h}{\pi R^2 H \rho c_P} t \tag{3.168}$$

after factoring out $\widetilde{T}_l$. Integration of Eq. (3.168) ensues as

$$\left. \frac{\ln\left\{a \ln R + \alpha - \frac{\mathcal{P}R^2}{4k} - (1 + b \ln R - \beta)\widetilde{T}\right\}}{-(1 + b \ln R - \beta)} \right|_{T_{l,0}}^{T_l} = \frac{h}{\pi R^2 H \rho c_P} t, \tag{3.169}$$

which gives rise to

(continued)

Problem 3.5 (continued)

$$-\frac{1}{1+b\ln R-\beta}\ln\frac{a\ln R+\alpha-\frac{\mathcal{P}R^2}{4k}-(1+b\ln R-\beta)T_l}{a\ln R+\alpha-\frac{\mathcal{P}R^2}{4k}-(1+b\ln R-\beta)T_{l,0}}=\frac{h}{\pi R^2 H\rho c_P}t \quad (3.170)$$

after replacing by upper and lower limits of integration; multiplication of both sides by $1 + b \ln R - \beta$ unfolds

$$\ln\frac{a\ln R+\alpha-\frac{\mathcal{P}R^2}{4k}-(1+b\ln R-\beta)T_l}{a\ln R+\alpha-\frac{\mathcal{P}R^2}{4k}-(1+b\ln R-\beta)T_{l,0}}=-\frac{h(1+b\ln R-\beta)}{\pi R^2 H\rho c_P}t, \quad (3.171)$$

which prompts, in turn,

$$\frac{a\ln R+\alpha-\frac{\mathcal{P}R^2}{4k}-(1+b\ln R-\beta)T_l}{a\ln R+\alpha-\frac{\mathcal{P}R^2}{4k}-(1+b\ln R-\beta)T_{l,0}}=\exp\left\{-\frac{h(1+b\ln R-\beta)}{\pi R^2 H\rho c_P}t\right\} \quad (3.172)$$

once exponentials have been taken of both sides. Elimination of denominators transforms Eq. (3.172) to

$$\begin{aligned}&a\ln R+\alpha-\frac{\mathcal{P}R^2}{4k}-(1+b\ln R-\beta)T_l\\&=\left(a\ln R+\alpha-\frac{\mathcal{P}R^2}{4k}-(1+b\ln R-\beta)T_{l,0}\right)\exp\left\{-\frac{h(1+b\ln R-\beta)}{\pi R^2 H\rho c_P}t\right\},\end{aligned} \quad (3.173)$$

or else

$$\begin{aligned}T_l&=\frac{a\ln R+\alpha-\frac{\mathcal{P}R^2}{4k}-\left(a\ln R+\alpha-\frac{\mathcal{P}R^2}{4k}-(1+b\ln R-\beta)T_{l,0}\right)\exp\left\{-\frac{h(1+b\ln R-\beta)}{\pi R^2 H\rho c_P}t\right\}}{1+b\ln R-\beta}\\&=\frac{\left(a\ln R+\alpha-\frac{\mathcal{P}R^2}{4k}\right)\left(1-\exp\left\{-\frac{h(1+b\ln R-\beta)}{\pi R^2 H\rho c_P}t\right\}\right)+(1+b\ln R-\beta)T_{l,0}\exp\left\{-\frac{h(1+b\ln R-\beta)}{\pi R^2 H\rho c_P}t\right\}}{1+b\ln R-\beta}\end{aligned} \quad (3.174)$$

upon lumping alike terms; a further cancellation of identical factors between numerator and denominator unfolds

(continued)

Problem 3.5 (continued)

$$T_l = T_{l,0} \exp\left\{-\frac{h(1+b\ln R-\beta)}{\pi R^2 H\rho c_P}t\right\} + \frac{a\ln R+\alpha-\frac{\mathcal{P}R^2}{4k}}{1+b\ln R-\beta}\left(1-\exp\left\{-\frac{h(1+b\ln R-\beta)}{\pi R^2 H\rho c_P}t\right\}\right), \tag{3.175}$$

which entails a reverse exponential-based, increasing pattern on time. The corresponding long-time horizontal asymptote satisfies

$$\lim_{t\to\infty} T_l = T_{l,0}\mathrm{e}^{-\infty} + \frac{a\ln R+\alpha-\frac{\mathcal{P}R^2}{4k}}{1+b\ln R-\beta}\left(1-\mathrm{e}^{-\infty}\right), \tag{3.176}$$

or merely

$$\lim_{t\to\infty} T_l = \frac{a\ln R+\alpha-\frac{\mathcal{P}R^2}{4k}}{1+b\ln R-\beta}; \tag{3.177}$$

Equation (3.177) describes the steady-state temperature reached by the medium. Despite the cumulative effect arising from the supplied heat upon the batch of liquid, said steady state will eventually be attained (without the risk of runaway sterilization) because the extra heat will actually be lost to the environment kept at a constant temperature T_0 as per Eq. (3.131), driven by a sufficiently steep slope $(dT/dr)|_{r=R+\delta}$, in agreement with Fourier's law.

3.2.2.1 Resistor

If the liquid medium is heated via an electrical resistor submerged therein, the electric power at stake, $\mathcal{P}$, is fully dissipated as thermal energy according to Joule's law, namely,

$$\mathcal{P} = I\,\Delta V, \tag{3.178}$$

where I denotes intensity of electrical current and ΔV denotes difference of electrical potential; after invoking Ohm's law, that is,

$$\Delta V = \mathcal{R}I \tag{3.179}$$

with $\mathcal{R}$ denoting electrical resistance, Eq. (3.178) will give rise to

$$\mathcal{P} = I\mathcal{R}I = \mathcal{R}I^2, \tag{3.180}$$

or instead

$$\mathcal{P} = \frac{\Delta V}{\mathcal{R}}\Delta V = \frac{(\Delta V)^2}{\mathcal{R}} \tag{3.181}$$

after solving Eq. (3.179) for I and then inserting the result in Eq. (3.178).

In this case, the enthalpic balance to liquid medium, assumed to be fully insulated from the surroundings, looks like

$$\mathcal{P} = V\rho c_P \frac{dT}{dt}, \tag{3.182}$$

with

$$T|_{t=0} = T_0 \tag{3.183}$$

serving as initial condition; division of both sides by $V\rho c_P$ converts Eq. (3.182) to

$$\frac{dT}{dt} = \frac{\mathcal{P}}{V\rho c_P}. \tag{3.184}$$

Integration of Eq. (3.184) is straightforward, namely,

$$\int_{T_0}^{T} d\tilde{T} = \frac{\mathcal{P}}{V\rho c_P}\int_0^t d\tilde{t}, \tag{3.185}$$

by resorting to separation of variables, under the (usual) assumption that $\mathcal{P}/V\rho c_P$ is essentially independent of temperature and time, and taking advantage of Eq. (3.183); trivial application of the fundamental theorem of integral calculus then unfolds

$$\left.\widetilde{T}\right|_{T_0}^{T} = \frac{\mathcal{P}}{V\rho c_P}\left.\widetilde{t}\right|_0^t, \tag{3.186}$$

which becomes merely

$$T - T_0 = \frac{\mathcal{P}}{V\rho c_P}t \tag{3.187}$$

or else

$$T = T_0 + \frac{\mathcal{P}}{V\rho c_P}t \tag{3.188}$$

following isolation of T. Division of both sides by T_0, complemented by multiplication and division of the last term by $k_{d,Y,0}$ transform Eq. (3.188) further to

$$\frac{T}{T_0} = 1 + \frac{\mathcal{P}}{V\rho c_P k_{d,Y,0} T_0} k_{d,Y,0}t; \tag{3.189}$$

Equation (3.189) is sketched in Fig. 3.5. Note the linear increase in temperature, without upper bound as time elapses; this is so because the rate of heat transfer is not set by a difference in temperature between heating fluid and liquid medium (which would gradually

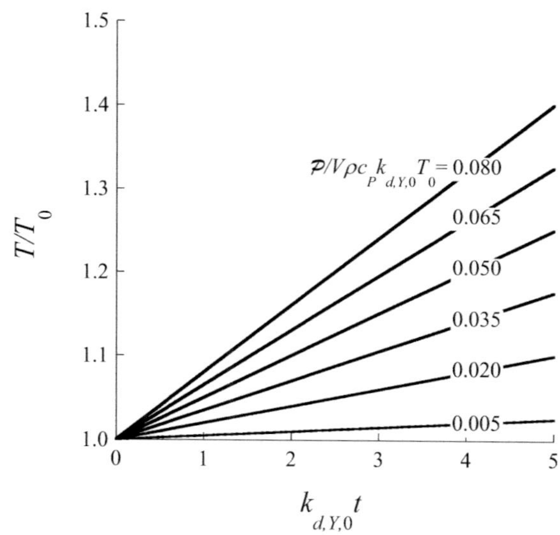

Fig. 3.5 Variation of dimensionless temperature, T/T_0, versus dimensionless time, $k_{d,Y,0}t$, for selected values of dimensionless parameter $\mathcal{P}/V\rho c_P k_{d,Y,0} T_0$

decrease in time), but by the electrical power made available (which is hypothesized to remain constant)—see Eq. (3.178), or Eqs. (3.180) or (3.181) for that matter.

Insertion of Eq. (3.189) converts Eq. (3.30) to

$$\frac{C_Y}{C_{Y,0}} = \exp\left\{ - \int_0^{k_{d,Y,0}t} \exp\left\{ \frac{E_{act,Y}}{\mathcal{R}T_0} \left(1 - \frac{1}{1 + \frac{\mathcal{P}}{V\rho c_P k_{d,Y,0} T_0} (k_{d,Y,0}\tilde{t})} \right) \right\} d(k_{d,Y,0}\tilde{t}) \right\}, \quad (3.190)$$

where $E_{act,Y}/\mathcal{R}T_0$ appears as germane dimensionless parameter, in addition to $\mathcal{P}/V\rho c_P k_{d,Y,0}T_0$; a graphical interpretation is conveyed by Fig. 3.6. Once again, the fractional viability of contaminating microorganisms decays slowly at early times, when temperature is still low (see Fig. 3.5); but decays faster and faster as temperature goes on growing up, with no upper limit for the magnitude of the corresponding (absolute) slope. This pattern is enhanced by a higher activation energy, via $E_{act,Y}/\mathcal{R}T_0$, since this is a measure of sensitivity to temperature; and a higher electrical power, or a lower volume or heat capacity, via $\mathcal{P}/V\rho c_P k_{d,Y,0}T_0$, since these are measures of rate of heat supply and inertia accompanying heat absorption, respectively.

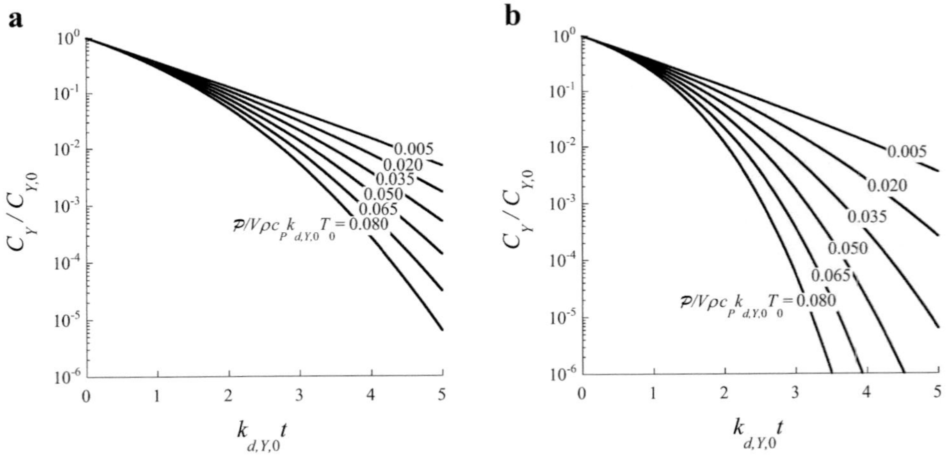

Fig. 3.6 Variation of normalized concentration of viable biomass, $C_Y/C_{Y,0}$, versus dimensionless time, $k_{d,Y,0}t$, for selected values of dimensionless parameters $\mathcal{P}/V\rho c_P k_{d,Y,0}T_0$, and (**a**) $E_{act,Y}/\mathcal{R}T_0 = 5$ and (**b**) $E_{act,Y}/\mathcal{R}T_0 = 10$

Problem 3.6

If time t_∞ is defined by

$$t_\infty \equiv t\Big|_{1-\frac{1}{1+\frac{\mathcal{P}}{V\rho c_P T_0}t}\approx 1-\frac{1}{1+\infty}=1-0=1}, \tag{3.191}$$

find the asymptotic expression that drives the $C_Y/C_Y|_{t=t_\infty}$-vs.-$k_{d,Y,0}(t-t_\infty)$ curve pertaining to resistor-driven sterilization, on a log/lin scale for $t \geq t_\infty$.

Solution

The integral in Eq. (3.190) may be exactly splitted as

$$\frac{C_Y|_{t=t_\infty}}{C_{Y,0}}\frac{C_Y}{C_Y|_{t=t_\infty}} = \exp\left\{-\left(\begin{array}{l}\displaystyle\int_0^{k_{d,Y,0}t_\infty}\exp\left\{\frac{E_{act,Y}}{\mathcal{R}T_0}\left(1-\frac{1}{1+\frac{\mathcal{P}}{V\rho c_P k_{d,Y,0}T_0}\left(k_{d,Y,0}\tilde{t}\right)}\right)\right\}d\left(k_{d,Y,0}\tilde{t}\right)\\ \displaystyle+\int_{k_{d,Y,0}t_\infty}^{k_{d,Y,0}t}\exp\left\{\frac{E_{act,Y}}{\mathcal{R}T_0}\left(1-\frac{1}{1+\frac{\mathcal{P}}{V\rho c_P k_{d,Y,0}T_0}\left(k_{d,Y,0}\tilde{t}\right)}\right)\right\}d\left(k_{d,Y,0}\tilde{t}\right)\end{array}\right)\right\} \tag{3.192}$$

along with multiplication and division of the left-hand side by $C_Y|_{t=t_\infty}$, since the exponential operator converts the sum of partial integrals to the product of their exponentials, which in turn map into biomass concentration fractions—with $C_Y|_{t=t_\infty}$ playing the role of normalizing factor within $[t_\infty, t]$, in much the same way $C_{Y,0}$ was used for normalization within $[0, t_\infty]$; Eq. (3.192) immediately implies

$$\begin{aligned}\frac{C_Y}{C_Y|_{t=t_\infty}} &= \exp\left\{-\int_{k_{d,Y,0}t_\infty}^{k_{d,Y,0}t}\exp\left\{\frac{E_{act,Y}}{\mathcal{R}T_0}\left(1-\frac{1}{1+\frac{\mathcal{P}}{V\rho c_P k_{d,Y,0}T_0}\left(k_{d,Y,0}\tilde{t}\right)}\right)\right\}d\left(k_{d,Y,0}\tilde{t}\right)\right\}\\ &\approx \exp\left\{-\int_{k_{d,Y,0}t_\infty}^{k_{d,Y,0}t}\exp\left\{\frac{E_{act,Y}}{\mathcal{R}T_0}\right\}d\left(k_{d,Y,0}\tilde{t}\right)\right\},\end{aligned} \tag{3.193}$$

in view of Eq. (3.191). The exponential in the kernel of Eq. (3.193) can be taken off as

(continued)

Problem 3.6 (continued)

$$\frac{C_Y}{C_Y|_{t=t_\infty}} = \exp\left\{ -\exp\left\{\frac{E_{act,Y}}{\mathcal{R}T_0}\right\} \int_{k_{d,Y,0}t_\infty}^{k_{d,Y,0}t} d(k_{d,Y,0}\tilde{t}) \right\}, \tag{3.194}$$

where trivial application of the fundamental theorem of integral calculus unfolds

$$\frac{C_Y}{C_Y|_{t=t_\infty}} = \exp\left\{ -\exp\left\{\frac{E_{act,Y}}{\mathcal{R}T_0}\right\} k_{d,Y,0}\tilde{t}\Big|_{k_{d,Y,0}t_\infty}^{k_{d,Y,0}t} \right\}; \tag{3.195}$$

Equation (3.195) breaks down to

$$\frac{C_Y}{C_Y|_{t=t_\infty}} = \exp\left\{ -\exp\left\{\frac{E_{act,Y}}{\mathcal{R}T_0}\right\} (k_{d,Y,0}t - k_{d,Y,0}t_\infty) \right\}, \tag{3.196}$$

after replacing by upper and lower limits of integration. Application of natural logarithms to both sides transforms Eq. (3.196) to

$$\ln \frac{C_Y}{C_Y|_{t=t_\infty}} = -\exp\left\{\frac{E_{act,Y}}{\mathcal{R}T_0}\right\} k_{d,Y,0}(t - t_\infty), \tag{3.197}$$

concomitant with factoring out of $k_{d,Y,0}$—and the left-hand side may then be reexpressed as a decimal logarithm, namely,

$$\log_{10} \frac{C_Y}{C_Y|_{t=t_\infty}} \ln 10 = -\exp\left\{\frac{E_{act,Y}}{\mathcal{R}T_0}\right\} k_{d,Y,0}(t - t_\infty); \tag{3.198}$$

upon division of both sides by ln 10, Eq. (3.198) becomes

$$\log_{10} \frac{C_Y}{C_Y|_{t=t_\infty}} = -\frac{\exp\left\{\frac{E_{act,Y}}{\mathcal{R}T_0}\right\}}{\ln 10} k_{d,Y,0}(t - t_\infty). \tag{3.199}$$

Inspection of Eq. (3.199) unfolds $-\exp\{E_{act,Y}/\mathcal{R}T_0\}/\ln 10$ as the slope, on a log/lin scale, of $\log_{10} C_Y/C_Y|_{t=t_\infty}$ when $k_{d,Y,0}(t - t_\infty)$ is used as independent variable; this justifies the extremely high slopes apparent in Fig. 3.6 at sufficiently long times, even though the graphical representation of Eq. (3.199) would result from Fig. 3.6 upon a downward vertical translation leading to a nil (common) vertical intercept of the curves therein.

3.2.2.2 Steam

Consider that high pressure saturated steam, at a (saturation) temperature sufficient to bring about sterilization, is sparged directly into the liquid medium. If the latter is at a lower temperature, the large rate of heat transfer driven by the large temperature difference results in fast condensation of steam. Such large rate arises from the highly turbulent conditions produced that support a very high heat transfer coefficient, and from the small size of the bubbles entailing a very large specific area for heat transfer; hence, the heat transfer process is, in fact, almost instantaneous when compared to the rate of supply of fresh steam.

Since the condensate of the original steam becomes a part of the (heated) liquid medium, the mass balance to the latter, considered (for simplicity) to exhibit properties similar to those of water, will look like

$$W\,dt = \rho\,dV, \tag{3.200}$$

where W denotes mass flow rate of high-pressure steam supplied that eventually condenses, ρ and V denote (constant) mass density and (variable) volume, respectively, of (liquid water) condensate, and t denotes time; division of both sides by dt promptly yields

$$W = \rho\frac{dV}{dt}, \tag{3.201}$$

where

$$V|_{t=0} = V_0 \tag{3.202}$$

will hereby serve as initial condition, with initial volume of liquid medium denoted by V_0. By the same token, an enthalpic balance will read

$$\Big(\lambda + c_P(T_{st} - T)\Big)W\,dt = \rho(V + dV)c_P\,dT, \tag{3.203}$$

with λ and T_{st} denoting (constant) latent heat of condensation and temperature of steam, respectively, and c_P denoting isobaric specific heat capacity of condensate; the left-hand side accordingly represents the latent and sensible heat that elementary mass $W\,dt$ of steam releases upon condensation at temperature T_{st}, coupled with cooling down from T_{st} to the current temperature of the liquid medium T, whereas the right-hand side accounts for the enthalpy taken up as sensible heat by the whole mass of liquid medium, $\rho(V + dV)$, to increase its temperature by dT. After regrouping terms in the left-hand side and factoring in dT in the right-hand side, Eq. (3.203) becomes

$$\left((\lambda + c_P T_{st}) - c_P T\right) W\,dt = \rho c_P (V\,dT + dV\,dT); \tag{3.204}$$

since $dV\,dT$ represents a second-order differential, one realizes that it must be much smaller than $V\,dT$ as first-order differential, and thus negligible when $dT \rightarrow 0$ (as is the case under scrutiny). Therefore, Eq. (3.204) simplifies to

$$W(\lambda + c_P T_{st})dt = c_P WT\,dt + \rho c_P V\,dT \tag{3.205}$$

after elimination of parenthesis and $\rho c_P\ dV\,dT$; division of both sides by dt and factoring out of c_P entail

$$W(\lambda + c_P T_{st}) = c_P\left(WT + \rho V\frac{dT}{dt}\right), \tag{3.206}$$

and a suitable initial condition reads

$$T|_{t=0} = T_0, \tag{3.207}$$

pertaining to the starting temperature of the liquid medium, T_0. Note that WT may be rewritten as $\rho(dV/dt)T$ in view of Eq. (3.201), so the right-hand side of Eq. (3.206) would read $\rho c_P(T(dV/dt) + V(dT/dt))$ once ρ is factored out; this is equivalent to $\rho c_P d(VT)/dt$ as per the rule of differentiation of a product of functions, or $d(\rho V c_P T)/dt$, where $\rho V c_P T$ represents the bulk manifestation of enthalpy received from the steam, upon condensation (affecting V) and cooling afterward (affecting T). Note also that a reference temperature, T^θ, for enthalpy proves redundant because the terms explicitly containing temperature can be rearranged as $Wc_P(T_{st} - T) \equiv Wc_P((T_{st} - T^\theta) - (T - T^\theta))$ and $dT \equiv d(T - T^\theta)$. On the other hand, condensation of steam will occur as long as a driving force exists for heat transfer between steam and liquid medium, that is, $T_{st} - T > 0$; therefore, Eq. (3.206) applies only for $T < T_{st}$ as descriptor of the initial temperature transient.

If a constant rate of supply of steam is hypothesized, then one may proceed to integration of Eq. (3.201) as

$$W\int_0^t d\tilde{t} = \rho\int_{V_0}^{V} d\tilde{V}, \tag{3.208}$$

where Eq. (3.202) was brought on board and advantage taken from the constancy of W and ρ; the fundamental theorem of integral calculus supports transformation to

$$W\tilde{t}\Big|_0^t = \rho\,\tilde{V}\Big|_{V_0}^V, \tag{3.209}$$

which readily produces

$$Wt = \rho(V - V_0) \tag{3.210}$$

as per the indicated upper and lower limits of integration, or else

$$V = V_0 + \frac{W}{\rho}t \tag{3.211}$$

following isolation of V. Combination with Eq. (3.211) converts Eq. (3.206) to

$$W(\lambda + c_P T_{st}) = c_P\left(WT + \rho\left(V_0 + \frac{W}{\rho}t\right)\frac{dT}{dt}\right), \tag{3.212}$$

where factoring in of ρ leaves

$$W(\lambda + c_P T_{st}) = c_P\left(WT + (\rho V_0 + Wt)\frac{dT}{dt}\right); \tag{3.213}$$

division of both sides by c_P, followed by subtraction of WT from both sides justify

$$(\rho V_0 + Wt)\frac{dT}{dt} = \frac{W(\lambda + c_P T_{st})}{c_P} - WT, \tag{3.214}$$

while factoring out W in the right-hand side yields

$$(\rho V_0 + Wt)\frac{dT}{dt} = W\left(\frac{\lambda}{c_P} + T_{st} - T\right). \tag{3.215}$$

Integration of Eq. (3.215) is now feasible via separation of variables, according to

$$\int_{T_0}^{T} \frac{d\tilde{T}}{\frac{\lambda}{c_P} + T_{st} - \tilde{T}} = \int_0^t \frac{W\,d\tilde{t}}{\rho V_0 + W\tilde{t}}, \tag{3.216}$$

with the aid of Eq. (3.207); the fundamental theorem of integral calculus may again be invoked to get

$$\left.\frac{\ln\left\{\frac{\lambda}{c_P}+T_{st}-\widetilde{T}\right\}}{-1}\right|_{T_0}^{T}=\ln\left\{\rho V_0+W\widetilde{t}\right\}\Big|_0^t, \tag{3.217}$$

which readily produces

$$\ln\frac{\frac{\lambda}{c_P}+T_{st}-T_0}{\frac{\lambda}{c_P}+T_{st}-T}=\ln\frac{\rho V_0+Wt}{\rho V_0} \tag{3.218}$$

with the reciprocal of the logarithmic function taken on account of the preceding minus sign. After applying exponentials to both sides, Eq. (3.218) becomes

$$\frac{\frac{\lambda}{c_P}+T_{st}-T_0}{\frac{\lambda}{c_P}+T_{st}-T}=\frac{\rho V_0+Wt}{\rho V_0}, \tag{3.219}$$

where reciprocals are, in turn, to be taken of both sides as

$$\frac{\frac{\lambda}{c_P}+T_{st}-T}{\frac{\lambda}{c_P}+T_{st}-T_0}=\frac{\rho V_0}{\rho V_0+Wt}; \tag{3.220}$$

multiplication of both sides by $\lambda/c_P + T_{st} - T_0$ unfolds

$$\frac{\lambda}{c_P}+T_{st}-T=\frac{\rho V_0\left(\frac{\lambda}{c_P}+T_{st}-T_0\right)}{\rho V_0+Wt}, \tag{3.221}$$

while solution for T gives

$$T=\frac{\lambda}{c_P}+T_{st}-\frac{\frac{\lambda}{c_P}+T_{st}-T_0}{1+\frac{W}{\rho V_0}t} \tag{3.222}$$

along with division of both numerator an denominator of the last term by ρV_0. Conversion of the right-hand side of Eq. (3.222) to a single term yields

$$T = \frac{\left(\frac{\lambda}{c_P} + T_{st}\right)\left(1 + \frac{W}{\rho V_0}t\right) - \left(\frac{\lambda}{c_P} + T_{st} - T_0\right)}{1 + \frac{W}{\rho V_0}t}, \tag{3.223}$$

where elimination of parenthesis leaves

$$T = \frac{\frac{\lambda}{c_P} + \frac{\lambda W}{\rho c_P V_0}t + T_{st} + \frac{W T_{st}}{\rho V_0}t - \frac{\lambda}{c_P} - T_{st} + T_0}{1 + \frac{W}{\rho V_0}t}; \tag{3.224}$$

after cancelling λ/c_P and T_{st} with the corresponding negatives, Eq. (3.224) simplifies as

$$T = \frac{\frac{\lambda W}{\rho c_P V_0}t + \frac{W T_{st}}{\rho V_0}t + T_0}{1 + \frac{W}{\rho V_0}t}, \tag{3.225}$$

with condensation of terms alike, complemented by factoring out of $W/\rho V_0$ finally producing

$$T = \frac{T_0 + \frac{W}{\rho V_0}\left(\frac{\lambda}{c_P} + T_{st}\right)t}{1 + \frac{W}{\rho V_0}t}. \tag{3.226}$$

A hyperbolic pattern has accordingly been obtained for $T \equiv T\{t\}$, as per the double-linear rational fraction apparent in Eq. (3.226); based on the order of magnitude of λ for water (ca. 2200 kJ kg^{-1}) and c_P for liquid water (c. 4 kJ kg^{-1}), as well as of steam temperature suitable for sterilization (ca. 395 K), one finds $\lambda/c_P \approx 2200/4 = 550 > 395 = T_{st}$. Upon division of both sides by T_0, Eq. (3.226) turns to

$$\frac{T}{T_0} = \frac{1 + \left(\frac{\lambda}{c_P T_0} + \frac{T_{st}}{T_0}\right)\frac{W}{\rho V_0}t}{1 + \frac{W}{\rho V_0}t}, \tag{3.227}$$

whereas further multiplication and division of t by $k_{d,Y,0}$ unfolds

$$\frac{T}{T_0} = \frac{1 + \left(\frac{\lambda}{c_P T_0} + \frac{T_{st}}{T_0}\right)\frac{W}{\rho k_{d,Y,0} V_0}k_{d,Y,0}t}{1 + \frac{W}{\rho k_{d,Y,0} V_0}k_{d,Y,0}t}; \tag{3.228}$$

the variation of dimensionless temperature T/T_0 with dimensionless time $k_{d,Y,0}t$ is plotted in Fig. 3.7, for selected values of $\lambda/c_P T_0$, T_{st}/T_0, and $W/\rho k_{d,Y,0} V_0$ as germane parameters. The curves resemble slight rectangular hyperbolae, with incipient curvature

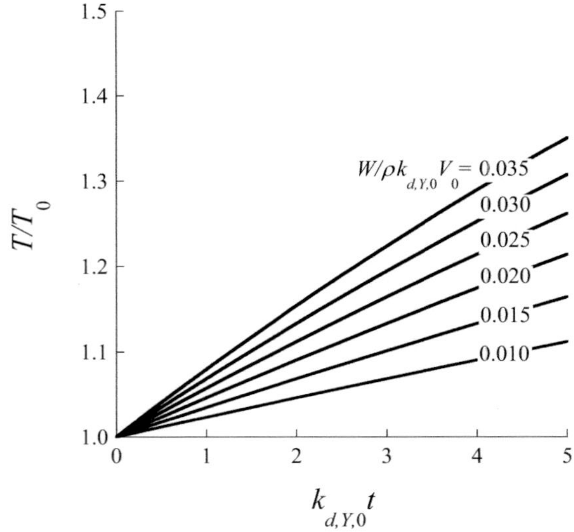

Fig. 3.7 Variation of dimensionless temperature, T/T_0, versus dimensionless time, $k_{d,Y,0}t$, for selected values of dimensionless parameters $W/\rho k_{d,Y,0}V_0$, $\lambda/c_P T_0 = 2$, and $T_{st}/T_0 = 1.35$

that becomes more notorious and displacement of curve upward as steam flow rate W increases, due to a larger inventory of heat supplied, or mass density ρ or (initial) volume V_0 decreases, due to a larger thermal inertia of the liquid system; both these trends give indeed rise to a larger $W/\rho k_{d,Y,0}V_0$.

After retrieving T/T_0 from Eq. (3.228), it becomes possible to redo Eq. (3.30) as

$$\frac{C_Y}{C_{Y,0}} = \exp\left\{ -\int_0^{k_{d,Y,0}t} \exp\left\{ \frac{E_{act,Y}}{\mathcal{R}T_0}\left(1 - \frac{1}{\dfrac{1+\left(\frac{\lambda}{c_P T_0}+\frac{T_{st}}{T_0}\right)\frac{W}{\rho k_{d,Y,0}V_0}k_{d,Y,0}t}{1+\frac{W}{\rho k_{d,Y,0}V_0}k_{d,Y,0}t}} \right) \right\} d\left(k_{d,Y,0}\tilde{t}\right) \right\}, \tag{3.229}$$

which may be algebraically rearranged to read

$$\frac{C_Y}{C_{Y,0}} = \exp\left\{ -\int_0^{k_{d,Y,0}t} \exp\left\{ \frac{E_{act,Y}}{\mathcal{R}T_0}\left(1 - \frac{1+\frac{W}{\rho k_{d,Y,0}V_0}\left(k_{d,Y,0}t\right)}{1+\left(\frac{\lambda}{c_P T_0}+\frac{T_{st}}{T_0}\right)\frac{W}{\rho k_{d,Y,0}V_0}\left(k_{d,Y,0}t\right)} \right) \right\} d\left(k_{d,Y,0}\tilde{t}\right) \right\} \tag{3.230}$$

through reduction in number of fraction levels. Equation (3.230) is illustrated in Fig. 3.8, for preselected values of dimensionless parameters $E_{act,Y}/\mathcal{R}T_0$, $W/\rho k_{d,Y,0}V_0$,

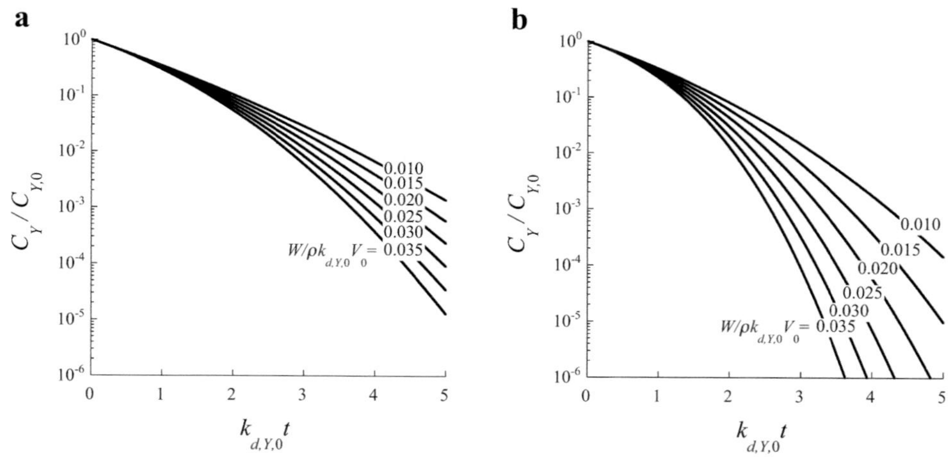

Fig. 3.8 Variation of normalized concentration of viable biomass, $C_Y/C_{Y,0}$, versus dimensionless time, $k_{d,Y,0}t$, for selected values of dimensionless parameters $W/\rho k_{d,Y,0}V_0$, $\lambda/c_PT_0 = 2$, $T_{st}/T_0 = 1.35$, and (**a**) $E_{act,Y}/\mathcal{R}T_0 = 5$ and (**b**) $E_{act,Y}/\mathcal{R}T_0 = 10$

λ/c_PT_0, and T_{st}/T_0. After a brief stage of linear decay with time (in the log/lin scale elected) as processing temperature remains low, $C_Y/C_{Y,0}$ starts decreasing faster than exponentially, visible as an increasing slope of the curve as a function of $k_{d,Y,0}t$; this trend is more visible at larger $W/\rho k_{d,Y,0}V_0$ and larger $E_{act,Y}/\mathcal{R}T_0$, as a consequence of a larger rate of supply of steam to be utilized in full (i.e., larger W) or a lower mass of liquid medium to be heated (i.e., lower ρV_0), or else a higher sensitivity of the contaminating microbiota to thermal processing (i.e., higher $E_{act,Y}$).

Problem 3.7

Consider use of steam to sterilize the liquid medium via direct sparging therein.

(a) Investigate and discuss the asymptotic trends of the hyperbolic evolution of $T \equiv T\{t\}$.

(b) Estimate the minimum height of the column of liquid required to guarantee that all steam supplied, at mass flow rate W, actually condenses.

Solution

(a) At startup, Eq. (3.226) leads obviously to

(continued)

Problem 3.7 (continued)

$$\lim_{t\to 0} T = \frac{T_0 + \frac{W}{\rho V_0}\left(\frac{\lambda}{c_P} + T_{st}\right)0}{1 + \frac{W}{\rho V_0}0} = \frac{T_0}{1} = T_0, \tag{3.231}$$

in agreement with Eq. (3.207). Conversely, Eq. (3.226) will be driven by

$$\lim_{t\to\infty} T = \lim_{t\to\infty} \frac{\frac{W}{\rho V_0}\left(\frac{\lambda}{c_P} + T_{st}\right)t}{\frac{W}{\rho V_0}t} = \lim_{t\to\infty}\left(\frac{\lambda}{c_P} + T_{st}\right) = T_{st} + \frac{\lambda}{c_P} \tag{3.232}$$

when time grows unbounded, because the second term in both numerator and denominator overrides the first one, along with dropping of common factors between numerator and denominator; Eq. (3.232) consubstantiates a physically unfeasible result, as the liquid medium temperature would then have exceeded the heating steam temperature (by $\lambda/c_P > 0$). This result violates the conditions of validity of Eq. (3.206), as emphasized previously; heating will indeed take place only until the driving force for heat transfer vanishes, meaning that

$$\frac{T_0 + \frac{W}{\rho V_0}\left(\frac{\lambda}{c_P} + T_{st}\right)t}{1 + \frac{W}{\rho V_0}t} \leq T_{st}, \tag{3.233}$$

based on Eq. (3.226), has to be enforced. Since $1 + Wt/\rho V_0 > 0$, Eq. (3.233) is equivalent to

$$T_0 + \frac{W}{\rho V_0}\left(\frac{\lambda}{c_P} + T_{st}\right)t \leq T_{st} + \frac{W}{\rho V_0}T_{st}t \tag{3.234}$$

after getting rid of denominator in the left-hand side and parenthesis in the right-hand side. Upon condensation of terms alike, Eq. (3.234) unfolds

$$\frac{W}{\rho V_0}\left(\frac{\lambda}{c_P} + T_{st} - T_{st}\right)t \leq T_{st} - T_0, \tag{3.235}$$

while elimination of symmetrical terms leaves

(continued)

Problem 3.7 (continued)

$$\frac{\lambda W}{\rho c_P V_0} t \leq T_{st} - T_0; \tag{3.236}$$

isolation of $\lambda W t$ finally yields

$$\lambda W t \leq \rho V_0 c_P (T_{st} - T_0), \tag{3.237}$$

again in view of $\rho c_P V_0 > 0$. Equation (3.237) indicates that heating takes place until the whole latent heat supplied, $\lambda W t$, reaches the whole sensible heat, $\rho V_0 c_P (T_{st} - T_0)$, necessary to bring the original liquid medium, with mass ρV_0 and specific heat capacity c_P, from its initial temperature, T_0, up to the steam temperature, T_{st}. Note that the sensible heat released after steam condensation will eventually be used back to heat the condensate to T_{st}, thus justifying why the initial mass of liquid medium rather than the final mass of liquid medium plus (water) condensate appears explicitly in Eq. (3.237).

(b) A mass balance to a single bubble of steam reads

$$-\frac{dm}{dt} = \frac{UA(T_{st} - T)}{\lambda}, \tag{3.238}$$

where U denotes overall heat transfer coefficient, and m and A denote mass and outer area, respectively, of bubble; the heat originated from steam condensation is in fact transferred to the liquid medium as per Fourier's law, while the corresponding portion of (liquid) condensate is no longer a part of the (gaseous) bubble. After recalling the definition of mass density of steam, ρ_{st}, that is,

$$\rho_{st} \equiv \frac{m}{V} \tag{3.239}$$

with V denoting bubble volume, Eq. (3.238) can be reformulated as

$$-\frac{d}{dt}(\rho_{st} V) = \frac{UA(T_{st} - T)}{\lambda}; \tag{3.240}$$

the rule of differentiation of a product then justifies

$$-V\frac{d\rho_{st}}{dt} - \rho_{st}\frac{dV}{dt} = \frac{UA(T_{st} - T)}{\lambda}. \tag{3.241}$$

Assume that the steam is supplied at the bottom of a tall bioreactor, and that bubbles collapse rapidly (so that they will not reach the top of the

(continued)

Problem 3.7 (continued)

column). If the variation in hydrostatic pressure along bubble path can be neglected (an assumption to be double-checked a posteriori) and the effect of interface curvature upon inner pressure will not be noticed until microscopic size is attained, then ρ_{st} will remain essentially constant; this permits simplification of Eq. (3.241) to

$$-\rho_{st}\frac{dV}{dt}=\frac{UA(T_{st}-T)}{\lambda}, \tag{3.242}$$

along with realization that T has no effect upon V because the bubble remains constituted solely by steam and steam condensation occurs at constant temperature T_{st}. The ideal gas equation may accordingly be invoked to support

$$PV=n\mathcal{R}T_{st} \tag{3.243}$$

pertaining to pure steam, where the definition of molecular weight M as

$$M\equiv\frac{m}{n}, \tag{3.244}$$

with n denoting number of moles, permits reformulation of Eq. (3.243) to

$$PV=\frac{m}{M}\mathcal{R}T_{st}; \tag{3.245}$$

exchange of V and M unfolds

$$PM=\frac{m}{V}\mathcal{R}T_{st}, \tag{3.246}$$

whereas Eq. (3.239) supports further transformation to

$$MP=\rho_{st}\mathcal{R}T_{st} \tag{3.247}$$

or else

$$\rho_{st}=\frac{MP}{\mathcal{R}T_{st}} \tag{3.248}$$

(continued)

Problem 3.7 (continued)

upon isolation of ρ_{st}. Insertion of Eq. (3.248) converts Eq. (3.242) to

$$-\frac{MP}{\mathcal{R}T_{st}}\frac{dV}{dt}=\frac{UA(T_{st}-T)}{\lambda}, \tag{3.249}$$

which may be algebraically rearranged to read

$$\frac{dV}{dt}=-\frac{\mathcal{R}UAT_{st}(T_{st}-T)}{\lambda MP}. \tag{3.250}$$

On the other hand, the outer area of a spherical bubble of radius R looks like

$$A=4\pi R^2, \tag{3.251}$$

and its volume satisfies

$$V=\frac{4}{3}\pi R^3; \tag{3.252}$$

differentiation of both sides of Eq. (3.252), with regard to R, then unfolds

$$dV=\frac{4}{3}\pi 3R^2 dR=4\pi R^2 dR, \tag{3.253}$$

along with lumping of numerical constants. Elimination of dV and A from Eq. (3.250), at the expense of Eqs. (3.253) and (3.251), respectively, gives rise to

$$4\pi R^2\frac{dR}{dt}=-\frac{\mathcal{R}U4\pi R^2 T_{st}(T_{st}-T)}{\lambda MP}, \tag{3.254}$$

so $4\pi R^2$ can be dropped off both sides to leave

$$\frac{dR}{dt}=-\frac{\mathcal{R}UT_{st}(T_{st}-T)}{\lambda MP} \tag{3.255}$$

that entails an essentially constant rate of change in bubble radius; Eq. (3.255) is subjected to

(continued)

Problem 3.7 (continued)

$$R|_{t=0} = R_0 \tag{3.256}$$

as initial condition. Integration of Eq. (3.255) ensues via separation of variables, namely,

$$\int_{R_0}^{R} d\tilde{R} = -\frac{\mathcal{R}UT_{st}(T_{st}-T)}{\lambda MP}\int_0^t d\tilde{t}, \tag{3.257}$$

written with the aid of Eq. (3.256); the fundamental theorem of integral calculus leads readily to

$$\tilde{R}\Big|_{R_0}^{R} = -\frac{\mathcal{R}UT_{st}(T_{st}-T)}{\lambda MP}\tilde{t}\Big|_0^t \tag{3.258}$$

or merely

$$R - R_0 = -\frac{\mathcal{R}UT_{st}(T_{st}-T)}{\lambda MP}t, \tag{3.259}$$

which becomes

$$R = R_0 - \frac{\mathcal{R}UT_{st}(T_{st}-T)}{\lambda MP}t \tag{3.260}$$

upon solution for R. Of particular interest here is the time, t_{cl}, after birth when the steam bubble collapses, obtainable from Eq. (3.260) upon setting $R = 0$ as

$$0 = R_0 - \frac{\mathcal{R}UT_{st}(T_{st}-T)}{\lambda MP}t_{cl}; \tag{3.261}$$

once solved for t_{cl}, Eq. (3.261) unfolds

$$t_{cl} = \frac{\lambda MP}{\mathcal{R}UT_{st}(T_{st}-T)}R_0. \tag{3.262}$$

Small bubbles (i.e., with diameter below ca. 1 cm) form and behave essentially as rigid spheres; they detach from the nozzle, of radius R_{nz}, when the buoyant forces are balanced by the contact forces associated with

(continued)

Problem 3.7 (continued)

surface tension, σ, at the steam/liquid interface along the perimeter of the nozzle tip. As will be seen in due course, this leads to

$$R_0 = \left(\frac{3}{2}\frac{\sigma R_{nz}}{\rho_l - \rho_{st}}\right)^{\frac{1}{3}}, \tag{3.263}$$

where ρ_l denotes mass density of liquid. Equation (3.263) can be combined with Eq. (3.262) to get

$$t_{cl} = \frac{\lambda MP}{\mathscr{R}UT_{st}(T_{st} - T)}\left(\frac{3}{2}\frac{\sigma R_{nz}}{\rho_l - \rho_{st}}\right)^{\frac{1}{3}}, \tag{3.264}$$

where insertion of Eq. (3.248) gives rise to

$$t_{cl} = \frac{\lambda MP}{\mathscr{R}UT_{st}(T_{st} - T)}\left(\frac{3}{2}\frac{\sigma R_{nz}}{\rho_l - \frac{MP}{\mathscr{R}T_{st}}}\right)^{\frac{1}{3}}. \tag{3.265}$$

In view of $\lambda = 2200$ kJ kg^{-1}, $M = 18$ g mol^{-1}, $\mathscr{R} = 8.314$ J K^{-1} mol^{-1}, $U = 15$ kW m^{-2} K^{-1}, $T_{st} = 121$ °C, $\sigma = 48$ mN m^{-1}, and $\rho_l = 940$ kg m^{-3}, Eq. (3.265) becomes

$$t_{cl} = \frac{2200\times10^3\cdot18\times10^{-3}P}{8.314\cdot15\times10^3(121+273)\left((121+273)-T\right)}\left(\frac{3}{2}\frac{48\times10^{-3}R_{nz}}{940-\frac{18\times10^{-3}P}{8.314(121+273)}}\right)^{\frac{1}{3}}, \tag{3.266}$$

where performance of calculations as indicated yields

$$t_{cl} = \frac{8.06\times10^{-4}P}{394 - T}\left(\frac{7.20\times10^{-2}R_{nz}}{940 - 5.49\times10^{-6}P}\right)^{\frac{1}{3}}; \tag{3.267}$$

inspection of Eq. (3.267) indicates that t_{cl} increases with P, R_{nz}, and T, so an upper estimate for t_{cl} may be obtained after arbitrarily setting $P = 2$ atm (corresponding to the maximum practical height of 10 m for the vessel), $R_{nz} = 1$ mm (corresponding to the maximum size normally encountered in steam individual diffusers, and leading to bubbles with final diameter of the order of 1 cm), and $T = 120$ °C (corresponding to a temperature very close to the target sterilization temperature of 121 °C), according to

(continued)

Problem 3.7 (continued)

$$t_{cl}=\frac{8.06\times10^{-4}\cdot2.02\times10^{5}}{394-(120+273)}\left(\frac{7.20\times10^{-2}\cdot0.5\times10^{-3}}{940-5.49\times10^{-6}\cdot2.02\times10^{5}}\right)^{\frac{1}{3}}=0.549\text{ s};\qquad(3.268)$$

in other words, the life expectancy of a steam bubble at birth is less than 1 s, so it will collapse quite rapidly. For the assumed bubble size and at the elevated temperature considered, surface tension and viscosity effects are not significant, so Davies and Taylor's law applies as

$$v=2\sqrt{gR};\qquad(3.269)$$

one promptly realizes that

$$v\leq2\sqrt{gR_0},\qquad(3.270)$$

because steam is being removed therefrom as condensate and thus $R\leq R_0$. The height spanned, h_{cl}, along the vertical motion of a bubble, during a period of duration t_{cl} following detachment from the nozzle will then be given by

$$h_{cl}=vt_{cl}\leq2\sqrt{gR_0}\,t_{cl},\qquad(3.271)$$

at the expense of Eq. (3.270), complemented by the definition of velocity, that is, $v\equiv h_{cl}/t_{cl}$; insertion of Eq. (3.263) converts Eq. (3.271) to

$$h_{cl}\leq2g^{\frac{1}{2}}\left(\left(\frac{3}{2}\frac{\sigma R_{nz}}{\rho_l-\rho_{st}}\right)^{\frac{1}{3}}\right)^{\frac{1}{2}}t_{cl}=2g^{\frac{1}{2}}\left(\frac{3}{2}\frac{\sigma R_{nz}}{\rho_l-\rho_{st}}\right)^{\frac{1}{6}}t_{cl},\qquad(3.272)$$

while Eq. (3.248) can be retrieved to write

$$h_{cl}\leq2g^{\frac{1}{2}}\left(\frac{3}{2}\frac{\sigma R_{nz}}{\rho_l-\frac{MP}{\mathcal{R}T_{st}}}\right)^{\frac{1}{6}}t_{cl}.\qquad(3.273)$$

Based on $g=9.8$ m s^{-2}, $\sigma=48$ mN m^{-1}, $R_{nz}=1$ mm, $\rho_l=940$ kg m^{-3}, $M=18$ g mol^{-1}, $P=2$ atm, $\mathcal{R}=8.314$ J mol^{-1} K^{-1}, and $T_{st}=121$ °C, further to Eq. (3.268), one obtains

(continued)

Problem 3.7 (continued)

$$h_{cl} \leq 2 \cdot 9.8^{\frac{1}{2}} \left(\frac{3}{2} \frac{48 \times 10^{-3} \cdot 1 \times 10^{-3}}{940 - \frac{18 \times 10^{-3} \cdot 2.02 \times 10^{5}}{8.314(121+273)}} \right)^{\frac{1}{6}} 0.549 = 0.224 \text{ m} \tag{3.274}$$

from Eq. (3.273); in other words, a steam bubble will require ca. 22 cm of water column to disappear. This length also serves as maximum distance travelled thereby, which is rather small compared to the height of 10 m assumed for the vessel; therefore, the change in hydrostatic pressure along bubble path is negligible (as it is expected not to exceed 2.3%), and the original hypothesis of a constant ρ_{st} is fully justified.

3.3 Temperature

As emphasized previously, the heating-up and cooling-down periods contribute somehow to sterilization, yet a major effect thereof comes from the holding time, carried out at a temperature clearly above the (average) temperature prevailing in either of such periods. To simplify calculations, such transients are often neglected though. Under these circumstances, one may resort to

$$\ln \frac{C_Y}{C_{Y,0}} = -k_{0,Y} e^{-\frac{E_{act,Y}}{\mathcal{R}T_{st}}} \int_0^t d\tilde{t} \tag{3.275}$$

in lieu of Eq. (3.26), where the constancy of sterilization temperature, T_{st}, during an adiabatic process was taken into account; this promptly simplifies to

$$\ln \frac{C_Y}{C_{Y,0}} = -k_{0,Y} e^{-\frac{E_{act,Y}}{\mathcal{R}T_{st}}} t, \tag{3.276}$$

following trivial application of the fundamental theorem of integral calculus. For a target reduction in concentration of viable microbial contaminant(s) from $C_{Y,0}$ to $C_{Y,tg}$, the corresponding sterilization time, t_{st}, will read

$$t_{st} = \frac{\ln \frac{C_{Y,tg}}{C_{Y,0}}}{-k_{0,Y} e^{-\frac{E_{act,Y}}{\mathcal{R}T_{st}}}} \tag{3.277}$$

stemming from Eq. (3.276), or else

$$t_{st} = \frac{\ln \frac{C_{Y,0}}{C_{Y,tg}}}{k_{0,Y}} e^{\frac{E_{act,Y}}{\mathcal{R}T_{st}}} \tag{3.278}$$

after taking the reciprocal of the argument of the logarithm to remove the minus sign in denominator.

Owing to its essential unselectivity, any thermal treatment affects both undesirable and desirable traits of the medium; hence, sterilization carried out at any temperature and for any period will necessarily damage nutrients to some extent. Assuming that the rate of destruction of a reference nutrient (usually the most heat-labile), Z, follows first-order kinetics as done with Eq. (3.18), one may write

$$\frac{dC_Z}{dt} = -k_{d,Z}C_Z, \tag{3.279}$$

where $k_{d,Z}$ denotes kinetic constant and C_Z denotes concentration of said nutrient;

$$C_Z|_{t=0} = C_{Z,0} \tag{3.280}$$

will accordingly serve as boundary condition, using Eq. (3.19) for template. If the preexponential factor and activation energy associated with nutrient decay are denoted by $k_{0,Z}$ and $E_{act,Z}$, respectively, then Arrhenius' law may be invoked to write

$$k_{d,Z} = k_{0,Z} e^{-\frac{E_{act,Z}}{\mathcal{R}T}} \tag{3.281}$$

under the same functional form used for Eq. (3.20), which prompts transformation of Eq. (3.279) to

$$\frac{dC_Z}{dt} = -k_{0,Z} e^{-\frac{E_{act,Z}}{\mathcal{R}T}} C_Z; \tag{3.282}$$

integration ensues via separation of variables, according to

$$\int_{C_{Z,0}}^{C_Z} \frac{d\widetilde{C}_Z}{\widetilde{C}_Z} = -\int_0^t k_{0,Z} e^{-\frac{E_{act,Z}}{\mathcal{R}T}} d\widetilde{t}, \tag{3.283}$$

where Eq. (3.280) was meanwhile brought on board. If attention is again to be focused only on the holding period characterized by (constant) temperature T_{st}, then Eq. (3.283) becomes

$$\int_{C_{Z,0}}^{C_Z} \frac{d\widetilde{C}_Z}{\widetilde{C}_Z} = -k_{0,Z} e^{-\frac{E_{act,Z}}{\mathcal{R}T_{st}}} \int_0^t d\widetilde{t}; \tag{3.284}$$

application of the fundamental theorem of integral calculus then yields

$$\ln \widetilde{C}_Z \Big|_{C_{Z,0}}^{C_Z} = -k_{0,Z} e^{-\frac{E_{act,Z}}{\mathcal{R}T_{st}}} \widetilde{t}\Big|_0^t, \tag{3.285}$$

which degenerates to

$$\ln \frac{C_Z}{C_{Z,0}} = -k_{0,Z} e^{-\frac{E_{act,Z}}{\mathcal{R}T_{st}}} t \tag{3.286}$$

after replacing the indicated functions by their upper and lower limits of integration. If t is set equal to t_{st}, then C_Z will attain some value $C_{Z,tg}$ that satisfies

$$\ln \frac{C_{Z,tg}}{C_{Z,0}} = -k_{0,Z} e^{-\frac{E_{act,Z}}{\mathcal{R}T_{st}}} t_{st}, \tag{3.287}$$

based on Eq. (3.286); after taking negatives of both sides, Eq. (3.287) turns to

$$\ln \frac{C_{Z,0}}{C_{Z,tg}} = k_{0,Z} e^{-\frac{E_{act,Z}}{\mathcal{R}T_{st}}} t_{st}, \tag{3.288}$$

and isolation of t_{st} ensues as

$$t_{st} = \frac{\ln \frac{C_{Z,0}}{C_{Z,tg}}}{k_{0,Z}} e^{\frac{E_{act,Z}}{\mathcal{R}T_{st}}}. \tag{3.289}$$

Elimination of t_{st} between Eqs. (3.278) and (3.289) produces

$$\frac{\ln \frac{C_{Y,0}}{C_{Y,tg}}}{k_{0,Y}} e^{\frac{E_{act,Y}}{\mathcal{R}T_{st}}} = \frac{\ln \frac{C_{Z,0}}{C_{Z,tg}}}{k_{0,Z}} e^{\frac{E_{act,Z}}{\mathcal{R}T_{st}}}, \tag{3.290}$$

where functions alike can be combined as

$$\frac{k_{0,Z}}{k_{0,Y}} \frac{\ln \frac{C_{Y,0}}{C_{Y,tg}}}{\ln \frac{C_{Z,0}}{C_{Z,tg}}} = e^{\frac{E_{act,Z}}{\mathcal{R}T_{st}} - \frac{E_{act,Y}}{\mathcal{R}T_{st}}}; \tag{3.291}$$

isolation of the logarithmic factors and factoring out of $-1/\mathcal{R}T_{st}$ finally lead to

$$\frac{\ln \frac{C_{Y,0}}{C_{Y,tg}}}{\ln \frac{C_{Z,0}}{C_{Z,tg}}} = \frac{k_{0,Y}}{k_{0,Z}} e^{-\frac{E_{act,Y} - E_{act,Z}}{\mathcal{R}T_{st}}}, \tag{3.292}$$

graphically depicted in Fig. 3.9 as ratio of $\ln\{C_{Y,0}/C_{Y,tg}\}$ to $\ln\{C_{Z,0}/C_{Z,tg}\}$ versus dimensionless sterilization temperature, $\mathcal{R}T_{st}/(E_{act,Y} - E_{act,Z})$, for selected values of dimensionless parameter $k_{0,Y}/k_{0,Z}$. Note the increasing behavior of $\ln\{C_{Y,0}/C_{Y,tg}\}/\ln\{C_{Z,0}/C_{Z,tg}\}$ with $\mathcal{R}T_{st}/(E_{act,Y} - E_{act,Z})$, as the latter increases in value within the positive range; and the reverse, when $\mathcal{R}T_{st}/(E_{act,Y} - E_{act,Z})$ increases in absolute value within the negative range. The variation with $\mathcal{R}T_{st}/(E_{act,Y} - E_{act,Z})$ becomes faster and faster under a larger $k_{0,Y}/k_{0,Z}$ within the positive range, but slower within the negative range; the dashed line in Fig. 3.9 indicates the threshold above which the relative decrease in microbial viability surpasses the

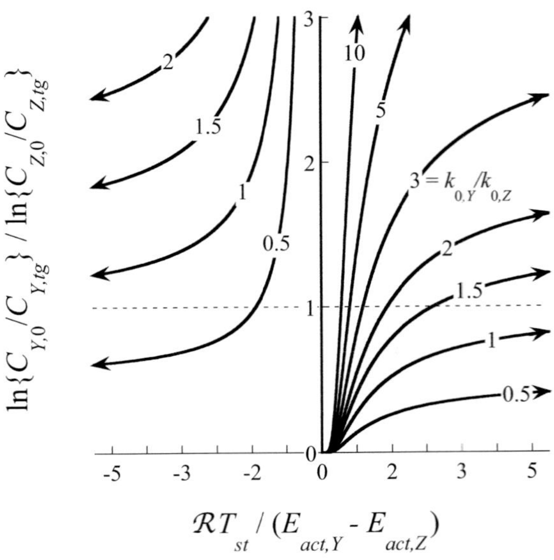

Fig. 3.9 Variation of ratio of logarithms of relative reductions in biomass viability and nutrient wholesomeness, $\ln\{C_{Y,0}/C_{Y,tg}\}/\ln\{C_{Z,0}/C_{Z,tg}\}$, versus dimensionless temperature, $\mathcal{R}T_{st}/(E_{act,Y} - E_{act,Z})$, for selected values of dimensionless parameter $k_{0,Y}/k_{0,Z}$, with direction of increasing $|E_{act,Y} - E_{act,Z}|$ indicated by arrows

corresponding relative decrease in nutrient decay, being reached at a lower and lower processing temperature as $k_{0,Y}/k_{0,Z}$ increases.

The monotony of the left-hand side of Eq. (3.292) with regard to T_{st} may be ascertained upon differentiation of both its sides, that is,

$$\frac{d}{dT_{st}}\left(\frac{\ln\frac{C_{Y,0}}{C_{Y,tg}}}{\ln\frac{C_{Z,0}}{C_{Z,tg}}}\right)=\frac{k_{0,Y}}{k_{0,Z}}\mathrm{e}^{-\frac{E_{act,Y}-E_{act,Z}}{\mathcal{R}T_{st}}}\left(-\frac{E_{act,Y}-E_{act,Z}}{\mathcal{R}}\left(-\frac{1}{T_{st}^{2}}\right)\right)\tag{3.293}$$

or, upon factoring in of -1 and factoring out of $1/\mathcal{R}T_{st}^{2}$,

$$\frac{d}{dT_{st}}\left(\frac{\ln\frac{C_{Y,0}}{C_{Y,tg}}}{\ln\frac{C_{Z,0}}{C_{Z,tg}}}\right)=\frac{k_{0,Y}}{k_{0,Z}\mathcal{R}T_{st}^{2}}\mathrm{e}^{-\frac{E_{act,Y}-E_{act,Z}}{\mathcal{R}T_{st}}}\left(E_{act,Y}-E_{act,Z}\right);\tag{3.294}$$

a multitude of available data pertaining to medium nutrients and microbial contaminants indicate that

$$E_{act,Y}>E_{act,Z}\tag{3.295}$$

in most cases, so Eq. (3.294) supports

$$\frac{d}{dT_{st}}\left(\frac{\ln\frac{C_{Y,0}}{C_{Y,tg}}}{\ln\frac{C_{Z,0}}{C_{Z,tg}}}\right)>0\tag{3.296}$$

since $k_{0,Y}/k_{0,Z}\mathcal{R}T_{st}^{2}>0$ and $\mathrm{e}^{-(E_{act,Y}-E_{act,Z})/\mathcal{R}T_{st}}>0$. Equation (3.296) implies that, for a given t_{st}, a higher T_{st} causes a larger increase in $\ln\{C_{Y,0}/C_{Y,tg}\}$ than in $\ln\{C_{Z,0}/C_{Z,tg}\}$ or, equivalently, a larger relative decrease in $C_{Y,tg}$ than in $C_{Z,tg}$, since $C_{Y,0}\geq C_{Y,tg}$ as per Eq. (3.278) with $k_{0,Y}t_{st}\mathrm{e}^{-E_{act,Y}/\mathcal{R}T_{st}}>0$, and likewise $C_{Z,0}\geq C_{Z,tg}$ as per Eq. (3.289) with $k_{0,Z}t_{st}\mathrm{e}^{-E_{act,Z}/\mathcal{R}T_{st}}>0$. Therefore, a higher sterilization temperature is comparatively less detrimental to nutrient(s) in the medium than to contaminating microorganisms; a critical temperature, T_{cr}, exists indeed for which the fractional decay in contaminating microorganisms equals the fractional decay in nutrients, that is,

$$T_{cr}\equiv T_{st}\Big|_{\ln\frac{C_{Y,0}}{C_{Y,tg}}=\ln\frac{C_{Z,0}}{C_{Z,tg}}};\tag{3.297}$$

upon combination with Eq. (3.292), one gets

$$1 = \frac{k_{0,Y}}{k_{0,Z}} e^{-\frac{E_{act,Y} - E_{act,Z}}{\mathscr{R}T_{cr}}}, \tag{3.298}$$

where multiplication of both sides by $k_{0,Z}/k_{0,Y}$ and removal of the minus sign preceding the argument of the exponential function unfold

$$e^{\frac{E_{act,Z} - E_{act,Y}}{\mathscr{R}T_{cr}}} = \frac{k_{0,Z}}{k_{0,Y}}. \tag{3.299}$$

Application of logarithms to both sides of Eq. (3.299) produces

$$\frac{E_{act,Z} - E_{act,Y}}{\mathscr{R}T_{cr}} = \ln \frac{k_{0,Z}}{k_{0,Y}}, \tag{3.300}$$

while isolation of T_{cr} leads to

$$T_{cr} = \frac{E_{act,Z} - E_{act,Y}}{\mathscr{R} \ln \frac{k_{0,Z}}{k_{0,Y}}} \tag{3.301}$$

or, after multiplication of both sides by $\mathscr{R}/(E_{act,Z} - E_{act,Y})$,

$$\frac{\mathscr{R}T_{cr}}{E_{act,Z} - E_{act,Y}} = \frac{1}{\ln \frac{k_{0,Z}}{k_{0,Y}}}; \tag{3.302}$$

Equation (3.302) is depicted in Fig. 3.10, using $k_{0,Z}/k_{0,Y}$ as (dimensionless) independent variable. Beware that Eq. (3.295), taken as valid for hypothesis, implies a negative numerator in the right-hand side of Eq. (3.301), so $k_{0,Z} < k_{0,Y}$ and thus $\ln\{k_{0,Z}/k_{0,Y}\} < 0$ are a requirement for existence of a physically meaningful T_{cr} (i.e., $T_{cr} > 0$); this means that only the curve in the bottom left portion of Fig. 3.10 is of practical interest, which, in addition, conveys a value within 0.01–0.02 over the very many orders of magnitude of $k_{0,Z}/k_{0,Y}$ found in practice. In fact, a steeper (monotonically) decreasing curve will necessarily intercept another monotonically decreasing curve if (and only if) the former departs from a higher ordinate; note, in this regard, that $\ln k_{0,Y} - E_{act,Y}/\mathscr{R}T$ and $\ln k_{0,Z} - E_{act,Z}/\mathscr{R}T$ serve as vertical intercepts for the $\ln\{\ln\{C_{Y,0}/C_Y\}\}$-vs.-$\ln t$ and $\ln\{\ln\{C_{Z,0}/C_Z\}\}$-vs.-$\ln t$ curves, respectively (see Eqs. (3.276) and (3.288)).

Values of $E_{act,Y}$ of many unwanted microorganisms are relatively high, usually within the range 220–290 kJ mol^{-1}, with spores being more resistant (and consequently described by lower $E_{act,Y}$) than their vegetative cell counterparts; such values are typical of

Fig. 3.10 Variation of dimensionless critical temperature, $\mathcal{R}T_{cr}/(E_{act,Z} - E_{act,Y})$, versus ratio of preexponential factors of nutrient to cell decay, $k_{0,Z}/k_{0,Y}$

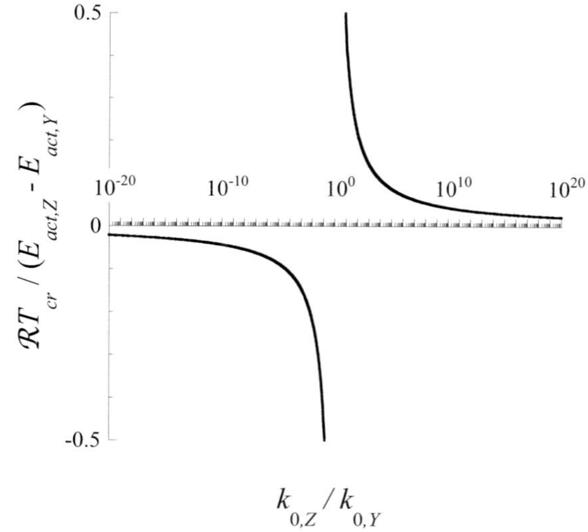

nonbiochemical reactions as expected, because denaturation of some enzyme in the cells that is critical for their viability is not an enzyme-mediated phenomenon. Values of $E_{act,Z}$ are quite lower, for example, 84–92 kJ mol^{-1} for destruction of vitamins and amino acids, and ca. 165 kJ mol^{-1} for proteins.

In practice, any processing temperature above T_{cr} will bring about more extensive decrease in viability of microbial contaminants than degradation of nutrient integrity, that is, $T_{st} > T_{cr}$ will imply $\ln\{C_{Y,0}/C_{Y,tg}\} > \ln\{C_{Z,0}/C_{Z,tg}\}$ as long as Eq. (3.295) is valid; the corresponding t_{st} will also be reduced as per T_{st} appearing in denominator of the argument of the exponential function in Eq. (3.278). Although in theory $t_{st} \to 0$ when T_{st} is sufficiently high, the safety precautions required to handle ultrahigh-pressure steam would set an upper value for T_{st}, normally within the range 121–138 °C, which corresponds to 2.0–3.4 atm for pressure range. Furthermore, it becomes harder and harder to guarantee that heat transfer will be sufficiently fast for compatibility with the minute t_{st} at stake, in agreement with Fourier's law (valid also for steam sparging); hence, it would be useless to go to excessively high T_{st}, when unnecessary, yet unavoidable overprocessing would take place anyway for any predefined target of lethality.

Problem 3.8

When designing a sterilization process, some microbial species and some nutrient are to be chosen in advance for reference, so that the corresponding $E_{act,Y}$ and $E_{act,Z}$, respectively, will be used to calculate T_{cr}.

(continued)

Problem 3.8 (continued)

(a) In a given nutrient medium containing vitamin C as ingredient and with expected contamination by *Bacillus stearothermophilus*, it has been found that the death process of the latter is characterized by $k_{0,Y} = 1.0 \times 10^{47}$ min^{-1} and $E_{act,Y} = 83.7$ kcal mol^{-1} K^{-1}, whereas the rate of decay of the vitamin is characterized by $k_{0,Z} = 6.6 \times 10^9$ min^{-1} and $E_{act,Z} = 20.8$ kcal mol^{-1} K^{-1}. Calculate the critical temperature for thermal processing and the time it will take to guarantee a reduction of viability by 12 orders of magnitude, and the corresponding decay in nutrient integrity, at the standard sterilization temperature.

(b) Investigate how a mixed microbial population, in a batch bioreactor, will respond to a thermal process, in terms of viability reduction at some temperature $T_{st} > T_{cr}$, when the first-order death constants at that temperature read $k_{d,Y,1}$ and $k_{d,Y,2}$ for the reference and second microbial species, respectively; vis-à-vis with the behavior of a single population of reference species. For simplicity, neglect interactions between cells of the two species.

Solution

(a) The critical temperature, above which relative cell decay becomes faster than relative nutrient decay, is conveyed by Eq. (3.301), which becomes

$$T_{cr} = \frac{8.7 \times 10^4 - 3.5 \times 10^5}{8.314 \ln \frac{1.1 \times 10^8}{1.7 \times 10^{45}}} = 369.4\,\text{K} = 96.3\,^\circ\text{C} \qquad (3.303)$$

after having inserted $k_{0,Y} = 1.0 \times 10^{47}/60 = 1.7 \times 10^{45}$ s^{-1}, $E_{act,Y} = 83.7 \times 10^3 \cdot 4.184 = 3.5 \times 10^5$ J mol^{-1} K^{-1}, $k_{0,Z} = 6.6 \times 10^9/60 = 1.1 \times 10^8$ s^{-1}, and $E_{act,Z} = 20.8 \times 10^3 \cdot 4.184 = 8.7 \times 10^4$ J mol^{-1} K^{-1}. For the standard sterilization temperature of 121 °C, the corresponding first-order death constants will specifically read

$$k_{d,Y} = 1.7 \times 10^{45} \exp\left\{ -\frac{3.5 \times 10^5}{8.314 \cdot (121 + 273.2)} \right\} = 0.071\,\text{s}^{-1} \qquad (3.304)$$

and

(continued)

Problem 3.8 (continued)

$$k_{d,Z} = 1.1 \times 10^8 \exp\left\{-\frac{8.7 \times 10^4}{8.314 \cdot (121 + 273.2)}\right\} = 0.00033\,\mathrm{s}^{-1}, \tag{3.305}$$

as per Eqs. (3.20) and (3.281), respectively; $k_{d,Z} \ll k_{d,Y}$ confirms suitability of such processing temperature, 121 > 96.3 °C as per Eq. (3.303), to bring about elimination of microbial viability more than two orders of magnitude faster than the accompanying reduction in available vitamin. After reformulating Eq. (3.33) to

$$t_{st} = \frac{\ln \frac{C_{Y,tg}}{C_{Y,0}}}{-\mathrm{e}^0 k_{d,Y}} = \frac{\ln \frac{C_{Y,0}}{C_{Y,tg}}}{k_{d,Y}} = \frac{\log_{10} \frac{C_{Y,0}}{C_{Y,tg}} \ln 10}{k_{d,Y}} \tag{3.306}$$

with $T = T_0$, insertion of Eq. (3.304) coupled with $C_{Y,0}/C_{Y,tg} = 10^{12}$ by hypothesis give rise to

$$t_{st} = \frac{\left(\log_{10} 10^{12}\right) \ln 10}{0.071} = 389\,\mathrm{s} = 6.5\,\mathrm{min}, \tag{3.307}$$

meaning that ca. 6.5 min of thermal treatment will attain the intended degree of lethality. By the same token, Eqs. (3.281) and (3.289) can be combined to obtain

$$t_{st} = \frac{\ln \frac{C_{Z,0}}{C_{Z,tg}}}{k_{0,Z}\,\mathrm{e}^{-\frac{E_{act,Z}}{\mathcal{R}T_{st}}}} \equiv \frac{\ln \frac{C_{Z,0}}{C_{Z,tg}}}{k_{d,Z}} = \frac{\log_{10} \frac{C_{Z,0}}{C_{Z,tg}} \ln 10}{k_{d,Z}} \tag{3.308}$$

or, after isolation of $C_{Z,0}/C_{Z,tg}$,

$$\frac{C_{Z,0}}{C_{Z,tg}} = 10^{\frac{k_{d,Z} t_{st}}{\ln 10}}; \tag{3.309}$$

combination with Eqs. (3.305) and (3.307) then unfolds

(continued)

Problem 3.8 (continued)

$$\frac{C_{Z,0}}{C_{Z,tg}} = 10^{\frac{0.00033 \cdot 389}{\ln 10}} = 1.14, \tag{3.310}$$

so a mere 1 − 1/1.14 = 12% of the initial vitamin will be lost, thus confirming that 121 °C for sterilization temperature is extremely powerful toward elimination of the contaminating microorganism at stake, yet quite soft in terms of degradation of the reference nutrient under scrutiny.

(b) The mass balance to the reference microbial species reads

$$\frac{dC_{Y,1}}{dt} = -k_{d,Y,1} C_{Y,1} \tag{3.311}$$

in agreement with Eq. (3.18), subjected to

$$C_{Y,1}|_{t=0} = C_{Y,1,0} \tag{3.312}$$

as initial condition also compatible with Eq. (3.19); by the same token, the evolution in viability of the second species satisfies

$$\frac{dC_{Y,2}}{dt} = -k_{d,Y,2} C_{Y,2}, \tag{3.313}$$

with

$$C_{Y,2}|_{t=0} = C_{Y,2,0} \tag{3.314}$$

serving as initial condition. Integration of Eq. (3.311), with the aid of Eq. (3.312), leads to

$$\int_{C_{Y,1,0}}^{C_{Y,1}} \frac{d\widetilde{C}_{Y,1}}{\widetilde{C}_{Y,1}} = -k_{d,Y,1} \int_0^t d\widetilde{t} \tag{3.315}$$

via separation of variables, which becomes

(continued)

Problem 3.8 (continued)

$$\ln \tilde{C}_{Y,1} \Big|_{C_{Y,1,0}}^{C_{Y,1}} = -k_{d,Y,1}\, \tilde{t} \Big|_{0}^{t} \tag{3.316}$$

after the fundamental theorem of integral calculus; Eq. (3.316) breaks down to

$$\ln \frac{C_{Y,1}}{C_{Y,1,0}} = -k_{d,Y,1} t, \tag{3.317}$$

or else

$$C_{Y,1} = C_{Y,1,0} e^{-k_{d,Y,1} t} \tag{3.318}$$

upon taking exponentials of both sides, and multiplying them by $C_{Y,1,0}$ afterward. In view of the analogy between Eqs. (3.311) and (3.313), and between Eqs. (3.312) and (3.314), one can resort to Eq. (3.318) as template to write

$$C_{Y,2} = C_{Y,2,0} e^{-k_{d,Y,2} t} \tag{3.319}$$

pertaining to the second species; ordered addition of Eqs. (3.318) and (3.319) gives rise to

$$C_{Y,1} + C_{Y,2} = C_{Y,1,0} e^{-k_{d,Y,1} t} + C_{Y,2,0} e^{-k_{d,Y,2} t}, \tag{3.320}$$

where division of both sides by $C_{Y,1,0} + C_{Y,2,0}$ unfolds

$$\frac{C_{Y,1} + C_{Y,2}}{C_{Y,1,0} + C_{Y,2,0}} = \frac{C_{Y,1,0}}{C_{Y,1,0} + C_{Y,2,0}} e^{-k_{d,Y,1} t} + \frac{C_{Y,2,0}}{C_{Y,1,0} + C_{Y,2,0}} e^{-k_{d,Y,2} t}. \tag{3.321}$$

After defining α as

$$\alpha \equiv \frac{C_{Y,1,0}}{C_{Y,1,0} + C_{Y,2,0}}, \tag{3.322}$$

thus representing the initial fraction of reference species in the overall microbial load, one promptly realizes that

(continued)

Problem 3.8 (continued)

$$1-\alpha=1-\frac{C_{Y,1,0}}{C_{Y,1,0}+C_{Y,2,0}}=\frac{C_{Y,1,0}+C_{Y,2,0}-C_{Y,1,0}}{C_{Y,1,0}+C_{Y,2,0}}=\frac{C_{Y,2,0}}{C_{Y,1,0}+C_{Y,2,0}} \quad (3.323)$$

upon lumping fractions and then dropping off symmetrical terms; insertion of Eqs. (3.322) and (3.323) permits reformulation of Eq. (3.321) to

$$\frac{C_{Y,1}+C_{Y,2}}{C_{Y,1,0}+C_{Y,2,0}}=\alpha e^{-k_{d,Y,1}t}+(1-\alpha)e^{-k_{d,Y,2}t}. \quad (3.324)$$

If the second species were absent, then Eq. (3.324) would break down to

$$\left.\frac{C_{Y,1}+C_{Y,2}}{C_{Y,1,0}+C_{Y,2,0}}\right|_{C_{Y,2,0}=0}=1e^{-k_{d,Y,1}t}+(1-1)e^{-k_{d,Y,2}t}=e^{-k_{d,Y,1}t}, \quad (3.325)$$

since $\alpha = C_{Y,0}/(C_{Y,0} + 0) = C_{Y,0}/C_{Y,0} = 1$ as per Eq. (3.322). If the ratio of fractional binary population survival to fractional single population survival, under identical conditions, Ω, is put forward as

$$\Omega \equiv \frac{\frac{C_{Y,1}+C_{Y,2}}{C_{Y,1,0}+C_{Y,2,0}}}{\left.\frac{C_{Y,1}+C_{Y,2}}{C_{Y,1,0}+C_{Y,2,0}}\right|_{C_{Y,2,0}=0}}, \quad (3.326)$$

then combination with Eqs. (3.324) and (3.325) will produce

$$\Omega=\frac{\alpha e^{-k_{d,Y,1}t}+(1-\alpha)e^{-k_{d,Y,2}t}}{e^{-k_{d,Y,1}t}}; \quad (3.327)$$

cancellation of common factors between numerator and denominator gives rise to

$$\Omega=\alpha+(1-\alpha)e^{k_{d,Y,1}t}e^{-k_{d,Y,2}t}=\alpha+(1-\alpha)e^{(k_{d,Y,1}-k_{d,Y,2})t}, \quad (3.328)$$

where t was meanwhile factored out, or, equivalently,

$$\Omega=\alpha+(1-\alpha)e^{\left(1-\frac{k_{d,Y,2}}{k_{d,Y,1}}\right)k_{d,Y,1}t} \quad (3.329)$$

after multiplication and division of the argument of the outstanding exponential function by $k_{d,Y,1}$. A graphical interpretation of Eq. (3.329) is provided in Fig. 3.11, using $k_{2,Y,1}t$ as dimensionless independent variable,

(continued)

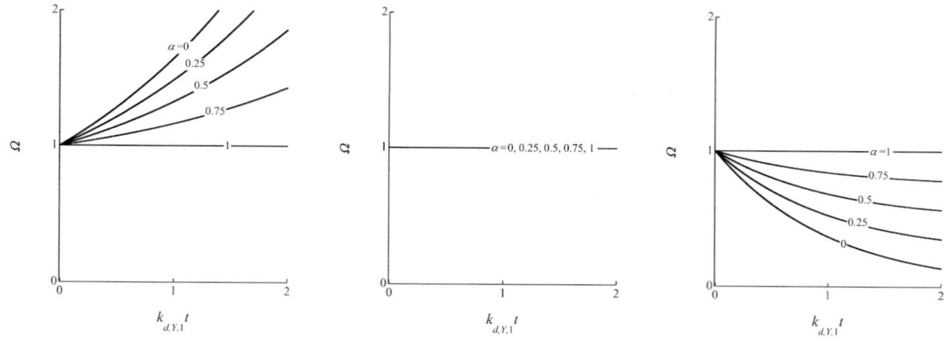

Fig. 3.11 Evolution of ratio of fractional binary population survival to fractional single population survival, Ω, versus dimensionless time, $k_{d,Y,1}t$, for selected values of dimensionless parameters α, and $k_{d,Y,2}/k_{d,Y,1} = 0.5$ (left), $k_{d,Y,2}/k_{d,Y,1} = 1$ (center), and $k_{d,Y,2}/k_{d,Y,1} = 2$ (right)

Problem 3.8 (continued)

for selected values of dimensionless parameters α and $k_{d,Y,2}/k_{d,Y,1}$. As expected,

$$\Omega|_{\alpha=1} = 1 + (1-1)e^{\left(1-\frac{k_{d,Y,2}}{k_{d,Y,1}}\right)k_{d,Y,1}t} = 1 + 0 = 1 \tag{3.330}$$

as per Eq. (3.329), thus justifying the horizontal lines in all three plots when $\alpha = 1$; whereas

$$\Omega\Big|_{\frac{k_{d,Y,2}}{k_{d,Y,1}}=1} = \alpha + (1-\alpha)e^{(1-1)k_{d,Y,1}t} = \alpha + (1-\alpha)e^{0} = \alpha + 1 - \alpha = 1, \tag{3.331}$$

based again on Eq. (3.329), describes the single line in the center plot. When $k_{d,Y,2}/k_{d,Y,1}$ exceeds unity, all lines lie below that labeled as $\Omega = 1$; the secondary species being more labile than the reference one makes indeed the overall population lose viability faster than the latter one if considered alone. Conversely, $k_{d,Y,2}/k_{d,Y,1} < 1$ forces all lines to above that corresponding to $\Omega = 1$, so overall lethality will fall behind that of the reference species when considered by itself. In both cases, deviation from the horizontal line of equation $\Omega = 1$ (and associated to $\alpha = 1$) increases as α decreases from 1 toward 0. This example helps justify why the most resistant species anticipated as contaminant of the medium (i.e., the one characterized by the lowest k_d) should be selected for reference species, when designing a sterilization treatment; the actual degree of lethality will accordingly be higher than

(continued)

Problem 3.8 (continued)

predicted for that species considered by itself, so a conservative estimate results.

3.4 Continuous Operation

Two basic modes of operation are normally pinpointed in bioprocesses—batch operation and continuous operation; certain processes, usually at lower scale, resort to the former, whereas large-scale operation is normally carried out continuouswise. A specific hybrid of thereof, known as fed-batch, is of particular relevance in bioprocessing; it entails intermittent or continuous provision of an inlet stream, devoid of inoculum and aimed at replenishing the medium with fresh substrate, in the absence of an outlet stream, so the medium volume increases, stepwise or continuously, as time elapses.

The underlying mass balances in batch operation encompass only two terms—production and accumulation; hence, all nutrients supplied will be directly converted to biomass and/or specific metabolites, at rates of depletion or accumulation that match the rates of consumption or production thereof, respectively. Therefore, the system remains under transient conditions, from the very beginning until the intended batch time has been reached. The amount of medium to be used is known in advance, and may be changed, at the end of a batch, to some other; this facilitates experimental checks on the effect of distinct environmental conditions, and justifies its preference at bench-scale operation.

Conversely, continuous operation goes through only an initial (rather than a sustained) transient; such transient is accordingly expected to elapse after some time, thus giving rise to a steady state. This represents indeed the major advantage of this mode of operation, as long as the conditions meant for steady-state operation have been preset criteriously; the outcome is thus not only fully predictable on account of being constant, but also the best ever possible on account of having been optimized in advance. However, the associated mass balances now rely on three terms—production, accumulation, and convection; the last magnifies the time period prior to reaching steady state, so part of the outlet stream may have to be discarded (or recycled) for lack of compliance with product specifications. Continuous operation is preferred at industrial settings because the large amounts of feedstock processed are too expensive to afford significant and persistent deviations from their preset/optimum conditions; however, it further requires stricter control measures, either implemented in feedback or feedforward fashion.

Depending on whether batch or continuous operation is sought, a different mode of sterilization is required for the liquid medium. In the former case, the whole batch must obviously be sterilized in advance; continuous operation may instead resort to either sterilization of inlet medium on a continuous basis, or continuous supply of sterile medium, pumped through the piping network under absolutely aseptic conditions, and originating from a large reservoir of medium where it has been sterilized in advance. Since small bioreactors are normally operated batchwise, they can be taken as a whole, that is, together with sensing probes and inlet/outlet fittings, into an autoclave of capacity sufficient to carry out sterilization. Larger bioreactors are obviously not compatible with this simple procedure, so sterilizing devices have to be built in the equipment itself, with steam inlets and condensate drains provided in situ; this approach is more expensive regarding capital investment, yet compatible with the much more expensive bioreactors at stake and the much larger amounts processed.

The situation is, nevertheless, disparate when supply of molecular oxygen is concerned—as required by aerobic respiration. In fact, the inventory of oxygen dissolved in water, even under saturation, would rapidly fall short of that required by microorganisms performing regular metabolism because of the poor solubility of this gas in water at room temperatures. Therefore, oxygen must be supplied to the broth on a continuous basis, irrespective of batch of continuous operation with regard to load of medium; the least expensive source of molecular oxygen is obviously atmospheric air itself. Although thermal sterilization of air is feasible, it would require an extra heat exchanger, characterized by a particularly large area because resistance of air to heat transfer is much higher than that of its liquid medium counterpart. On the other hand, air is much less viscous than liquid water; hence, pumping of air through a porous membrane is technically quite easy, since the associated pressure drops due to attrition are acceptable even for pores narrow enough to halt passage of microorganisms (and even viruses).

Problem 3.9

Consider a bioreactor possessing an outer jacket of heat transfer area A_l, through which condensing steam at temperature T_s, and characterized by latent heat of condensation λ_s, will be flown; filled with a volume V_l of liquid at temperature T_l, characterized by mass density ρ_l and isobaric-specific heat capacity $c_{P,l}$; and renewed at a sufficiently high velocity so that the bed of solid, compact particles at temperature T_b inside it, characterized by overall volume V_b, overall outer heat transfer area A_b, mass density ρ_b, and isobaric-specific heat capacity $c_{P,b}$, remains fluidized. The aforementioned flow of steam is meant to sterilize those particles, before immobilization proceeds upon an inlet flow of inoculum. Estimate the evolution in temperature of those solid particles after condensing steam starts to be supplied, with given overall heat transfer coefficient between steam and liquid, U_{sl}, and overall heat transfer coefficient between liquid and bed particles, U_{lb}, under the hypothesis of a large thermal conductivity of the solid phase.

(continued)

Problem 3.9 (continued)

Solution

An enthalpy balance to the liquid takes the form

$$U_{sl}A_l(T_s - T_l) = \rho_l V_l c_{P,l}\frac{dT_l}{dt}, \tag{3.332}$$

subjected to

$$T_l|_{t=0} = T_{l,0} \tag{3.333}$$

as initial condition, whereas a similar balance to the solid phase reads

$$U_{lb}A_b(T_l - T_b) = \rho_b V_b c_{P,b}\frac{dT_b}{dt} \tag{3.334}$$

subjected to

$$T_b|_{t=0} = T_{b,0} \tag{3.335}$$

for initial condition; no thermal gradients were considered inside the liquid as a consequence of its being well stirred, neither inside the solid for its exhibiting a high thermal conductivity. Before steam is supplied, the liquid lies at temperature $T_{l,0}$ as per Eq. (3.333), and steam in the jacket at temperature $T_{s,0} = T_{l,0}$; hence,

$$U_{sl}A_l(T_{s,0} - T_{l,0}) = 0, \tag{3.336}$$

based on Eq. (3.332), trivially describes the prevailing steady state. Ordered subtraction of Eq. (3.336) from Eq. (3.332) yields

$$U_{sl}A_l\Big(T_s - T_{s,0} - (T_l - T_{l,0})\Big) = \rho_l V_l c_{P,l}\frac{d(T_l - T_{l,0})}{dt}, \tag{3.337}$$

where advantage was taken of the constancy of $T_{l,0}$. By the same token, the initial steady state encompassing the solid phase abides to

$$U_{lb}A_b(T_{l,0} - T_{b,0}) = 0 \tag{3.338}$$

stemming from Eq. (3.334), with $T_{b,0} = T_{l,0}$ denoting initial temperature of solid phase; ordered subtraction of Eq. (3.338) from Eq. (3.334) permits replacement of the latter by

(continued)

Problem 3.9 (continued)

$$U_{lb}A_b\left(T_l - T_{l,0} - (T_b - T_{b,0})\right) = \rho_b V_b c_{P,b} \frac{d(T_b - T_{b,0})}{dt}, \tag{3.339}$$

again based on realization that $d(T_b - T_{b,0})$ coincides with dT_b for $T_{b,0}$ being constant. For simplicity of algebraic manipulation, deviation variables relative to the corresponding initial steady states will be introduced, namely,

$$\widehat{T}_s \equiv T_s - T_{s,0}, \tag{3.340}$$

$$\widehat{T}_l \equiv T_l - T_{l,0}, \tag{3.341}$$

and

$$\widehat{T}_b \equiv T_b - T_{b,0}; \tag{3.342}$$

Equations (3.340)–(3.342) permit replacement of Eqs. (3.337) and (3.339) by

$$\rho_l V_l c_{P,l} \frac{d\widehat{T}_l}{dt} = U_{sl}A_l\left(\widehat{T}_s - \widehat{T}_l\right) \tag{3.343}$$

and

$$\rho_b V_b c_{P,b} \frac{d\widehat{T}_b}{dt} = U_{lb}A_b\left(\widehat{T}_l - \widehat{T}_b\right), \tag{3.344}$$

respectively, and

$$\widehat{T}_l\big|_{t=0} = 0 \tag{3.345}$$

and

$$\widehat{T}_b\big|_{t=0} = 0 \tag{3.346}$$

similarly serve as aliases to Eqs. (3.333) and (3.335), respectively, with the aid of Eqs. (3.341) and (3.342). Application of Laplace's transforms to Eq. (3.343) supports

(continued)

Problem 3.9 (continued)

$$\rho_l V_l c_{P,l}\left(s\bar{\bar{\hat{T}}}_l - T_l|_{t=0}\right) = U_{sl}A_l\left(\bar{\bar{\hat{T}}}_s - \bar{\bar{\hat{T}}}_l\right), \tag{3.347}$$

which simplifies to

$$\rho_l V_l c_{P,l} s\bar{\bar{\hat{T}}}_l = U_{sl}A_l\bar{\bar{\hat{T}}}_s - U_{sl}A_l\bar{\bar{\hat{T}}}_l \tag{3.348}$$

in view of Eq. (3.345) and upon removal of the outstanding parenthesis; by the same token, Eq. (3.344) will give rise to

$$\rho_b V_b c_{P,b}\left(s\bar{\bar{\hat{T}}}_b - \hat{T}_b|_{t=0}\right) = U_{lb}A_b\left(\bar{\bar{\hat{T}}}_l - \bar{\bar{\hat{T}}}_b\right), \tag{3.349}$$

or else

$$\rho_b V_b c_{P,b} s\bar{\bar{\hat{T}}}_b = U_{lb}A_b\bar{\bar{\hat{T}}}_l - U_{lb}A_b\bar{\bar{\hat{T}}}_b \tag{3.350}$$

once Eq. (3.346) is brought on board and parenthesis in the right-hand side taken off. After factoring out $\bar{\bar{\hat{T}}}_l$, Eq. (3.348) becomes

$$(U_{sl}A_l + \rho_l V_l c_{P,l} s)\bar{\bar{\hat{T}}}_l = U_{sl}A_l\bar{\bar{\hat{T}}}_s, \tag{3.351}$$

and Eq. (3.350) similarly yields

$$(U_{lb}A_b + \rho_b V_b c_{P,b} s)\bar{\bar{\hat{T}}}_b = U_{lb}A_b\bar{\bar{\hat{T}}}_l \tag{3.352}$$

after factoring out $\bar{\bar{\hat{T}}}_b$; isolation of $\bar{\bar{\hat{T}}}_l$ in Eq. (3.351) produces

$$\bar{\bar{\hat{T}}}_l = \frac{U_{sl}A_l}{U_{sl}A_l + \rho_l V_l c_{P,l} s}\bar{\bar{\hat{T}}}_s, \tag{3.353}$$

whereas Eq. (3.352) unfolds

$$\bar{\bar{\hat{T}}}_b = \frac{U_{lb}A_b}{U_{lb}A_b + \rho_b V_b c_{P,b} s}\bar{\bar{\hat{T}}}_l \tag{3.354}$$

(continued)

Problem 3.9 (continued)

after solving for $\overline{\overline{T}}_b$. Division of both numerator and denominator by $\rho_l V_l c_{P,l}$ converts Eq. (3.353) to

$$\overline{\overline{T}}_l = \frac{\frac{U_{sl}A_l}{\rho_l V_l c_{P,l}}}{\frac{U_{sl}A_l}{\rho_l V_l c_{P,l}} + s} \overline{\overline{T}}_s, \tag{3.355}$$

and Eq. (3.354) likewise generates

$$\overline{\overline{T}}_b = \frac{\frac{U_{lb}A_b}{\rho_b V_b c_{P,b}}}{\frac{U_{lb}A_b}{\rho_b V_b c_{P,b}} + s} \overline{\overline{T}}_l \tag{3.356}$$

upon division of both numerator and denominator by $\rho_b V_b c_{P,b}$; insertion of Eq. (3.355) then transforms Eq. (3.356) to

$$\overline{\overline{T}}_b = \frac{\frac{U_{lb}A_b}{\rho_b V_b c_{P,b}}}{\frac{U_{lb}A_b}{\rho_b V_b c_{P,b}} + s} \frac{\frac{U_{sl}A_l}{\rho_l V_l c_{P,l}}}{\frac{U_{sl}A_l}{\rho_l V_l c_{P,l}} + s} \overline{\overline{T}}_s. \tag{3.357}$$

The factors preceding $\overline{\overline{T}}_s$ in Eq. (3.357) can be rewritten as

$$\frac{\frac{U_{lb}A_b}{\rho_b V_b c_{P,b}}}{\frac{U_{lb}A_b}{\rho_b V_b c_{P,b}} + s} \frac{\frac{U_{sl}A_l}{\rho_l V_l c_{P,l}}}{\frac{U_{sl}A_l}{\rho_l V_l c_{P,l}} + s} = \frac{\alpha}{\frac{U_{lb}A_b}{\rho_b V_b c_{P,b}} + s} + \frac{\beta}{\frac{U_{sl}A_l}{\rho_l V_l c_{P,l}} + s} \tag{3.358}$$

after Heaviside's decomposition in partial fractions, where α and β denote constants to be determined; elimination of denominators from both sides leaves

$$\frac{U_{lb}A_b}{\rho_b V_b c_{P,b}} \frac{U_{sl}A_l}{\rho_l V_l c_{P,l}} = \alpha\left(\frac{U_{sl}A_l}{\rho_l V_l c_{P,l}} + s\right) + \beta\left(\frac{U_{lb}A_b}{\rho_b V_b c_{P,b}} + s\right). \tag{3.359}$$

Equation (3.359) should be satisfied, no matter the value taken by s; in particular, it should be valid for $s = -U_{sl}A_l/\rho_l V_l c_{P,l}$, that is,

(continued)

Problem 3.9 (continued)

$$\frac{U_{lb}A_b}{\rho_b V_b c_{P,b}} \frac{U_{sl}A_l}{\rho_l V_l c_{P,l}} = \alpha\left(\frac{U_{sl}A_l}{\rho_l V_l c_{P,l}} - \frac{U_{sl}A_l}{\rho_l V_l c_{P,l}}\right) + \beta\left(\frac{U_{lb}A_b}{\rho_b V_b c_{P,b}} - \frac{U_{sl}A_l}{\rho_l V_l c_{P,l}}\right)$$
$$= \beta\left(\frac{U_{lb}A_b}{\rho_b V_b c_{P,b}} - \frac{U_{sl}A_l}{\rho_l V_l c_{P,l}}\right), \tag{3.360}$$

while setting s equal to $-U_{lb}A_b/\rho_b V_b c_{P,b}$ converts Eq. (3.359) to

$$\frac{U_{lb}A_b}{\rho_b V_b c_{P,b}} \frac{U_{sl}A_l}{\rho_l V_l c_{P,l}} = \alpha\left(\frac{U_{sl}A_l}{\rho_l V_l c_{P,l}} - \frac{U_{lb}A_b}{\rho_b V_b c_{P,b}}\right) + \beta\left(\frac{U_{lb}A_b}{\rho_b V_b c_{P,b}} - \frac{U_{lb}A_b}{\rho_b V_b c_{P,b}}\right)$$
$$= \alpha\left(\frac{U_{sl}A_l}{\rho_l V_l c_{P,l}} - \frac{U_{lb}A_b}{\rho_b V_b c_{P,b}}\right). \tag{3.361}$$

Isolation of β from Eq. (3.360) leads to

$$\beta = \frac{\frac{U_{lb}A_b}{\rho_b V_b c_{P,b}} \frac{U_{sl}A_l}{\rho_l V_l c_{P,l}}}{\frac{U_{lb}A_b}{\rho_b V_b c_{P,b}} - \frac{U_{sl}A_l}{\rho_l V_l c_{P,l}}} = -\frac{U_{sl}A_l U_{lb}A_b}{U_{sl}A_l \rho_b V_b c_{P,b} - U_{lb}A_b \rho_l V_l c_{P,l}}, \tag{3.362}$$

and Eq. (3.361) likewise supports

$$\alpha = \frac{\frac{U_{lb}A_b}{\rho_b V_b c_{P,b}} \frac{U_{sl}A_l}{\rho_l V_l c_{P,l}}}{\frac{U_{sl}A_l}{\rho_l V_l c_{P,l}} - \frac{U_{lb}A_b}{\rho_b V_b c_{P,b}}} = \frac{U_{sl}A_l U_{lb}A_b}{U_{sl}A_l \rho_b V_b c_{P,b} - U_{lb}A_b \rho_l V_l c_{P,l}}, \tag{3.363}$$

both along with reduction in number of fraction levels; combination with Eqs. (3.362) and (3.363) converts Eq. (3.358) to

$$\frac{\frac{U_{lb}A_b}{\rho_b V_b c_{P,b}}}{\frac{U_{lb}A_b}{\rho_b V_b c_{P,b}} + s} \frac{\frac{U_{sl}A_l}{\rho_l V_l c_{P,l}}}{\frac{U_{sl}A_l}{\rho_l V_l c_{P,l}} + s} = \frac{\frac{U_{sl}A_l U_{lb}A_b}{U_{sl}A_l \rho_b V_b c_{P,b} - U_{lb}A_b \rho_l V_l c_{P,l}}}{\frac{U_{lb}A_b}{\rho_b V_b c_{P,b}} + s} - \frac{\frac{U_{sl}A_l U_{lb}A_b}{U_{sl}A_l \rho_b V_b c_{P,b} - U_{lb}A_b \rho_l V_l c_{P,l}}}{\frac{U_{sl}A_l}{\rho_l V_l c_{P,l}} + s}, \tag{3.364}$$

and Eq. (3.357) consequently becomes

$$\bar{\hat{T}}_b = \left(\frac{\frac{U_{sl}A_l U_{lb}A_b}{U_{sl}A_l \rho_b V_b c_{P,b} - U_{lb}A_b \rho_l V_l c_{P,l}}}{\frac{U_{lb}A_b}{\rho_b V_b c_{P,b}} + s} - \frac{\frac{U_{sl}A_l U_{lb}A_b}{U_{sl}A_l \rho_b V_b c_{P,b} - U_{lb}A_b \rho_l V_l c_{P,l}}}{\frac{U_{sl}A_l}{\rho_l V_l c_{P,l}} + s}\right) \bar{\hat{T}}_s \tag{3.365}$$

where the common numerator will be factored out as

(continued)

Problem 3.9 (continued)

$$\overline{\overline{T}}_b = \frac{U_{sl}A_l U_{lb}A_b}{U_{sl}A_l\rho_b V_b c_{P,b} - U_{lb}A_b\rho_l V_l c_{P,l}}\left(\frac{1}{\frac{U_{lb}A_b}{\rho_b V_b c_{P,b}} + s} - \frac{1}{\frac{U_{sl}A_l}{\rho_l V_l c_{P,l}} + s}\right)\widehat{\overline{T}}_s. \tag{3.366}$$

For the suggested step input of steam at condensing temperature, say, $T_{s,in}$, that is,

$$T_s = T_{s,in}H\{t\}, \tag{3.367}$$

the corresponding deviation variable to $T_{s,0}$ will look like

$$\widehat{T}_s = (T_{s,in} - T_{s,0})H\{t\} \tag{3.368}$$

in agreement with Eq. (3.340); its counterpart in Laplace's domain then reads

$$\widehat{\overline{T}}_s = \frac{T_{s,in} - T_{s,0}}{s}, \tag{3.369}$$

so Eq. (3.366) will degenerate to

$$\widehat{\overline{T}}_b = \frac{U_{sl}A_l U_{lb}A_b}{U_{sl}A_l\rho_b V_b c_{P,b} - U_{lb}A_b\rho_l V_l c_{P,l}}\left(\frac{1}{\frac{U_{lb}A_b}{\rho_b V_b c_{P,b}} + s} - \frac{1}{\frac{U_{sl}A_l}{\rho_l V_l c_{P,l}} + s}\right)\frac{T_{s,in} - T_{s,0}}{s}. \tag{3.370}$$

After inverse Laplace's transforms are taken of both sides, Eq. (3.370) gives rise to

$$\widehat{T}_b\{t\} \equiv \mathcal{L}^{-1}\left\{\widehat{\overline{T}}_b\right\} = \frac{U_{sl}A_l U_{lb}A_b(T_{s,in} - T_{s,0})}{U_{sl}A_l\rho_b V_b c_{P,b} - U_{lb}A_b\rho_l V_l c_{P,l}}\mathcal{L}^{-1}\left\{\frac{1}{s}\left(\frac{1}{\frac{U_{lb}A_b}{\rho_b V_b c_{P,b}} + s} - \frac{1}{\frac{U_{sl}A_l}{\rho_l V_l c_{P,l}} + s}\right)\right\} \tag{3.371}$$

in view of the linearity of said operator; the theorem of integration in the time domain justifies

$$\widehat{T}_b\{t\} = \frac{U_{sl}A_l U_{lb}A_b(T_{s,in} - T_{s,0})}{U_{sl}A_l\rho_b V_b c_{P,b} - U_{lb}A_b\rho_l V_l c_{P,l}}\int_0^t \mathcal{L}^{-1}\left\{\frac{1}{\frac{U_{lb}A_b}{\rho_b V_b c_{P,b}} + s} - \frac{1}{\frac{U_{sl}A_l}{\rho_l V_l c_{P,l}} + s}\right\}d\tilde{t}, \tag{3.372}$$

or else

(continued)

Problem 3.9 (continued)

$$\widehat{T}_b\{t\}=\frac{U_{sl}A_lU_{lb}A_b(T_{s,in}-T_{s,0})}{U_{sl}A_l\rho_bV_bc_{P,b}-U_{lb}A_b\rho_lV_lc_{P,l}}\int_0^t\left(\mathcal{L}^{-1}\left\{\frac{1}{\frac{U_{lb}A_b}{\rho_bV_bc_{P,b}}+s}\right\}-\mathcal{L}^{-1}\left\{\frac{1}{\frac{U_{sl}A_l}{\rho_lV_lc_{P,l}}+s}\right\}\right)d\tilde{t} \tag{3.373}$$

again owing to the aforementioned linearity. After splitting the outstanding integral, Eq. (3.373) supports

$$\widehat{T}_b\{t\}=\frac{U_{sl}A_lU_{lb}A_b(T_{s,in}-T_{s,0})}{U_{sl}A_l\rho_bV_bc_{P,b}-U_{lb}A_b\rho_lV_lc_{P,l}}\left(\int_0^t\mathcal{L}^{-1}\left\{\frac{1}{\frac{U_{lb}A_b}{\rho_bV_bc_{P,b}}+s}\right\}d\tilde{t}-\int_0^t\mathcal{L}^{-1}\left\{\frac{1}{\frac{U_{sl}A_l}{\rho_lV_lc_{P,l}}+s}\right\}d\tilde{t}\right); \tag{3.374}$$

the translation theorem in Laplace's domain will now be invoked to justify

$$\widehat{T}_b\{t\}=\frac{U_{sl}A_lU_{lb}A_b(T_{s,in}-T_{s,0})}{U_{sl}A_l\rho_bV_bc_{P,b}-U_{lb}A_b\rho_lV_lc_{P,l}}\times\left(\int_0^t\exp\left\{-\frac{U_{lb}A_b}{\rho_bV_bc_{P,b}}\tilde{t}\right\}\mathcal{L}^{-1}\left\{\frac{1}{s}\right\}d\tilde{t}-\int_0^t\exp\left\{-\frac{U_{sl}A_l}{\rho_lV_lc_{P,l}}\tilde{t}\right\}\mathcal{L}^{-1}\left\{\frac{1}{s}\right\}d\tilde{t}\right), \tag{3.375}$$

which is equivalent to

$$\widehat{T}_b\{t\}=\frac{U_{sl}A_lU_{lb}A_b(T_{s,in}-T_{s,0})}{U_{sl}A_l\rho_bV_bc_{P,b}-U_{lb}A_b\rho_lV_lc_{P,l}}\times\left(\int_0^t\exp\left\{-\frac{U_{lb}A_b}{\rho_bV_bc_{P,b}}\tilde{t}\right\}H\{\tilde{t}\}d\tilde{t}-\int_0^t\exp\left\{-\frac{U_{sl}A_l}{\rho_lV_lc_{P,l}}\tilde{t}\right\}H\{\tilde{t}\}d\tilde{t}\right) \tag{3.376}$$

or else

(continued)

Problem 3.9 (continued)

$$\widehat{T}_b\{t\} = \frac{U_{sl}A_l U_{lb}A_b (T_{s,in} - T_{s,0})}{U_{sl}A_l \rho_b V_b c_{P,b} - U_{lb}A_b \rho_l V_l c_{P,l}} \left(\left. \frac{\exp\left\{ -\frac{U_{lb}A_b}{\rho_b V_b c_{P,b}} \tilde{t} \right\}}{-\frac{U_{lb}A_b}{\rho_b V_b c_{P,b}}} \right|_0^t - \left. \frac{\exp\left\{ -\frac{U_{sl}A_l}{\rho_l V_l c_{P,l}} \tilde{t} \right\}}{-\frac{U_{sl}A_l}{\rho_l V_l c_{P,l}}} \right|_0^t \right) H\{t\} \tag{3.377}$$

as per the fundamental theorem of integral calculus. After replacing by upper and lower limits of integration, Eq. (3.377) becomes

$$\widehat{T}_b\{t\} = \frac{U_{sl}A_l U_{lb}A_b (T_{s,in} - T_{s,0})}{U_{sl}A_l \rho_b V_b c_{P,b} - U_{lb}A_b \rho_l V_l c_{P,l}} \left(\frac{e^0 - \exp\left\{ -\frac{U_{lb}A_b}{\rho_b V_b c_{P,b}} t \right\}}{\frac{U_{lb}A_b}{\rho_b V_b c_{P,b}}} - \frac{e^0 - \exp\left\{ -\frac{U_{sl}A_l}{\rho_l V_l c_{P,l}} t \right\}}{\frac{U_{sl}A_l}{\rho_l V_l c_{P,l}}} \right) H\{t\}, \tag{3.378}$$

where reduction in number of fraction levels leaves

$$\widehat{T}_b\{t\} = \frac{U_{sl}A_l U_{lb}A_b (T_{s,in} - T_{s,0})}{U_{sl}A_l \rho_b V_b c_{P,b} - U_{lb}A_b \rho_l V_l c_{P,l}} \left(\begin{array}{c} \frac{\rho_b V_b c_{P,b}}{U_{lb}A_b} \left(1 - \exp\left\{ -\frac{U_{lb}A_b}{\rho_b V_b c_{P,b}} t \right\} \right) \\ -\frac{\rho_l V_l c_{P,l}}{U_{sl}A_l} \left(1 - \exp\left\{ -\frac{U_{sl}A_l}{\rho_l V_l c_{P,l}} t \right\} \right) \end{array} \right) H\{t\}; \tag{3.379}$$

factoring in of $U_{sl}A_l U_{lb}A_b$ converts Eq. (3.379) to

$$\widehat{T}_b\{t\} = \frac{T_{s,in} - T_{s,0}}{U_{sl}A_l \rho_b V_b c_{P,b} - U_{lb}A_b \rho_l V_l c_{P,l}} \left(\begin{array}{c} U_{sl}A_l \rho_b V_b c_{P,b} \left(1 - \exp\left\{ -\frac{U_{lb}A_b}{\rho_b V_b c_{P,b}} t \right\} \right) \\ -U_{lb}A_b \rho_l V_l c_{P,l} \left(1 - \exp\left\{ -\frac{U_{sl}A_l}{\rho_l V_l c_{P,l}} t \right\} \right) \end{array} \right) H\{t\}, \tag{3.380}$$

where elimination of inner parentheses unfolds

(continued)

Problem 3.9 (continued)

$$\widehat{T}_b\{t\}=\frac{T_{s,in}-T_{s,0}}{U_{sl}A_l\rho_b V_b c_{P,b}-U_{lb}A_b\rho_l V_l c_{P,l}}\begin{pmatrix} U_{sl}A_l\rho_b V_b c_{P,b}-U_{sl}A_l\rho_b V_b c_{P,b}\exp\left\{-\frac{U_{lb}A_b}{\rho_b V_b c_{P,b}}t\right\} \\ -U_{lb}A_b\rho_l V_l c_{P,l}+U_{lb}A_b\rho_l V_l c_{P,l}\exp\left\{-\frac{U_{sl}A_l}{\rho_l V_l c_{P,l}}t\right\}\end{pmatrix}H\{t\}. \tag{3.381}$$

After having terms reordered, Eq. (3.381) gives rise to

$$\widehat{T}_b\{t\}=\frac{T_{s,in}-T_{s,0}}{U_{sl}A_l\rho_b V_b c_{P,b}-U_{lb}A_b\rho_l V_l c_{P,l}} \times\begin{pmatrix} U_{sl}A_l\rho_b V_b c_{P,b}-U_{lb}A_b\rho_l V_l c_{P,l}-U_{sl}A_l\rho_b V_b c_{P,b}\exp\left\{-\frac{U_{lb}A_b}{\rho_b V_b c_{P,b}}t\right\} \\ +U_{lb}A_b\rho_l V_l c_{P,l}\exp\left\{-\frac{U_{sl}A_l}{\rho_l V_l c_{P,l}}t\right\}\end{pmatrix}H\{t\}, \tag{3.382}$$

so factoring in of $U_{sl}A_l\rho_b c_{P,b}-U_{lb}A_b\rho_l c_{P,l}$ in denominator is in order, namely,

$$\widehat{T}_b\{t\}=(T_{s,in}-T_{s,0}) \times\left(1-\frac{U_{sl}A_l\rho_b V_b c_{P,b}\exp\left\{-\frac{U_{lb}A_b}{\rho_b V_b c_{P,b}}t\right\}-U_{lb}A_b\rho_l V_l c_{P,l}\exp\left\{-\frac{U_{sl}A_l}{\rho_l V_l c_{P,l}}t\right\}}{U_{sl}A_l\rho_b V_b c_{P,b}-U_{lb}A_b\rho_l V_l c_{P,l}}\right)H\{t\}; \tag{3.383}$$

Equation (3.383) degenerates to

$$\frac{T_b-T_{b,0}}{T_{s,in}-T_{s,0}}=1-\frac{U_{sl}A_l\rho_b V_b c_{P,b}\exp\left\{-\frac{U_{lb}A_b}{\rho_b V_b c_{P,b}}t\right\}-U_{lb}A_b\rho_l V_l c_{P,l}\exp\left\{-\frac{U_{sl}A_l}{\rho_l V_l c_{P,l}}t\right\}}{U_{sl}A_l\rho_b V_b c_{P,b}-U_{lb}A_b\rho_l V_l c_{P,l}}, \tag{3.384}$$

after dividing both sides by $T_{s,in}-T_{s,0}$ and recalling Eq. (3.342). For the sake of graphical interpretation, it is convenient to divide both numerator and denominator in the right-hand side of Eq. (3.384) by $U_{sl}A_l\rho_b V_b c_{P,b}$ to obtain

(continued)

Problem 3.9 (continued)

$$\frac{T_b - T_{b,0}}{T_{s,in} - T_{s,0}} = 1 - \frac{\exp\left\{-\frac{U_{lb}A_b}{\rho_b V_b c_{P,b}}t\right\} - \frac{U_{lb}A_b\rho_l V_l c_{P,l}}{U_{sl}A_l\rho_b V_b c_{P,b}}\exp\left\{-\frac{U_{sl}A_l}{\rho_l V_l c_{P,l}}t\right\}}{1 - \frac{U_{lb}A_b\rho_l V_l c_{P,l}}{U_{sl}A_l\rho_b V_b c_{P,b}}}, \tag{3.385}$$

while multiplying and dividing the argument of the second exponential function by $U_{lb}A_b/\rho_b V_b c_{P,b}$ produce

$$\frac{T_b - T_{b,0}}{T_{s,in} - T_{s,0}} = 1 - \frac{\exp\left\{-\frac{U_{lb}A_b}{\rho_b V_b c_{P,b}}t\right\} - \frac{U_{lb}A_b\rho_l V_l c_{P,l}}{U_{sl}A_l\rho_b V_b c_{P,b}}\exp\left\{-\frac{U_{sl}A_l}{\rho_l V_l c_{P,l}\frac{U_{lb}A_b}{\rho_b V_b c_{P,b}}}\frac{U_{lb}A_b}{\rho_b V_b c_{P,b}}t\right\}}{1 - \frac{U_{lb}A_b\rho_l V_l c_{P,l}}{U_{sl}A_l\rho_b V_b c_{P,b}}}; \tag{3.386}$$

a final reduction in number of fraction levels leaves

$$\frac{T_b - T_{b,0}}{T_{s,in} - T_{s,0}} = 1 - \frac{\exp\left\{-\frac{U_{lb}A_b}{\rho_b V_b c_{P,b}}t\right\} - \frac{U_{lb}A_b\rho_l V_l c_{P,l}}{U_{sl}A_l\rho_b V_b c_{P,b}}\exp\left\{-\frac{U_{sl}A_l\rho_b V_b c_{P,b}}{U_{lb}A_b\rho_l V_l c_{P,l}}\frac{U_{lb}A_b}{\rho_b V_b c_{P,b}}t\right\}}{1 - \frac{U_{lb}A_b\rho_l V_l c_{P,l}}{U_{sl}A_l\rho_b V_b c_{P,b}}} \tag{3.387}$$

or, equivalently,

$$\frac{T_b - T_{b,0}}{T_{s,in} - T_{s,0}} = 1 - \frac{\exp\left\{-\frac{U_{lb}A_b}{\rho_b V_b c_{P,b}}t\right\} - \frac{U_{lb}A_b\rho_l V_l c_{P,l}}{U_{sl}A_l\rho_b V_b c_{P,b}}\exp\left\{-\frac{\frac{U_{lb}A_b}{\rho_b V_b c_{P,b}}t}{\frac{U_{lb}A_b\rho_l V_l c_{P,l}}{U_{sl}A_l\rho_b V_b c_{P,b}}}\right\}}{1 - \frac{U_{lb}A_b\rho_l V_l c_{P,l}}{U_{sl}A_l\rho_b V_b c_{P,b}}} \tag{3.388}$$

after taking the reciprocal of the reciprocal in the argument of the second exponential function. Inspection of the functional form of Eq. (3.388) unfolds (dimensionless) dependent variable $(T_b - T_{b,0})/(T_{s,in} - T_{s,0})$ being a function of (dimensionless) independent variable $U_{lb}A_b t/\rho_b V_b c_{P,b}$, with $U_{lb}A_b\rho_l V_l c_{P,l}/U_{sl}A_l\rho_b V_b c_{P,b}$ appearing as single parameter; the corresponding plot is labeled as Fig. 3.12. As expected, when heat transfer between liquid and solid phases is very fast, the characteristic parameter approaches zero; hence, a first-order behavior is found, characterized by a rectangular hyperbola. Conversely, a higher thermal resistance to transfer between said phases relative to that between steam and liquid phases (i.e., $1/U_{sl} > 1/U_{lb}$), described by a larger $U_{lb}A_b\rho_l V_l c_{P,l}/U_{sl}A_l\rho_b V_b c_{P,b}$, produces a second-order behavior, with a stronger and stronger sigmoidicity. Similar patterns are observed when $A_l \gg A_b$, $\rho_b \gg \rho_l$,

(continued)

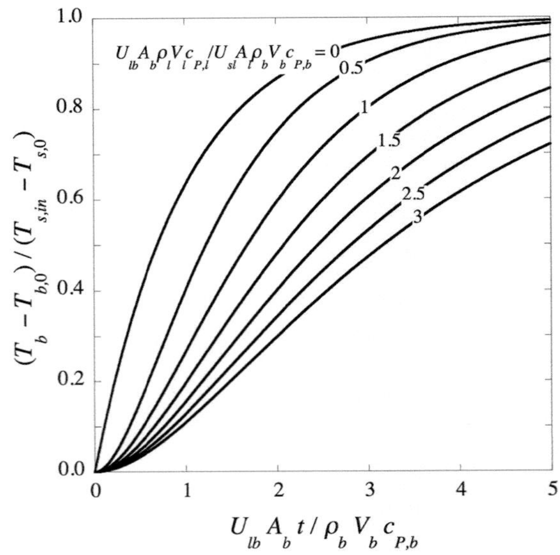

Fig. 3.12 Evolution of normalized temperature of beads, $(T_b - T_{b,0})/(T_{s,in} - T_{s,0})$, versus dimensionless time, $U_{lb}A_b t/\rho_b V_b c_{P,b}$, in fluidized bed

Problem 3.9 (continued)

$V_b \gg V_l$, or $c_{P,b} \gg c_{P,l}$, and $A_l < A_b$, $\rho_b < \rho_l$, $V_b < V_l$, or $c_{P,b} < c_{P,l}$, for first- and second-order behavior, respectively; remember that $U_{lb}A_b$ and $U_{sl}A_l$ are (kinetic) measures of ease to transfer heat, whereas $\rho_l V_l c_{P,l}$ and $\rho_b V_b c_{P,b}$ are (thermodynamic) measures of capacity to store (sensible) heat.

3.4.1 Thermal Processing

Continuous sterilization, especially when high-temperature, short-exposure time processes are employed, can significantly reduce thermal damage to the medium when compared to its batch counterpart, for a given degree of destruction of (contaminant) cells. Furthermore, the continuous process is known to save 20–25% steam in current industrial practice, and the time required for heating up and cooling down is considerably reduced.

A typical equipment layout for continuous sterilization is depicted in Fig. 3.13, pertaining to heat exchanger-based processing. Raw medium entering the system is first preheated by hot, sterile medium through a heat exchanger to reduce subsequent steam requirement for heating, while simultaneously cools the sterile medium; it is then brought to sterilization temperature by fresh steam in a second heat exchanger. The sterilization temperature is held in the holding tube, a well-insulated device that assures adiabatic conditions; the sterile medium is then cooled by heat exchange with incoming cool medium, before entering the bioreactor. One disadvantage is that two heat exchanger

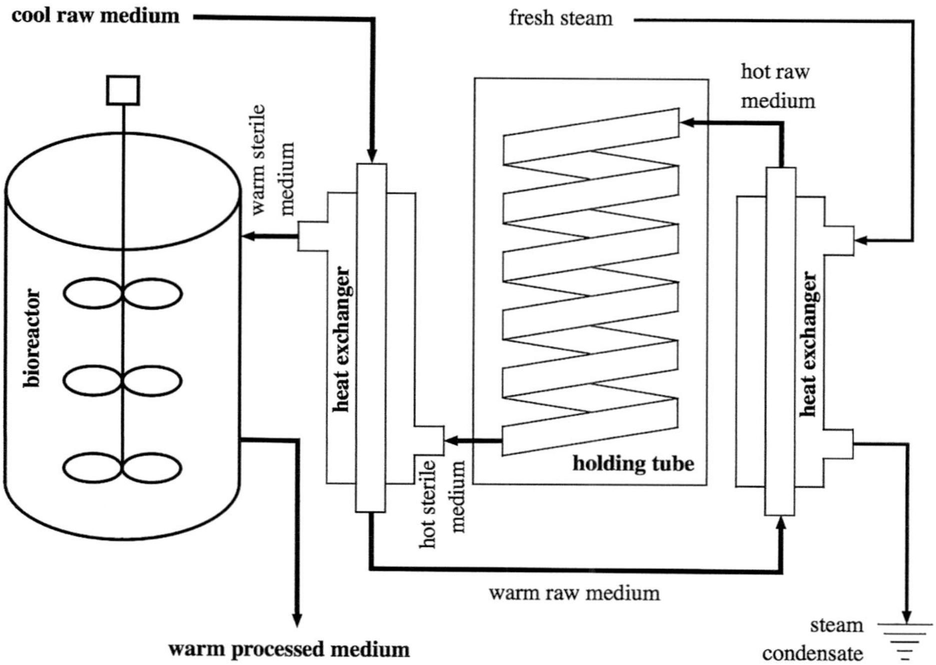

Fig. 3.13 Schematic representation of system for continuous sterilization of liquid medium in bioreactor, with heat indirectly supplied by steam via a heat exchanger

units are required; susceptibility to fouling of their inner surfaces may indeed reduce efficiency of heat transfer between cleanings.

A more efficient alternative resorts to direct injection of steam, as illustrated in Fig. 3.14. Raw medium enters the system and is first preheated by hot, sterile medium in a heat exchanger, so as to reduce the enthalpic need for heating while cooling the sterile medium to the proper mild temperature intended inside the bioreactor. Steam is then injected directly into the warm medium as it flows through a pipe; the temperature of the medium then rises suddenly to the intended sterilization temperature. The time of exposure to this target temperature depends on the space time associated with the holding tube; the sterile medium is then cooled down very fast by flash cooling, achieved by passing said medium through an expansion valve into a vacuum chamber, with residual (warm) steam recovered and disposed off as condensate. Further cooling of the medium takes place in the heat exchanger, with lost heat being employed to preheat the incoming medium; the warm sterile medium is finally ready for utilization in the bioreactor.

An injection device is less expensive to build and operate than a heat exchanger-based system; furthermore, the transient associated with heating up is much shorter, and so is the transient arising from flash expansion afterward. Therefore, lethality becomes correlated with essentially the time spent in the holding tube, thus facilitating operation design and

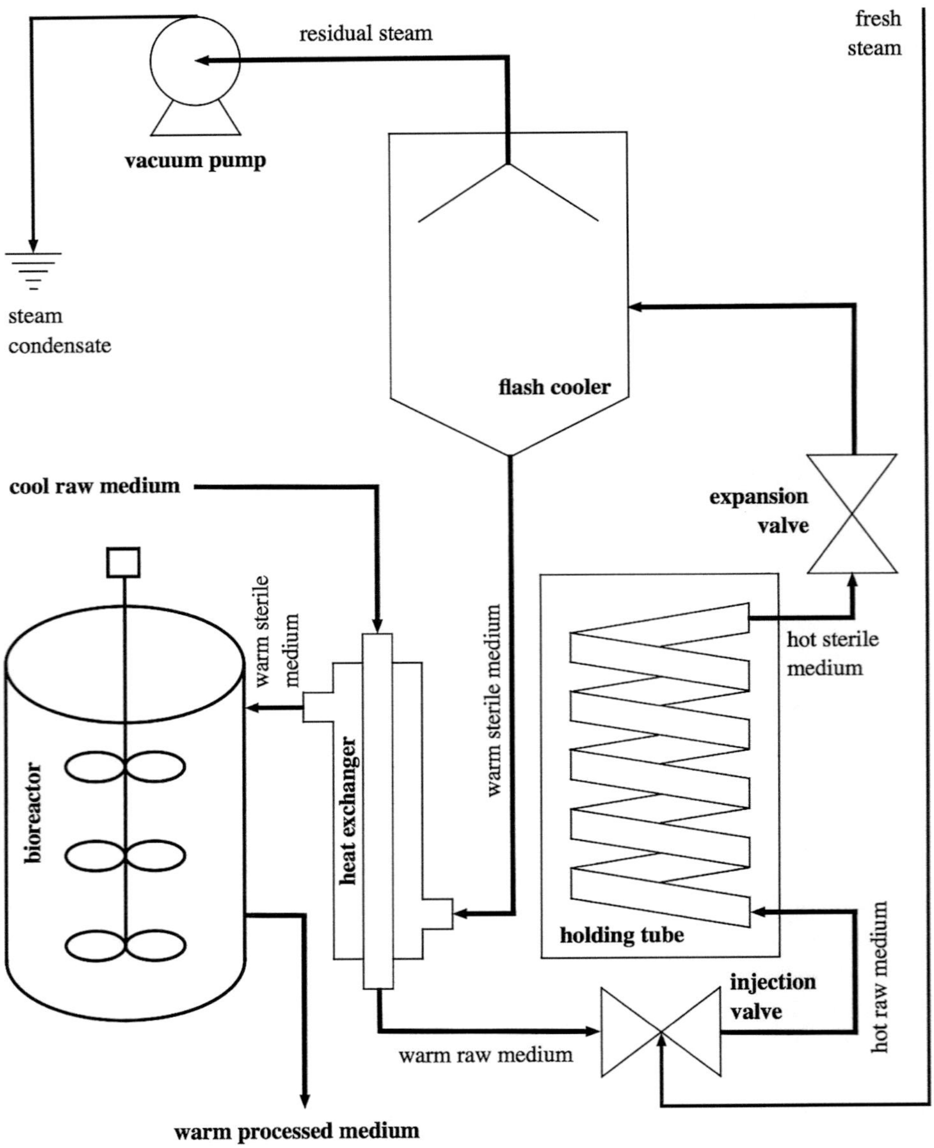

Fig. 3.14 Schematic representation of system for continuous sterilization of liquid medium in bioreactor, with heat directly supplied by steam via sparging therein

control. Conversely, the dilution of the medium as a result of steam condensation makes it necessary to eliminate afterward the associated condensate by flash cooling—inherently more expensive than plain heat transfer, due to the cost of vacuum pumping; in addition, efficiency is unfavorably affected by any foam developed during the process. On the other hand, the very fast rate of heating may cause problems with particulate media, because temperature in the core of such particles will not have the time to equilibrate with temperature prevailing in the bulk liquid itself. Therefore, the shorter contact times when steam is directly injected raise the risk that particles will not be heated thoroughly and for a sufficient time, so they might fail full sterilization; this means that clear media are preferred for this strategy of sterilization.

An important variable affecting performance of continuous sterilizers is the prevailing nature of mixing, especially within the holding tube; ideally, all fluid elements entering the system should spend the same time at the target sterilization temperature; otherwise, lethality conditions cannot be assured in all fluid elements. For instance, if fluid nearer the entrance of the holding tube mixes with fluid ahead, a risk arises of contamination transfer to the outlet of the sterilizer. Although plug flow should describe the hydrodynamic pattern within the holding tube, a continuous distribution of residence times characterized by a finite variance is actually the rule; this would be simulated with great accuracy by the model based on a finite number of equal stirred flow tanks in series. Turbulent flow, characterized by Reynolds' number above ca. 10^3, leads to a velocity profile much closer to plug flow than the laminar regime; however, axial dispersion ought to be taken into account for a conservative simulation, since the larger such dispersion, the poorer the performance of the sterilizer.

In view of the above discussion, consider a cascade of N ideal stirred tanks, sharing the same space time τ_i; as usual, microorganisms are hypothesized to undergo first-order thermal death, characterized by unit stoichiometry. If the volumetric flow rate of either inlet stream to or outlet stream from the i-th stirred tank of volume V_i is denoted by Q, then a steady-state mass balance to biomass will read

$$QC_{Y,i-1} - V_i k_d C_{Y,i} = QC_{Y,i}, \tag{3.389}$$

where $C_{Y,i\text{-}1}$ and $C_{Y,i}$ denote (viable) concentration of contaminant microorganisms, at inlet and outlet streams, respectively. After dividing both sides by Q, Eq. (3.389) reduces to

$$C_{Y,i-1} - k_d \tau_i C_{Y,i} = C_{Y,i}, \tag{3.390}$$

where space time of i-th unit, τ_i, obeys

$$\tau_i \equiv \frac{V_i}{Q}. \tag{3.391}$$

Isolation of $C_{Y,i}$ in Eq. (3.390) unfolds

$$C_{Y,i} = \frac{C_{Y,i-1}}{1 + k_d \tau_i}, \tag{3.392}$$

or else

$$\frac{C_{Y,i}}{C_{Y,i-1}} = \frac{1}{1 + k_d \tau_i} \quad ; \quad i = 1, 2, \ldots, N \tag{3.393}$$

upon division of both sides by $C_{Y,i\text{-}1}$. The ratio of $C_{Y,N}$ to $C_{Y,0}$ can be expressed as a telescopic (sequential) product, namely,

$$\frac{C_{Y,N}}{C_{Y,0}} = \frac{C_{Y,N}}{C_{Y,N-1}} \frac{C_{Y,N-1}}{C_{Y,N-2}} \cdots \frac{C_{Y,2}}{C_{Y,1}} \frac{C_{Y,1}}{C_{Y,0}}; \tag{3.394}$$

hence, one may insert Eq. (3.393) pertaining to each value of i to obtain

$$\frac{C_{Y,N}}{C_{Y,0}} = \frac{1}{1 + k_d \tau_N} \frac{1}{1 + k_d \tau_{N-1}} \cdots \frac{1}{1 + k_d \tau_2} \frac{1}{1 + k_d \tau_1}, \tag{3.395}$$

where a condensed notation is possible as

$$\frac{C_{Y,N}}{C_{Y,0}} = \prod_{i=1}^{N} \frac{1}{1 + k_d \tau_i}. \tag{3.396}$$

The N units in the cascade are identical to each other by hypothesis, that is,

$$\tau_i \equiv \frac{\tau}{N} \quad ; \quad i = 1, 2, \ldots, N \tag{3.397}$$

where τ denotes space time of overall cascade, defined, in turn, by

$$\tau \equiv \frac{V}{Q} \tag{3.398}$$

with overall volume, V, given by

$$V = \sum_{i=1}^{N} V_i, \tag{3.399}$$

and V_i denoting volume of i-th tank; insertion of Eq. (3.397) converts Eq. (3.396) to

$$\frac{C_{Y,N}}{C_{Y,0}} = \prod_{i=1}^{N} \frac{1}{1 + k_d \frac{\tau}{N}}, \tag{3.400}$$

where the mathematical definition of power permits further simplification to

$$\frac{C_{Y,N}}{C_{Y,0}} = \left(\frac{1}{1 + k_d \frac{\tau}{N}} \right)^N = \frac{1}{\left(1 + \frac{k_d \tau}{N}\right)^N}. \tag{3.401}$$

Inspection of lumped parameter $k_d t$ in Eq. (3.401) unfolds a dimensionless quantity, usually termed Damkohler's number, Da, abiding to

$$Da \equiv k_d \tau; \tag{3.402}$$

if rewritten as

$$Da = \frac{\tau}{\frac{1}{k_d}}, \tag{3.403}$$

Damkohler's number may be viewed as ratio of timescale characteristic of convection through the system, τ, to timescale characteristic of death process, $1/k_d$. In view of Eq. (3.402), one may rephrase Eq. (3.401) to

$$\frac{C_{Y,N}}{C_{Y,0}} = \frac{1}{\left(1 + \frac{Da}{N}\right)^N}; \tag{3.404}$$

Equation (3.404) conveys an estimate for the average lethality of the holding tube, when its hydrodynamic pattern is modelled by a cascade of N identical stirred flow tanks, as illustrated in Fig. 3.15. Parameter N correlates directly to dispersion within the liquid in transit. The lower its value, the higher the dispersion of space times of fluid elements relative to the average space time of fluid; the dispersion is critical toward performance of sterilization. In fact, a lower N causes the $C_{Y,N}/C_{Y,0}$ curve flatten at a smaller Da, so lethality of the thermal treatment may be compromised. Although not self-evident from inspection of the curves in Fig. 3.15 (due to the logarithmic scale used therein), a common threshold will eventually be reached, according to

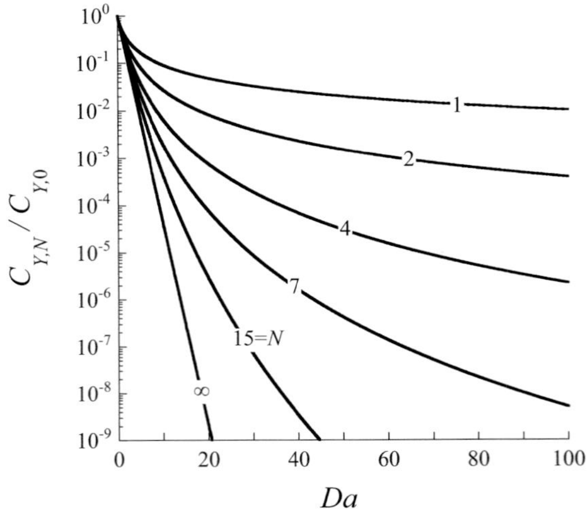

Fig. 3.15 Variation of normalized concentration of viable biomass at outlet of holding tube, $C_{Y,N}/C_{Y,0}$, versus Damkohler's number, Da, for selected values of dimensionless parameter N, under equal-sized stirred flow tank cascade model

$$\lim_{Da\to\infty}\frac{C_{Y.N}}{C_{Y,0}}=\frac{1}{\left(1+\frac{\infty}{N}\right)^N}=\frac{1}{(1+\infty)^N}=\frac{1}{\infty^N}=\frac{1}{\infty}=0 \tag{3.405}$$

stemming from Eq. (3.404); in other words, the horizontal axis serves as unique long-Da, horizontal asymptote.

Lethality is maximum when no dispersion is present, corresponding to $N\to\infty$, and thus equivalent to plug flow; under these circumstances, Eq. (3.404) will be driven by

$$\lim_{N\to\infty}\frac{C_{Y,N}}{C_{Y,0}}=\lim_{N\to\infty}\frac{1}{\left(1+\frac{Da}{N}\right)^{\frac{N}{Da}Da}}=\lim_{N\to\infty}\frac{1}{\left(\left(1+\frac{Da}{N}\right)^{\frac{N}{Da}}\right)^{Da}}=\lim_{\frac{N}{Da}\to\infty}\frac{1}{\left(\left(1+\frac{1}{\frac{N}{Da}}\right)^{\frac{N}{Da}}\right)^{Da}}, \tag{3.406}$$

upon multiplication and division of the exponent by Da, coupled with realization that $N/Da\to\infty$ when $N\to\infty$ for a finite Da; the classical theorems on limits support transformation to

$$\lim_{N\to\infty}\frac{C_{Y,N}}{C_{Y,0}}=\frac{1}{\lim\limits_{\frac{N}{Da}\to\infty}\left(\left(1+\frac{1}{\frac{N}{Da}}\right)^{\frac{N}{Da}}\right)^{Da}}=\frac{1}{\left(\lim\limits_{\frac{N}{Da}\to\infty}\left(1+\frac{1}{\frac{N}{Da}}\right)^{\frac{N}{Da}}\right)^{Da}}, \tag{3.407}$$

where the denominator is but the definition of Neper's number, that is,

$$\lim_{N \to \infty} \frac{C_{Y,N}}{C_{Y,0}} = \frac{1}{(\mathrm{e})^{Da}} = \frac{1}{\mathrm{e}^{Da}} \tag{3.408}$$

or, in simpler notation,

$$\lim_{N \to \infty} \frac{C_{Y,N}}{C_{Y,0}} = \mathrm{e}^{-Da}. \tag{3.409}$$

Equation (3.409) provides an estimate of the maximum degree of lethality attained for a given Da; if such degree is not severe enough, then Da has to be further increased, namely, by extending space time, τ, or enhancing intrinsic death constant k_d, in agreement with Eq. (3.402)—the latter being accessible via operation at a higher temperature (see Eq. (3.20)).

The actual efficiency of sterilization, η_{st}, may accordingly be computed as

$$\eta_{st} \equiv \frac{\frac{C_{Y,0}}{C_{Y,N}}}{\lim\limits_{N \to \infty} \frac{C_{Y,0}}{C_{Y,N}}} = \frac{\lim\limits_{N \to \infty} \frac{C_{Y,N}}{C_{Y,0}}}{\frac{C_{Y,N}}{C_{Y,0}}}; \tag{3.410}$$

insertion of Eqs. (3.404) and (3.409) converts Eq. (3.410) to

$$\eta_{st} = \frac{\mathrm{e}^{-Da}}{\frac{1}{\left(1+\frac{Da}{N}\right)^N}}, \tag{3.411}$$

where straightforward algebraic rearrangement leaves

$$\eta_{st} = \frac{\left(1 + \frac{Da}{N}\right)^N}{\mathrm{e}^{Da}} \tag{3.412}$$

as illustrated in Fig. 3.16. Inspection of this figure indicates that performance of sterilization declines significantly as N decreases, thus reflecting the unfavorable effect of dispersion; a smaller N worsens indeed the degree of lethality relative to what would be expected under true plug flow, for any given space time. Therefore, underprocessing of

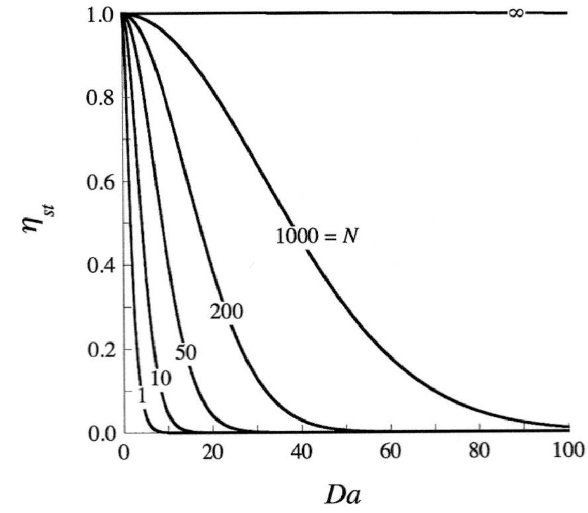

Fig. 3.16 Variation of efficiency of sterilization, η_{st}, versus Damkohler's number, Da, for selected values of dimensionless parameter N, under equal-sized stirred flow tank cascade model

fluid elements remaining in the vessel for a residence time below τ is never balanced by overprocessing of the corresponding fluid elements remaining in the vessel longer than τ; the nonlinear nature of the (integrated) first-order decay conveyed by Eq. (3.26) accounts for this realization, despite a given τ enforcing the former fraction of fluid elements to match the latter as per the intrinsic linearity of the mass balance.

The above model is widely applicable provided that flow in the holding tube is essentially turbulent, since equivalence to a series of well-mixed units was assumed from the very beginning; when mixing is poor, namely because of a higher viscosity of the medium, or velocity is low, the associated Reynolds' number will be low as well, and laminar flow will prevail characterized by (almost) no back mixing. Under these circumstances, the equal-sized stirred tank cascade model fails to convey reasonable predictions, and an alternative method, departing from its unique parabolic velocity profile, is preferred; as derived in due course, said profile looks like

$$v = \alpha\left(1 - \left(\frac{r}{R}\right)^2\right), \tag{3.413}$$

where r denotes radial coordinate, R denotes tube inner radius, and α denotes a proportionality constant. The average velocity, $\overline{v}$, is calculated as

$$\overline{v} \equiv \frac{\int_0^R v\{r\} 2\pi r \, dr}{\int_0^R 2\pi r \, dr}, \tag{3.414}$$

where 2π will be taken off the kernel and then dropped from both numerator and denominator to get

$$\overline{v} = \frac{2\pi \int_0^R v\{r\} r\, dr}{2\pi \int_0^R r\, dr} = \frac{\int_0^R v\{r\} r\, dr}{\int_0^R r\, dr}; \tag{3.415}$$

insertion of Eq. (3.413) then converts Eq. (3.415) to

$$\overline{v} = \frac{\int_0^R \alpha \left(1 - \left(\frac{r}{R}\right)^2\right) r\, dr}{\int_0^R r\, dr}, \tag{3.416}$$

with division of both numerator and denominator twice by R unfolding

$$\overline{v} = \frac{\int_0^1 \alpha \left(1 - \left(\frac{r}{R}\right)^2\right) \frac{r}{R} d\left(\frac{r}{R}\right)}{\int_0^1 \frac{r}{R} d\left(\frac{r}{R}\right)}, \tag{3.417}$$

along with replacement of lower and upper limits according to

$$\left.\frac{r}{R}\right|_{r=0} = 0 \tag{3.418}$$

and

$$\left.\frac{r}{R}\right|_{r=R} = 1, \tag{3.419}$$

respectively. After taking α off the kernel and factoring in r/R, Eq. (3.417) becomes

$$\overline{v} = \frac{\alpha \int_0^1 \left(\frac{r}{R} - \left(\frac{r}{R} \right)^3 \right) d\left(\frac{r}{R} \right)}{\int_0^1 \frac{r}{R} d\left(\frac{r}{R} \right)}, \tag{3.420}$$

where application of the fundamental theorem of integral calculus is in order, namely,

$$\overline{v} = \alpha \frac{\left(\frac{\left(\frac{r}{R} \right)^2}{2} - \frac{\left(\frac{r}{R} \right)^4}{4} \right) \Bigg|_0^1}{\frac{\left(\frac{r}{R} \right)^2}{2} \Bigg|_0^1}; \tag{3.421}$$

following replacement by upper and lower limits of integration, Eq. (3.421) yields

$$\overline{v} = \alpha \frac{\left(\frac{1^2}{2} - \frac{1^4}{4} \right) - \left(\frac{0^2}{2} - \frac{0^4}{4} \right)}{\frac{1^2}{2} - \frac{0^2}{2}} = \alpha \frac{\frac{1}{2} - \frac{1}{4}}{\frac{1}{2}} = \alpha \frac{\frac{1}{4}}{\frac{1}{2}} = \frac{1}{2} \alpha \tag{3.422}$$

along with straightforward algebraic manipulation, which may be used to rewrite Eq. (3.413) as

$$v = 2\overline{v} \left(1 - \left(\frac{r}{R} \right)^2 \right) \tag{3.423}$$

since $\alpha = 2\overline{v}$ after Eq. (3.422). No mixing whatsoever exists in true laminar flow, so one should consider the first-order process of microbial death as carried out in each annular element of fluid, of thickness dr and length L, as an independent plug flow reactor, according to

$$\frac{dC_Y}{d\tau} = -k_d C_Y, \tag{3.424}$$

which mimics Eq. (3.18) and reflects equivalence between a batch reactor by time t and a plug flow reactor of space time $\tau \equiv t$; the corresponding boundary condition reads

$$C_Y|_{\tau=0} = C_{Y,0}, \tag{3.425}$$

an alias to Eq. (3.19) once $t = 0$ for initial state has accordingly been replaced by $\tau = 0$ for inlet stream. Integration of Eq. (3.424) ensues, via separation of variables, as

$$\int_{C_{Y,0}}^{C_Y} \frac{d\widetilde{C}_Y}{\widetilde{C}_Y} = -k_d \int_0^{\tau} d\widetilde{\tau} \tag{3.426}$$

with the aid of Eq. (3.425) and the constancy of k_d, thus promptly leading to

$$\ln \widetilde{C}_Y \Big|_{C_{Y,0}}^{C_Y} = -k_d\, \widetilde{\tau}\Big|_0^{\tau}; \tag{3.427}$$

Equation (3.427) is equivalent to just

$$\ln \frac{C_Y}{C_{Y,0}} = -k_d \tau. \tag{3.428}$$

After taking exponentials of both sides, Eq. (3.428) gives rise to

$$\frac{C_Y}{C_{Y,0}} = \mathrm{e}^{-k_d \tau}; \tag{3.429}$$

Equation (3.429) applies to every space time, defined hereafter more conveniently as

$$\tau\{r\} \equiv \frac{L}{v\{r\}} \tag{3.430}$$

in lieu of Eq. (3.398), so Eq. (3.429) will be rephrased as

$$\frac{C_Y\{r\}}{C_{Y,0}} = \mathrm{e}^{-\frac{k_d L}{v\{r\}}}. \tag{3.431}$$

The (average) concentration of viable microorganisms, at outlet of the holding tube, can thus be expressed as

$$C_{Y,out} \equiv \frac{\int\limits_0^R C_Y\{r\}v\{r\}2\pi r\,dr}{\int\limits_0^R v\{r\}2\pi r\,dr}, \tag{3.432}$$

where Eq. (3.414) may be retrieved to write

$$C_{Y,out} = \frac{\int\limits_0^R 2\pi C_Y\{r\}v\{r\}\,r\,dr}{\overline{v}\int\limits_0^R 2\pi r\,dr}; \tag{3.433}$$

upon taking 2π or π off the kernel in numerator or denominator, respectively, and dropping π afterward, Eq. (3.433) simplifies to

$$C_{Y,out} = \frac{2\pi\int\limits_0^R C_Y\{r\}v\{r\}r\,dr}{\pi\overline{v}\int\limits_0^R 2r\,dr} = \frac{2\int\limits_0^R C_Y\{r\}v\{r\}r\,dr}{\overline{v}\int\limits_0^R 2r\,dr}. \tag{3.434}$$

The fundamental theorem of integral calculus may be invoked to redo the denominator of Eq. (3.434) as

$$C_{Y,out} = \frac{2\int\limits_0^R C_Y\{r\}v\{r\}r\,dr}{\overline{v}\,r^2\Big|_0^R} \tag{3.435}$$

or, equivalently,

$$C_{Y,out} = \frac{2\int\limits_0^R C_Y\{r\}v\{r\}r\,dr}{\overline{v}R^2}; \tag{3.436}$$

following combination with Eq. (3.431), one obtains

$$C_{Y,out} = \frac{2}{\overline{v}} \frac{\int_0^R C_{Y,0} e^{-\frac{k_d L}{v\{r\}}} v\{r\}\, r\, dr}{R^2}. \tag{3.437}$$

Division of both numerator and denominator in the right-hand side of Eq. (3.437) by R^2, while taking $C_{Y,0}$ off the kernel gives then rise to

$$C_{Y,out} = \frac{2}{\overline{v}} C_{Y,0} \int_0^1 e^{-\frac{k_d L}{v\left\{\frac{r}{R}\right\}}} v\left\{\frac{r}{R}\right\} \frac{r}{R} d\left(\frac{r}{R}\right), \tag{3.438}$$

where integration limits have again been updated with the aid of Eqs. (3.418) and (3.419); division of both sides by $C_{Y,0}$ supports conversion to

$$\frac{C_{Y,out}}{C_{Y,0}} = \frac{2}{\overline{v}} \int_0^1 e^{-\frac{k_d L}{v\left\{\frac{r}{R}\right\}}} v\left\{\frac{r}{R}\right\} \frac{r}{R} d\left(\frac{r}{R}\right). \tag{3.439}$$

On the other hand, differentiation of Eq. (3.423) unfolds

$$dv = 2\overline{v}(-1)2\frac{r}{R} d\left(\frac{r}{R}\right) = -4\overline{v}\frac{r}{R} d\left(\frac{r}{R}\right), \tag{3.440}$$

which can be redone as

$$\frac{r}{R} d\left(\frac{r}{R}\right) = -\frac{dv}{4\overline{v}}; \tag{3.441}$$

by the same token,

$$v\big|_{\frac{r}{R}=0} = 2\overline{v} \tag{3.442}$$

and

$$v\big|_{\frac{r}{R}=1} = 0, \tag{3.443}$$

stem from Eq. (3.423). Insertion of Eqs. (3.441)–(3.443) transforms Eq. (3.439) to

$$\frac{C_{Y,out}}{C_{Y,0}} = \frac{2}{\overline{v}} \int_{2\overline{v}}^{0} e^{-\frac{k_d L}{v}} v \left(-\frac{dv}{4\overline{v}} \right), \tag{3.444}$$

or else

$$\frac{C_{Y,out}}{C_{Y,0}} = -\frac{1}{2\overline{v}^2} \int_{2\overline{v}}^{0} e^{-\frac{k_d L}{v}} v \, dv \tag{3.445}$$

after taking constant factors off the kernel; reversal of the integration limits is convenient, that is,

$$\frac{C_{Y,out}}{C_{Y,0}} = \frac{1}{2\overline{v}^2} \int_{0}^{2\overline{v}} e^{-\frac{k_d L}{v}} v \, dv \tag{3.446}$$

at the expense of the preceding minus sign.

In attempts to calculate the integral in Eq. (3.446), one may to advantage move $\overline{v}^2$ into the kernel in split form as

$$\frac{C_{Y,out}}{C_{Y,0}} = 2 \int_{0}^{1} e^{-\frac{k_d L}{v}} \frac{v}{2\overline{v}} d\left(\frac{v}{2\overline{v}} \right), \tag{3.447}$$

where the constancy thereof was taken advantage of to place $2\overline{v}$ under the differential sign and concomitantly update integration limits; a further multiplication and division of the argument of the exponential function by $2\overline{v}$ produces

$$\frac{C_{Y,out}}{C_{Y,0}} = 2 \int_{0}^{1} e^{-\frac{1}{2}\frac{k_d L}{\overline{v}}\frac{2\overline{v}}{v}} \frac{v}{2\overline{v}} d\left(\frac{v}{2\overline{v}} \right), \tag{3.448}$$

where Eqs. (3.402), revisited as

$$Da \equiv k_d \frac{L}{\overline{v}} = \frac{k_d L}{\overline{v}}, \tag{3.449}$$

can be utilized to rewrite Eq. (3.448) as

$$\frac{C_{Y,out}}{C_{Y,0}} = 2\int_0^1 e^{-\frac{1}{2}Da\frac{2\bar{v}}{v}} \frac{v}{2\bar{v}} d\left(\frac{v}{2\bar{v}}\right). \tag{3.450}$$

The next step is defining an auxiliary variable ζ, according to

$$\zeta \equiv \frac{2\bar{v}}{v}, \tag{3.451}$$

or else

$$\frac{v}{2\bar{v}} = \frac{1}{\zeta} \tag{3.452}$$

after taking reciprocals of both sides; differentiation of both sides is now in order, namely,

$$d\left(\frac{v}{2\bar{v}}\right) = -\frac{d\zeta}{\zeta^2}, \tag{3.453}$$

while its asymptotic values look like

$$\zeta\Big|_{\frac{v}{2\bar{v}}=0} = \infty \tag{3.454}$$

and

$$\zeta_{\big|\frac{v}{2\bar{v}}=1} = 1, \tag{3.455}$$

stemming from Eq. (3.451). In view of Eqs. (3.451)–(3.455), it is possible to reformulate Eq. (3.450) to

$$\frac{C_{Y,out}}{C_{Y,0}} = 2\int_\infty^1 e^{-\frac{1}{2}Da\,\zeta} \frac{1}{\zeta}\left(-\frac{d\zeta}{\zeta^2}\right) = -2\int_\infty^1 e^{-\frac{1}{2}Da\,\zeta} \frac{d\zeta}{\zeta^3}, \tag{3.456}$$

along with condensation of factors alike and removal of -2 from the kernel; the preceding minus can now be taken off on account of exchanging the integration limits, that is,

$$\frac{C_{Y,out}}{C_{Y,0}} = 2\int_1^\infty \frac{e^{-\frac{Da}{2}\zeta}}{\zeta^3}\, d\zeta. \tag{3.457}$$

The exact solution to Eq. (3.457) reads

$$\frac{C_{Y,out}}{C_{Y,0}} = 2E_3\left\{\frac{Da}{2}\right\}, \tag{3.458}$$

where E_3 denotes exponential-integral function of third order; in general,

$$E_N\{a\} \equiv \int_1^\infty \frac{e^{-a\xi}}{\xi^N} d\xi \tag{3.459}$$

represents its N-th order counterpart. It is normally easier to work with $E_1\{a\}$, so one may take advantage of the general recurrence relationship

$$E_{N+1}\{a\} = \frac{1}{N}\left(e^{-a} - aE_N\{a\}\right); \tag{3.460}$$

in particular,

$$E_3\{a\} = \frac{1}{2}\left(e^{-a} - aE_2\{a\}\right) \tag{3.461}$$

holds for $N = 2$, where a second application of Eq. (3.460) unfolds

$$E_2\{a\} = \frac{1}{1}\left(e^{-a} - aE_1\{a\}\right) = e^{-a} - aE_1\{a\}. \tag{3.462}$$

Insertion of Eq. (3.462) converts Eq. (3.461) to

$$E_3\{a\} = \frac{1}{2}\left(e^{-a} - a\left(e^{-a} - aE_1\{a\}\right)\right) = \frac{1}{2}\left(e^{-a} - ae^{-a} + a^2E_1\{a\}\right) \tag{3.463}$$

together with elimination of inner parenthesis, or else

$$E_3\{a\} = \frac{1}{2}\left((1-a)e^{-a} + a^2E_1\{a\}\right) \tag{3.464}$$

once e^{-a} is factored out; a series expansion exists for $E_1\{a\}$, that is,

$$E_1\{a\} = -\gamma - \ln a - \sum_{i=1}^{\infty} \frac{(-1)^i a^i}{i \cdot i!}, \tag{3.465}$$

with γ denoting Euler and Mascheroni's constant, an irrational number defined by

$$\gamma \equiv \lim_{M \to \infty} \left(\sum_{k=1}^{M} \frac{1}{k} - \ln M \right), \tag{3.466}$$

and given in practice by

$$\gamma = 0.577216\ldots. \tag{3.467}$$

A final combination of Eqs. (3.464) and (3.465) unfolds

$$E_3\{a\} = \frac{1}{2} \left((1 - a) e^{-a} - a^2 \left(\gamma + \ln a + \sum_{i=1}^{\infty} \frac{(-1)^i a^i}{i \cdot i!} \right) \right); \tag{3.468}$$

in the present case, $a = Da/2$ as per comparative inspection of Eqs. (3.457)–(3.459), so Eq. (3.468) will take the form

$$E_3\left\{\frac{Da}{2}\right\} = \frac{1}{2} \left(\left(1 - \frac{Da}{2}\right) e^{-\frac{Da}{2}} - \left(\frac{Da}{2}\right)^2 \left(\gamma + \ln \frac{Da}{2} + \sum_{i=1}^{\infty} \frac{(-1)^i}{i \cdot i!} \left(\frac{Da}{2}\right)^i \right) \right), \tag{3.469}$$

which may be plugged in Eq. (3.458) to finally reach

$$\frac{C_{Y,out}}{C_{Y,0}} = \left(1 - \frac{Da}{2}\right) e^{-\frac{Da}{2}} - \left(\frac{Da}{2}\right)^2 \left(\gamma + \ln \frac{Da}{2} + \sum_{i=1}^{\infty} \frac{(-1)^i}{i \cdot i!} \left(\frac{Da}{2}\right)^i \right). \tag{3.470}$$

Equation (3.470) is illustrated in Fig. 3.17. The almost linear pattern observed in this figure, laid on a log/lin scale, resembles the curve labeled as $N = \infty$ in Fig. 3.15 that corresponds to true plug flow; however, it is not so steep, meaning that it is more difficult to attain a preset degree of sterility under laminar than plug flow. The series in Eq. (3.470) converges very slowly at large Da, in which case one should to advantage resort to

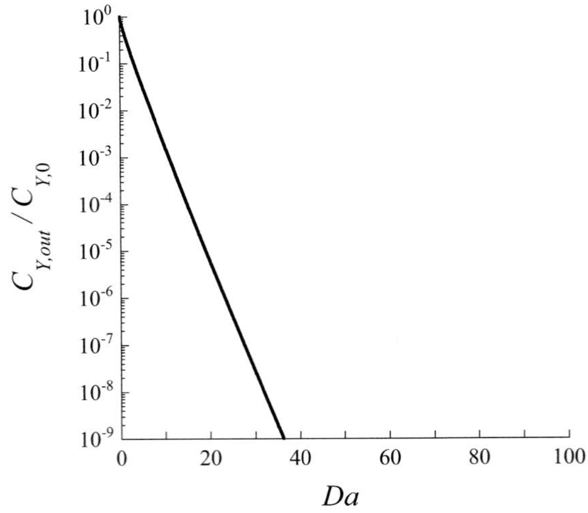

Fig. 3.17 Variation of normalized concentration of viable biomass at outlet of holding tube, $C_{Y,out}/C_{Y,0}$, versus Damkohler's number, Da, under laminar flow model

$$\frac{C_{Y,out}}{C_{Y,0}} = \left(1 - \frac{Da}{2}\right)\mathrm{e}^{-\frac{Da}{2}} + \left(\frac{Da}{2}\right)^2 \left(\frac{\mathrm{e}^{-\frac{Da}{2}}}{\frac{Da}{2}} \sum_{i=0}^{\infty} (-1)^i i! \left(\frac{Da}{2}\right)^{-i}\right), \tag{3.471}$$

as asymptotic series.

If true plug flow prevailed, then $v \to \overline{v}$, so Eq. (3.431) would produce

$$\lim_{v \to \overline{v}} \frac{C_Y\{r\}}{C_{Y,0}} = \mathrm{e}^{-\frac{k_d L}{\overline{v}}}; \tag{3.472}$$

owing to its constancy, $\mathrm{e}^{-k_d L/\overline{v}}$ would then be eligible to leave the kernel in Eq. (3.437) and thus generate

$$\lim_{v \to \overline{v}} C_{Y,out} = \frac{2}{\overline{v}} \frac{\int_0^R C_{Y,0}\mathrm{e}^{-\frac{k_d L}{\overline{v}}}\overline{v}\, r\, dr}{R^2} = \frac{2C_{Y,0}\mathrm{e}^{-\frac{k_d L}{\overline{v}}}\overline{v}}{\overline{v}} \frac{\int_0^R r\, dr}{R^2}. \tag{3.473}$$

After dropping common factors between numerator and denominator, Eq. (3.473) becomes

$$\lim_{v \to \bar{v}} C_{Y,out} = 2e^{-\frac{k_d L}{\bar{v}}} C_{Y,0} \frac{\int_0^R r\,dr}{R^2}, \tag{3.474}$$

while application of the fundamental theorem of integral calculus justifies

$$\lim_{v \to \bar{v}} C_{Y,out} = 2e^{-\frac{k_d L}{\bar{v}}} C_{Y,0} \frac{\left.\frac{r^2}{2}\right|_0^R}{R^2} = 2e^{-\frac{k_d L}{\bar{v}}} C_{Y,0} \frac{\frac{R^2}{2}}{R^2}; \tag{3.475}$$

a final dropping of 2 and R^2 from both numerator and denominator leaves

$$\lim_{v \to \bar{v}} C_{Y,out} = e^{-\frac{k_d L}{\bar{v}}} C_{Y,0}, \tag{3.476}$$

or else

$$\lim_{v \to \bar{v}} \frac{C_{Y,out}}{C_{Y,0}} = e^{-Da} \tag{3.477}$$

upon bringing Eq. (3.449) on board and dividing both sides by $C_{Y,0}$. Equation (3.477) coincides with Eq. (3.409), as expected, because both pertain to plug flow.

After revisiting Eq. (3.410) as

$$\eta_{st} \equiv \frac{\frac{C_{Y,0}}{C_{Y,out}}}{\lim\limits_{v \to \bar{v}} \frac{C_{Y,0}}{C_{Y,out}}} = \frac{\lim\limits_{v \to \bar{v}} \frac{C_{Y,out}}{C_{Y,0}}}{\frac{C_{Y,out}}{C_{Y,0}}} \tag{3.478}$$

for definition of efficiency of sterilization, one gets

$$\eta_{st} = \frac{e^{-Da}}{\left(1 - \frac{Da}{2}\right)e^{-\frac{Da}{2}} - \left(\frac{Da}{2}\right)^2 \left(\gamma + \ln \frac{Da}{2} + \sum\limits_{i=1}^{\infty} \frac{(-1)^i}{i \cdot i!} \left(\frac{Da}{2}\right)^i\right)} \tag{3.479}$$

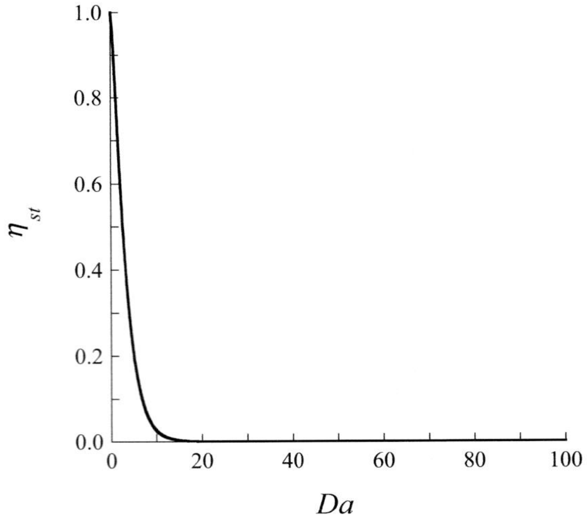

Fig. 3.18 Variation of efficiency of sterilization, η_{st}, versus Damkohler's number, Da, under laminar flow model

with the aid of Eqs. (3.470) and (3.477), which reduces to

$$\eta_{st} = \frac{1}{\left(1 - \frac{Da}{2}\right)e^{\frac{Da}{2}} - \left(\frac{Da}{2}\right)^2 e^{Da}\left(\gamma + \ln\frac{Da}{2} + \sum_{i=1}^{\infty}\frac{(-1)^i}{i \cdot i!}\left(\frac{Da}{2}\right)^i\right)}, \tag{3.480}$$

after multiplying both numerator and denominator by e^{Da}; Eq. (3.480) is depicted in Fig. 3.18. The degree of sublethality during continuous sterilization of a fluid undergoing laminar flow follows trends similar to those pinpointed n Fig. 3.16, notoriously close to $N = 10$ for the equal-sized stirred flow tank cascade model; this implies similar degrees of dispersion between these two models.

Problem 3.10

An alternate description of continuous sterilization, when the residence time distribution within the holding tube abides to the equal-sized stirred tank cascade model, is possible from a mass balance to a putative tracer injection under unsteady-state conditions.

(a) Obtain the residence time distribution, $E\{t\}$, of said cascade of N stirred tanks, knowing that

(continued)

Problem 3.10 (continued)

$$E\{t\} \equiv \frac{C_T\{t\}}{\int\limits_0^\infty C_T\{t\}\, dt}, \tag{3.481}$$

where $C_T\{t\}$ denotes concentration of tracer at exit of holding tube, following instantaneous injection of a finite amount thereof at its entrance.

(b) Calculate the variance of $E\{t\}$, and conclude on the dependence of dispersion on N. Remember that the i-th absolute moment of the residence time distribution, μ_i, is given by

$$\mu_i\{E\} = (-1)^i \left.\frac{d^i G\{s\}}{ds^i}\right|_{s=0} \quad ; \quad i = 1, 2, \ldots, \tag{3.482}$$

also known as van der Laan's theorem, and that

$$\sigma^2 = \mu_2 - {\mu_1}^2 \tag{3.483}$$

applies to the second relative moment (or variance), σ^2.

(c) Estimate the fraction of viable microorganisms if the total segregation model, given by

$$C_{Y,ts} = \int\limits_0^\infty C_Y\{t\} E\{t\} dt, \tag{3.484}$$

is combined with the result obtained in *a*).

(d) Discuss the dispersion of the residence time distribution of a vessel undergoing laminar flow.

Solution

(a) The residence time distribution can be derived directly from the theoretical model under scrutiny, because the individual units that make up the cascade are ideal stirred flow tanks; if some tracer were injected at the inlet to the i-th unit, then a balance thereto would read

(continued)

Problem 3.10 (continued)

$$QC_{T,i-1} = QC_{T,i} + V_i \frac{dC_{T,i}}{dt}, \quad (3.485)$$

where $C_{T,i\text{-}1}$ and $C_{T,i}$ denote tracer concentration at inlet and outlet streams, respectively, V_i denotes volume of each tank, and Q denotes constant volumetric fow rate through the whole cascade. Upon division of both sides by Q, Eq. (3.485) becomes

$$C_{T,i-1} = C_{T,i} + \tau_i \frac{dC_{T,i}}{dt}, \quad (3.486)$$

where Eq. (3.391) was meanwhile retrieved;

$$C_{T,i}\big|_{t=0^-} = 0 \quad (3.487)$$

will serve as initial condition, corresponding to absence of tracer prior to its injection. Application of Laplace's transforms to both sides converts Eq. (3.486) to

$$\overline{C}_{T,i-1} = \overline{C}_{T,i} + \tau_i \left(s\overline{C}_{T,i} - C_{T,i}\big|_{t=0^-} \right), \quad (3.488)$$

where combination with Eq. (3.487) allows simplification to

$$\overline{C}_{T,i-1} = \overline{C}_{T,i} + \tau_i s \overline{C}_{T,i}; \quad (3.489)$$

upon condensation of terms alike, Eq. (3.489) becomes

$$\overline{C}_{T,i-1} = (1 + \tau_i s)\overline{C}_{T,i}, \quad (3.490)$$

which degenerates to

$$\frac{\overline{C}_{T,i}}{\overline{C}_{T,i-1}} = \frac{1}{1 + \tau_i s} \quad ; \quad i = 1, 2, \ldots, N \quad (3.491)$$

following division of both sides by $(1 + \tau_i s)\overline{C}_{T,i-1}$. The corresponding ratio of $\overline{C}_{T,N}$ to $\overline{C}_{T,0}$ may be splitted as

(continued)

Problem 3.10 (continued)

$$\frac{\overline{C}_{T,N}}{\overline{C}_{T,0}} = \frac{\overline{C}_{T,N}}{\overline{C}_{T,N-1}} \frac{\overline{C}_{T,N-1}}{\overline{C}_{T,N-2}} \cdots \frac{\overline{C}_{T,2}}{\overline{C}_{T,1}} \frac{\overline{C}_{T,1}}{\overline{C}_{T,0}} \tag{3.492}$$

upon multiplication and division by $\overline{C}_{T,N-1}$, $\overline{C}_{T,N-2}$, …, $\overline{C}_{T,2}$, and $\overline{C}_{T,1}$, where Eq. (3.491) supports transformation to

$$\frac{\overline{C}_{T,N}}{\overline{C}_{T,0}} = \frac{1}{1+\tau_N s} \frac{1}{1+\tau_{N-1} s} \cdots \frac{1}{1+\tau_2 s} \frac{1}{1+\tau_1 s}; \tag{3.493}$$

in view of Eq. (3.397), one obtains

$$\frac{\overline{C}_{T,N}}{\overline{C}_{T,0}} = \frac{1}{1+\frac{\tau}{N} s} \frac{1}{1+\frac{\tau}{N} s} \cdots \frac{1}{1+\frac{\tau}{N} s} \frac{1}{1+\frac{\tau}{N} s} = \prod_{i=1}^{N} \frac{1}{1+\frac{\tau}{N} s}, \tag{3.494}$$

where definition of power and operational rule pertaining to its reciprocal yield

$$\frac{\overline{C}_{T,N}}{\overline{C}_{T,0}} = \left(\frac{1}{1+\frac{\tau}{N} s}\right)^N = \frac{1}{\left(1+\frac{\tau}{N} s\right)^N}. \tag{3.495}$$

The ratio of outlet response to inlet challenge, in Laplace's domain, is but the transfer function, $G\{s\}$, of the system; in the present case, it looks like

$$G\{s\} = \frac{1}{\left(1+\frac{\tau}{N} s\right)^N} \tag{3.496}$$

for the whole cascade, in agreement with Eq. (3.495), and pertaining to tracer concentration. If tracer is instantaneously injected into the inlet stream to the first unit, then $C_{T,0}$ will abide to

$$C_{T,0}\{t\} = C_T^0 \delta\{t\} \tag{3.497}$$

that unfolds

(continued)

Problem 3.10 (continued)

$$\overline{C}_{T,0}=C_T^0 \tag{3.498}$$

in Laplace's domain; here, C_T^0 denotes a constant and $\delta\{t\}$ denotes Dirac's function. Combination with Eq. (3.498) simplifies Eq. (3.495) to

$$\frac{\overline{C}_{T,N}}{C_T^0}=\frac{1}{\left(1+\frac{\tau}{N}s\right)^N}, \tag{3.499}$$

where $\overline{C}_{T,N}$ can be promptly isolated as

$$\overline{C}_{T,N}=\frac{C_T^0}{\left(1+\frac{\tau}{N}s\right)^N}. \tag{3.500}$$

The evolution of tracer concentration, in the outlet stream of the cascade of stirred flow tanks, will then be ascertained upon application of inverse Laplace's transforms to Eq. (3.500) as

$$\begin{aligned} C_{T,N}\{t\} &\equiv \mathcal{L}^{-1}\left(\overline{C}_{T,N}\right)=\mathcal{L}^{-1}\left(\frac{C_T^0}{\left(1+\frac{\tau}{N}s\right)^N}\right) \\ &=\mathcal{L}^{-1}\left(\frac{\left(\frac{N}{\tau}\right)^N C_T^0}{(N-1)!}\frac{(N-1)!}{\left(s+\frac{N}{\tau}\right)^N}\right) \end{aligned} \tag{3.501}$$

with numerator and denominator of the argument thereof meanwhile multiplied by $(N/\tau)^N/(N-1)!$; in view of the linearity of Laplace's transform, Eq. (3.501) will be redone to

$$C_{T,N}\{t\}=\frac{\left(\frac{N}{\tau}\right)^N C_T^0}{(N-1)!}\mathcal{L}^{-1}\left(\frac{(N-1)!}{\left(s+\frac{N}{\tau}\right)^N}\right). \tag{3.502}$$

The translation theorem in Laplace's domain may now be invoked to convert Eq. (3.502) to

(continued)

Problem 3.10 (continued)

$$C_{T,N}\{t\} = \frac{\left(\frac{N}{\tau}\right)^N C_T^0}{(N-1)!} \mathrm{e}^{-\frac{N}{\tau}t} \mathcal{L}^{-1}\left(\frac{(N-1)!}{s^N}\right), \tag{3.503}$$

where the outstanding Laplace's transform will finally produce

$$C_{T,N}\{t\} = \frac{\left(\frac{N}{\tau}\right)^N C_T^0}{(N-1)!} \mathrm{e}^{-\frac{N}{\tau}t} t^{N-1}. \tag{3.504}$$

Since the residence time distribution satisfies Eq. (3.481), one may retrieve Eq. (3.504) as

$$E\{t\} = \frac{\frac{\left(\frac{N}{\tau}\right)^N C_T^0}{(N-1)!} t^{N-1} \mathrm{e}^{-\frac{N}{\tau}t}}{\int_0^\infty \frac{\left(\frac{N}{\tau}\right)^N C_T^0}{(N-1)!} t^{N-1} \mathrm{e}^{-\frac{N}{\tau}t} dt}, \tag{3.505}$$

or else

$$E\{t\} = \frac{\frac{\left(\frac{N}{\tau}\right)^N C_T^0}{(N-1)!} t^{N-1} \mathrm{e}^{-\frac{N}{\tau}t}}{\frac{\left(\frac{N}{\tau}\right)^N C_T^0}{(N-1)!} \int_0^\infty t^{N-1} \mathrm{e}^{-\frac{N}{\tau}t} dt} = \frac{t^{N-1} \mathrm{e}^{-\frac{N}{\tau}t}}{\int_0^\infty t^{N-1} \mathrm{e}^{-\frac{N}{\tau}t} dt} \tag{3.506}$$

after taking constant factors off the kernel, followed by dropping $(N/\tau)^N C_T^0/(N-1)!$ from both numerator and denominator; integration proves easier upon defining an auxiliary variable ξ as

$$\zeta \equiv \frac{N}{\tau} t, \tag{3.507}$$

as long as it supports

(continued)

Problem 3.10 (continued)

$$\int_0^\infty t^{N-1} e^{-\frac{N}{\tau}t} dt = \int_0^\infty \left(\frac{\zeta\tau}{N}\right)^{N-1} e^{-\zeta} d\left(\frac{\zeta\tau}{N}\right) = \int_0^\infty \left(\frac{\tau}{N}\right)^{N-1} \zeta^{N-1} e^{-\zeta} \frac{\tau}{N} d\zeta = \left(\frac{\tau}{N}\right)^N \int_0^\infty \zeta^{N-1} e^{-\zeta} d\zeta \tag{3.508}$$

with constant factors meanwhile collapsed, while the lower and upper limits of integration are kept, that is, $\zeta|_{t\to 0} = 0$ and $\zeta|_{t\to\infty} = \infty$, in agreement with Eq. (3.507). After recalling the general mathematical result

$$\lim_{\zeta\to\infty} \frac{\zeta^{N-1}}{e^\zeta} = 0 \tag{3.509}$$

valid for every (integer) $N > 1$, one may proceed to integration by parts of the outstanding integral in Eq. (3.508) as

$$\int_0^\infty \zeta^{N-1} e^{-\zeta} d\zeta = \zeta^{N-1} \frac{e^{-\zeta}}{-1}\Bigg|_0^\infty - \int_0^\infty (N-1)\zeta^{N-2} \frac{e^{-\zeta}}{-1} d\zeta$$
$$= \zeta^{N-1} e^{-\zeta}\Big|_0 - \zeta^{N-1} e^{-\zeta}\Big|_\infty + (N-1)\int_0^\infty \zeta^{N-2} e^{-\zeta} d\zeta; \tag{3.510}$$

in view of Eq. (3.509), it is possible to redo Eq. (3.510) as

$$\int_0^\infty \zeta^{N-1} e^{-\zeta} d\zeta = 0^{N-1} e^0 - \lim_{\zeta\to\infty} \frac{\zeta^{N-1}}{e^\zeta} + (N-1)\int_0^\infty \zeta^{N-2} e^{-\zeta} d\zeta$$
$$= 0 \cdot 1 - 0 + (N-1)\int_0^\infty \zeta^{N-2} e^{-\zeta} d\zeta, \tag{3.511}$$

and thus

(continued)

Problem 3.10 (continued)

$$\int_0^\infty \zeta^{N-1}e^{-\zeta}d\zeta = (N-1)\int_0^\infty \zeta^{N-2}e^{-\zeta}d\zeta. \tag{3.512}$$

Equation (3.512) consubstantiates a recurrence relationship, and may be applied to every single (integer) exponent of ζ, say, $N-2, N-3, \ldots, 2, 1$, to eventually obtain

$$\int_0^\infty \zeta^{N-1}e^{-\zeta}d\zeta = (N-1)(N-2)\int_0^\infty \zeta^{N-3}e^{-\zeta}d\zeta = \ldots = (N-1)(N-2)\ldots 1\int_0^\infty e^{-\zeta}d\zeta; \tag{3.513}$$

after bringing the definition of factorial and the fundamental theorem of integral calculus on board, Eq. (3.513) becomes

$$\int_0^\infty \zeta^{N-1}e^{-\zeta}d\zeta = (N-1)!\left.\frac{e^{-\zeta}}{-1}\right|_0^\infty = (N-1)!\left(e^0 - e^{-\infty}\right) = (N-1)!(1-0), \tag{3.514}$$

which readily degenerates to

$$\int_0^\infty \zeta^{N-1}e^{-\zeta}d\zeta = (N-1)!. \tag{3.515}$$

Insertion of Eq. (3.515) converts Eq. (3.508) to

$$\int_0^\infty t^{N-1}e^{-\frac{N}{\tau}t}dt = \left(\frac{\tau}{N}\right)^N (N-1)!, \tag{3.516}$$

which may then be inserted in Eq. (3.506) to produce

(continued)

Problem 3.10 (continued)

$$E\{t\} = \frac{t^{N-1} e^{-\frac{N}{\tau}t}}{\left(\frac{\tau}{N}\right)^N (N-1)!}; \tag{3.517}$$

upon condensation of factors alike, Eq. (3.517) yields

$$E\{t\} = \frac{\left(\frac{N}{\tau}\right)^N t^{N-1} e^{-\frac{N}{\tau}t}}{(N-1)!} = \frac{\frac{N}{\tau}}{(N-1)!}\left(\frac{N}{\tau}t\right)^{N-1} e^{-\frac{N}{\tau}t} \tag{3.518}$$

that plays the role of residence time distribution of the vessel. A graphical interpretation of Eq. (3.518) is provided in Fig. 3.19, after having multiplied both sides by τ, that is,

$$\tau E\{t\} = \frac{N}{(N-1)!}\left(N\frac{t}{\tau}\right)^{N-1} e^{-N\frac{t}{\tau}} \tag{3.519}$$

meant to render it dimensionless; (residence) time was made dimensionless as well, in the form t/τ, so N appears isolated as germane parameter in Eq. (3.519). Inspection of this figure indicates that dispersion becomes

(continued)

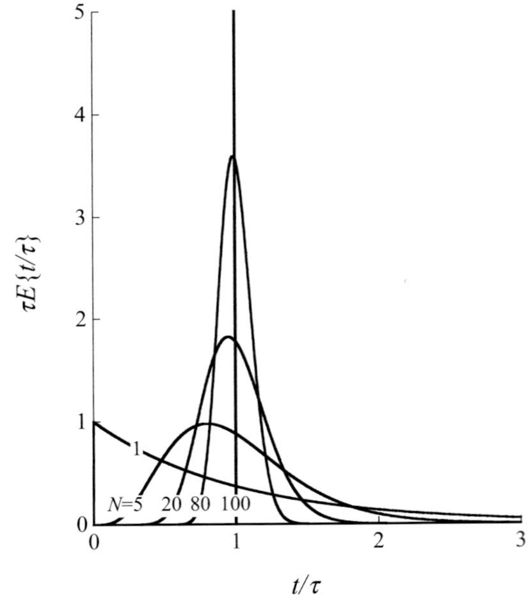

Fig. 3.19 Dimensionless residence time distribution, τE, of dimensionless residence times, t/τ, associated to equal-sized stirred flow tank model, for selected number of tanks, N

Problem 3.10 (continued)

smaller and smaller as N increases; when $N \rightarrow \infty$, all liquid elements will indeed share the very same residence time τ, that is,

$$\lim_{N \rightarrow \infty} E\{t\} = \frac{1}{\tau}\delta\left\{\frac{t}{\tau} - 1\right\}, \tag{3.520}$$

which describes a plug flow vessel with τ for space time.

(b) Insertion of Eq. (3.496) converts Eq. (3.482) to

$$\mu_1 = (-1)\frac{d}{ds}\left(\frac{1}{\left(1 + \frac{\tau}{N}s\right)^N}\right)\Bigg|_{s=0} = -\frac{d}{ds}\left(\left(1 + \frac{\tau}{N}s\right)^{-N}\right)\Bigg|_{s=0} \tag{3.521}$$

when $i = 1$; the classical rules of differentiation then justify

$$\mu_1 = -\left(-N\left(1 + \frac{\tau}{N}s\right)^{-N-1}\frac{\tau}{N}\right)\Bigg|_{s=0}. \tag{3.522}$$

After setting s equal to zero as indicated, followed by lumping of factors alike, Eq. (3.522) simplifies to

$$\mu_1 = N\left(1 + \frac{\tau}{N}0\right)^{-N-1}\frac{\tau}{N} = N1^{-N-1}\frac{\tau}{N} = N1\frac{\tau}{N} = N\frac{\tau}{N} = \tau, \tag{3.523}$$

with the expected value (or μ_1) coincident with space time; this result is anticipated because the mechanistic definition of space time is the mean (or average) residence time. By the same token, the second moment reads

$$\mu_2 = (-1)^2\frac{d^2G\{s\}}{ds^2}\Bigg|_{s=0} \equiv \frac{d}{ds}\left(\frac{dG\{s\}}{ds}\right)\Bigg|_{s=0} \tag{3.524}$$

stemming from Eq. (3.482) with $i = 2$, where insertion of Eq. (3.522) unfolds

$$\mu_2 = \frac{d}{ds}\left(-N\left(1 + \frac{\tau}{N}s\right)^{-N-1}\frac{\tau}{N}\right)\Bigg|_{s=0}; \tag{3.525}$$

differentiation of the germane power leads to

(continued)

Problem 3.10 (continued)

$$\mu_2 = \left(-N(-N-1)\left(1+\frac{\tau}{N}s\right)^{-N-1-1}\frac{\tau}{N}\frac{\tau}{N}\right)\Bigg|_{s=0}, \tag{3.526}$$

so $s=0$ will promptly produce

$$\mu_2 = N(N+1)\left(1+\frac{\tau}{N}0\right)^{-N-2}\left(\frac{\tau}{N}\right)^2 = N(N+1)1^{-N-2}\left(\frac{\tau}{N}\right)^2$$
$$= \frac{N(N+1)}{N^2}1\tau^2 = \frac{N+1}{N}\tau^2 = \left(1+\frac{1}{N}\right)\tau^2 \tag{3.527}$$

upon a sequence of self-explanatory algebraic steps. On the other hand, the degree of dispersion of a distribution is statistically measured by its variance, accessible via Eq. (3.483); in view of Eqs. (3.523) and (3.527), one obtains

$$\sigma^2 = \left(1+\frac{1}{N}\right)\tau^2 - \tau^2 = \left(1+\frac{1}{N}-1\right)\tau^2 \tag{3.528}$$

for Eq. (3.483) with τ^2 meanwhile factored out, which breaks down to

$$\sigma^2 = \frac{\tau^2}{N} \tag{3.529}$$

after dropping symmetrical terms. Inspection of Eq. (3.529) indicates that dispersion is inversely proportional to N, for a given τ—and, consequently,

$$\lim_{N\to\infty} \sigma^2 = \frac{\tau^2}{\infty} = 0; \tag{3.530}$$

hence, plug flow is characterized by nil dispersion (as claimed before), compatible with the asymptotic behavior apparent in Fig. 3.19.

(c) The total segregation model requires knowledge of the extent of viability loss in any ideal plug flow reactor, with the same first-order death kinetics at stake; the associated (differential) mass balance reads

$$QC_Y|_V - k_d C_Y\, dV = QC_Y|_{V+dV}, \tag{3.531}$$

where V denotes volumetric coordinate. After dividing both sides by Q, and differentiating both sides of Eq. (3.398) as

(continued)

Problem 3.10 (continued)

$$dt = \frac{dV}{Q} \tag{3.532}$$

once τ has been swapped for t (which will denote space time hereafter), one can reformulate Eq. (3.531) as

$$C_Y|_t - k_d C_Y \, dt = C_Y|_{t+dt}. \tag{3.533}$$

After similar terms are collapsed and both sides divided by dt, Eq. (3.533) becomes

$$-k_d C_Y = \frac{C_Y|_{t+dt} - C_Y|_t}{dt} \tag{3.534}$$

or, equivalently,

$$\frac{dC_Y}{dt} = -k_d C_Y \tag{3.535}$$

as per the definition of derivative; as expected, Eq. (3.535) mimics Eq. (3.18), due to the functional equivalence of a plug flow reactor along space time to a stirred batch reactor along time. After choosing

$$C_Y|_{t=0} = C_{Y,0} \tag{3.536}$$

for boundary condition, integration of Eq. (3.535) ensues via separation of variables as

$$\int_{C_{Y,0}}^{C_Y} \frac{d\widetilde{C}_Y}{\widetilde{C}_Y} = -k_d \int_0^t d\widetilde{t}; \tag{3.537}$$

the fundamental theorem of integral calculus then justifies

$$\ln \widetilde{C}_Y \Big|_{C_{Y,0}}^{C_Y} = -k_d \, \widetilde{t} \Big|_0^t \tag{3.538}$$

or, equivalently,

(continued)

Problem 3.10 (continued)

$$\ln \frac{C_Y}{C_{Y,0}} = -k_d t, \tag{3.539}$$

where exponentials will be taken of both sides to get

$$C_Y = C_{Y,0} \mathrm{e}^{-k_d t} \tag{3.540}$$

along with multiplication thereof by $C_{Y,0}$. One may now revisit Eq. (3.484) as

$$C_{Y,ts} = \int_0^\infty C_{Y,0} \mathrm{e}^{-k_d t} \frac{\frac{N}{\tau}}{(N-1)!} \left(\frac{N}{\tau} t\right)^{N-1} \mathrm{e}^{-\frac{N}{\tau} t} dt, \tag{3.541}$$

upon insertion of Eqs. (3.518) and (3.540); after taking $1/(N-1)!$ and $C_{Y,0}$ off the kernel, multiplying and dividing the argument of the first exponential function by N/τ, and taking (constant) N/τ under the differential operator, Eq. (3.541) becomes

$$\begin{aligned} C_{Y,ts} &= \frac{C_{Y,0}}{(N-1)!} \int_0^\infty \mathrm{e}^{-\frac{k_d \tau}{N} \frac{N}{\tau} t} \left(\frac{N}{\tau} t\right)^{N-1} \mathrm{e}^{-\frac{N}{\tau} t} d\left(\frac{N}{\tau} t\right) \\ &= \frac{C_{Y,0}}{(N-1)!} \int_0^\infty \mathrm{e}^{-\frac{N}{\tau} t \left(1 + \frac{k_d \tau}{N}\right)} \left(\frac{N}{\tau} t\right)^{N-1} d\left(\frac{N}{\tau} t\right), \end{aligned} \tag{3.542}$$

where exponential functions were meanwhile lumped. After recalling Eqs. (3.402) and using Eq. (3.507) as template to define an auxiliary (dummy) variable of integration, one may rewrite Eq. (3.542) as

$$C_{Y,ts} = \frac{C_{Y,0}}{(N-1)!} \int_0^\infty \xi^{N-1} \mathrm{e}^{-\left(1 + \frac{Da}{N}\right) \xi} d\xi. \tag{3.543}$$

To facilitate manipulation of the integral in Eq. (3.543), it is convenient to define another auxiliary variable ζ as

$$\zeta \equiv \left(1 + \frac{Da}{N}\right) \xi, \tag{3.544}$$

since one readily obtains

(continued)

Problem 3.10 (continued)

$$\int_0^\infty \xi^{N-1} e^{-\left(1+\frac{Da}{N}\right)\xi} d\xi = \int_0^\infty \left(\frac{\zeta}{1+\frac{Da}{N}}\right)^{N-1} e^{-\zeta} d\left(\frac{\zeta}{1+\frac{Da}{N}}\right) \tag{3.545}$$

where $\left(1/(1+Da/N)\right)^{N-1}$ and $1/(1+Da/N)$ are to be taken off the kernel as

$$\int_0^\infty \xi^{N-1} e^{-\left(1+\frac{Da}{N}\right)\xi} d\xi = \left(\frac{1}{1+\frac{Da}{N}}\right)^N \int_0^\infty \zeta^{N-1} e^{-\zeta} d\zeta. \tag{3.546}$$

The integral left in the right-hand side of Eq. (3.546) is identical to that labeled as Eq. (3.515), so one may take advantage of an analogue to Eq. (3.516) to write

$$\int_0^\infty \xi^{N-1} e^{-\left(1+\frac{Da}{N}\right)\xi} d\xi = \left(\frac{1}{1+\frac{Da}{N}}\right)^N (N-1)! = \frac{(N-1)!}{\left(1+\frac{Da}{N}\right)^N}; \tag{3.547}$$

hence, Eq. (3.543) will read

$$C_{Y,ts} = \frac{C_{Y,0}}{(N-1)!} \frac{(N-1)!}{\left(1+\frac{Da}{N}\right)^N} = \frac{C_{Y,0}}{\left(1+\frac{Da}{N}\right)^N} \tag{3.548}$$

upon cancellation of $(N-1)!$ between numerator and denominator or, following division of both sides by $C_{Y,0}$,

$$\frac{C_{Y,ts}}{C_{Y,0}} = \frac{1}{\left(1+\frac{Da}{N}\right)^N}. \tag{3.549}$$

Equation (3.549) coincides with Eq. (3.404); therefore, the same result was reached following two independent routes. This coincidence was possible because cell death was taken as a first-order process, and it is known that the total segregation model produces the same results as the maximum

(continued)

Problem 3.10 (continued)

mixedness model, alias to the ideal stirred flow conditions assumed when modelling each unit of the cascade of equal-sized stirred flow tanks.

(d) When fluid goes through a cylindrical vessel under laminar flow, the portion occupied by an annulus of inner radius r, length L, and (elementary) thickness dr will exhibit some (uniform) velocity $v\{r\}$, and will progress unmixed until leaving the vessel after a residence time t, given by

$$t = \frac{L}{v\{r\}} \tag{3.550}$$

that mimics Eq. (3.430) after renaming τ as t, and likewise

$$\tau = \frac{L}{\overline{v}} \tag{3.551}$$

applying to space time; here, $\overline{v}$ denotes average velocity throughout the whole circular cross section of radius R. In view of the parabolic velocity profile and the inherent cylindrical symmetry of the system, no other radial portions of fluid will experience that residence time. Therefore, the fraction of fluid with residence time within $[t, t + dt]$, given by $E\{t\}dt$ as per the definition of residence time distribution, can be expressed in terms of a radial distribution of volumetric flow rates, $\Theta\{r\}$, according to

$$E\{t\}dt = \Theta\{r\}dr; \tag{3.552}$$

Equation (3.552) guarantees equivalence between the two continuous statistical distributions, since they represent the same fluid. In view of the above,

$$\Theta\{r\} \equiv \frac{2\pi r v\{r\}}{\int_0^R 2\pi r v\{r\}dr}; \tag{3.553}$$

one may then take 2π off the kernel in denominator and drop it from both numerator and denominator afterward to get

(continued)

Problem 3.10 (continued)

$$\Theta\{r\} = \frac{2\pi r v\{r\}}{2\pi \int\limits_0^R r v\{r\} dr} = \frac{r v\{r\}}{\int\limits_0^R r v\{r\} dr}. \tag{3.554}$$

Insertion of Eq. (3.423) gives rise to

$$\Theta\{r\} = \frac{r 2\overline{v}\left(1 - \left(\frac{r}{R}\right)^2\right)}{\int\limits_0^R r 2\overline{v}\left(1 - \left(\frac{r}{R}\right)^2\right) dr} = \frac{2\overline{v}\left(r - \frac{r^3}{R^2}\right)}{2\overline{v} \int\limits_0^R \left(r - \frac{r^3}{R^2}\right) dr}, \tag{3.555}$$

where $2\overline{v}$ was taken off the kernel for its constancy and r was factored in for convenience; Eq. (3.555) breaks down to

$$\Theta\{r\} = \frac{r - \frac{r^3}{R^2}}{\int\limits_0^R \left(r - \frac{r^3}{R^2}\right) dr}, \tag{3.556}$$

after cancelling out common factors between numerator and denominator. Performance of the stated integration converts Eq. (3.556) to

$$\Theta\{r\} = \frac{r - \frac{r^3}{R^2}}{\left.\left(\frac{r^2}{2} - \frac{1}{R^2}\frac{r^4}{4}\right)\right|_0^R} = \frac{r - \frac{r^3}{R^2}}{\frac{R^2}{2} - \frac{1}{R^2}\frac{R^4}{4}} \tag{3.557}$$

with the aid of the fundamental theorem of integral calculus, where condensation of factors alike unfolds

$$\Theta\{r\} = \frac{r - \frac{r^3}{R^2}}{\frac{R^2}{2} - \frac{R^2}{4}} = \frac{r - \frac{r^3}{R^2}}{\frac{R^2}{4}}; \tag{3.558}$$

an extra multiplication of both numerator and denominator by $4/R^2$ gives rise to

(continued)

Problem 3.10 (continued)

$$\Theta\{r\} = \frac{4}{R^2}\left(r - \frac{r^3}{R^2}\right). \tag{3.559}$$

Insertion of Eq. (3.559) converts Eq. (3.552) to

$$E\{t\}dt = \frac{4}{R^2}\left(r - \frac{r^3}{R^2}\right)dr, \tag{3.560}$$

or else

$$E\{t\}\frac{dt}{dr} = \frac{4}{R^2}r\left(1 - \frac{r^2}{R^2}\right) \tag{3.561}$$

after factoring out r and dividing both sides by dr. On the other hand, combination of Eqs. (3.423) and (3.550) produces

$$t = \frac{L}{2\overline{v}\left(1 - \left(\frac{r}{R}\right)^2\right)}; \tag{3.562}$$

differentiation of both sides with regard to r ensues as

$$\frac{dt}{dr} = \frac{L}{2\overline{v}}\left(-\frac{-2\frac{r}{R}\frac{1}{R}}{\left(1 - \left(\frac{r}{R}\right)^2\right)^2}\right) = \frac{L}{\overline{v}R^2}\frac{r}{\left(1 - \left(\frac{r}{R}\right)^2\right)^2}, \tag{3.563}$$

as per the derivative of the reciprocal of a composite function, complemented by condensation of factors alike; combination with Eq. (3.563) transforms Eq. (3.561) to

$$E\{t\}\frac{L}{\overline{v}R^2}\frac{r}{\left(1 - \left(\frac{r}{R}\right)^2\right)^2} = \frac{4}{R^2}r\left(1 - \frac{r^2}{R^2}\right), \tag{3.564}$$

where r/R^2 can cancel off between sides to just leave

(continued)

Problem 3.10 (continued)

$$E\{t\}\frac{L}{\overline{\overline{v}}}\frac{1}{\left(1-\left(\frac{r}{R}\right)^2\right)^2}=4\left(1-\left(\frac{r}{R}\right)^2\right). \tag{3.565}$$

Isolation of $E\{t\}$ in E. (3.565) ensues as

$$E\{t\}=\frac{4}{\frac{L}{\overline{\overline{v}}}}\left(1-\left(\frac{r}{R}\right)^2\right)^3, \tag{3.566}$$

where Eq. (3.551) permits simplification to

$$E\{t\}=\frac{4}{\tau}\left(1-\left(\frac{r}{R}\right)^2\right)^3; \tag{3.567}$$

combination of Eqs. (3.423) and (3.550) generates

$$t=\frac{L}{2\overline{v}\left(1-\left(\frac{r}{R}\right)^2\right)} \tag{3.568}$$

or, equivalently,

$$1-\left(\frac{r}{R}\right)^2=\frac{\frac{L}{\overline{\overline{v}}}}{2t} \tag{3.569}$$

upon swapping t and $1-(r/R)^2$; Eq. (3.569) reduces further to

$$1-\left(\frac{r}{R}\right)^2=\frac{\tau}{2t}, \tag{3.570}$$

after bringing Eq. (3.551) on board again. In view of Eq. (3.570), one may reformulate Eq. (3.567) as

$$E\{t\}=\frac{4}{\tau}\left(\frac{\tau}{2t}\right)^3, \tag{3.571}$$

where similar powers will be collapsed to generate

$$E\{t\}=\frac{4}{\tau}\frac{\tau^3}{8t^3}=\frac{\tau^2}{2t^3}; \tag{3.572}$$

since the maximum velocity of fluid is attained on the axis, that is,

(continued)

Problem 3.10 (continued)

$$v\{r\} \leq v\{r\}|_{r=0} = 2\overline{v} \tag{3.573}$$

consistent with Eq. (3.423), the corresponding residence times are constrained by

$$t \geq \frac{L}{2\overline{v}} = \frac{\frac{L}{\overline{v}}}{2} \tag{3.574}$$

in agreement with Eq. (3.550), or else

$$t \geq \frac{\tau}{2} \tag{3.575}$$

in view of Eq. (3.551). Therefore, Eq. (3.572) should be completed as

$$E\{t\} = \frac{\tau^2}{2t^3} \mathrm{H}\left\{t - \frac{\tau}{2}\right\} \tag{3.576}$$

as full mathematical descriptor of the residence time distribution under scrutiny; Heaviside's function enforces indeed $E\{t\} = 0$ when $t < \tau/2$ (see Eq. (3.575)). After multiplying both sides by τ, Eq. (3.576) yields

$$\tau E\{t\} = \frac{\tau^3}{2t^3} \mathrm{H}\left\{t - \frac{\tau}{2}\right\}, \tag{3.577}$$

where lumping similar powers and dividing the argument of Heaviside's function by $\tau \neq 0$ lead to

$$\tau E\{t\} = \frac{1}{2} \frac{\mathrm{H}\left\{\frac{t}{\tau} - \frac{1}{2}\right\}}{\left(\frac{t}{\tau}\right)^3}; \tag{3.578}$$

hence, the independent variable appears naturally in dimensionless form, that is, t/τ, as happens with the dependent dimensionless variable, τE, as plotted in Fig. 3.20. The shape of the residence time distribution resembles that corresponding to a single stirred flow vessel (see Fig. 3.19) as discussed in *a*), upon rightward compression by $\tau/2$, which also makes the curve steeper in the vicinity of unity. The mean of t, defined as

(continued)

Fig. 3.20 Dimensionless residence time distribution, τE, of dimensionless residence times, t/τ, associated to laminar flow model

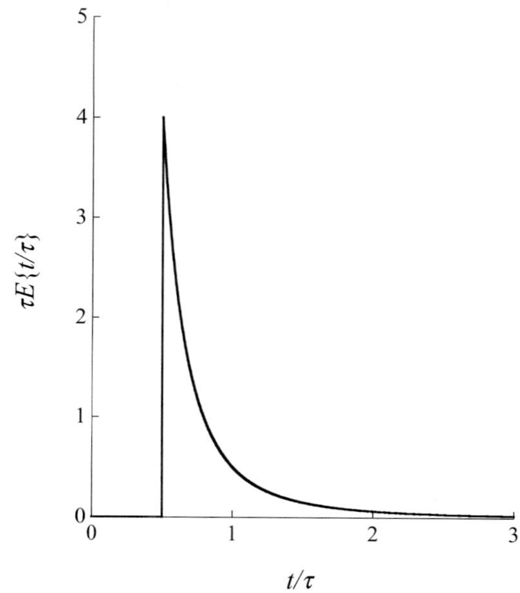

Problem 3.10 (continued)

$$\mu_1 \equiv \int_0^\infty tE\{t\}dt, \tag{3.579}$$

can be rewritten as

$$\mu_1 = \int_0^\infty t\frac{\tau^2}{2t^3}\mathrm{H}\left\{t - \frac{\tau}{2}\right\}dt \tag{3.580}$$

after having inserted Eq. (3.576); constants can be taken off the kernel, and the lower limit of integration readjusted on account of $\mathrm{H}\{t - \tau/2\}$ appearing as factor, to eventually get

$$\mu_1 = \frac{\tau^2}{2}\int_{\frac{\tau}{2}}^\infty t^{-2}dt. \tag{3.581}$$

The fundamental theorem of integral calculus has it that

(continued)

Problem 3.10 (continued)

$$\mu_1 = \frac{\tau^2}{2}\frac{t^{-1}}{-1}\bigg|_{\frac{\tau}{2}}^{\infty} = -\frac{\tau^2}{2}\left(\frac{1}{\infty} - \frac{1}{\frac{\tau}{2}}\right) = \frac{\tau^2}{2}\left(\frac{2}{\tau} - 0\right), \tag{3.582}$$

or merely

$$\mu_1 = \tau \tag{3.583}$$

after lumping factors alike; this is anticipated as per the (operational) definition of space time as mean residence time. The variance may become accessible also via its definition, namely,

$$\sigma^2 \equiv \int_0^{\infty} (t - \mu_1)^2 E\{t\} dt, \tag{3.584}$$

which gives rise to

$$\sigma^2 = \int_0^{\infty} (t - \tau)^2 \frac{\tau^2}{2t^3} \mathrm{H}\left\{t - \frac{\tau}{2}\right\} dt \tag{3.585}$$

after taking Eqs. (3.576) and (3.583) on board; Eq. (3.585) can be replaced by

$$\sigma^2 = \frac{\tau^2}{2} \int_{\frac{\tau}{2}}^{\infty} \frac{(t - \tau)^2}{t^3} dt \tag{3.586}$$

due to the constancy of $\tau^2/2$, together with the nil value of $\mathrm{H}\{t - \tau/2\}$ under the integral sign when $0 \le t < \tau/2$. Newton's binomial theorem permits expansion of Eq. (3.586) as

$$\sigma^2 = \frac{\tau^2}{2} \int_{\frac{\tau}{2}}^{\infty} \frac{t^2 - 2t\tau + \tau^2}{t^3} dt = \frac{\tau^2}{2} \int_{\frac{\tau}{2}}^{\infty} \left(\frac{1}{t} - \frac{2\tau}{t^2} + \frac{\tau^2}{t^3}\right) dt, \tag{3.587}$$

where the kernel was meanwhile split as three independent fractions; decomposition of the integral is in order, namely,

(continued)

Problem 3.10 (continued)

$$\sigma^2 = \frac{\tau^2}{2}\left(\int_{\frac{\tau}{2}}^{\infty}\frac{dt}{t} - 2\tau\int_{\frac{\tau}{2}}^{\infty}\frac{dt}{t^2} + \tau^2\int_{\frac{\tau}{2}}^{\infty}\frac{dt}{t^3}\right) \tag{3.588}$$

again with constant factors taken off the kernel, where the fundamental theorem of integral calculus can be invoked again to justify

$$\sigma^2 = \frac{\tau^2}{2}\left(\ln t\Big|_{\frac{\tau}{2}}^{\infty} - 2\tau\frac{t^{-1}}{-1}\Big|_{\frac{\tau}{2}}^{\infty} + \tau^2\frac{t^{-2}}{-2}\Big|_{\frac{\tau}{2}}^{\infty}\right). \tag{3.589}$$

Substitution by upper and lower limits of integration, followed by a sequence of straightforward steps of algebraic manipulation convert Eq. (3.589) to

$$\begin{aligned}\sigma^2 &= \frac{\tau^2}{2}\left(\ln\frac{\infty}{\frac{\tau}{2}} + 2\tau\left(\frac{1}{\infty} - \frac{1}{\frac{\tau}{2}}\right) - \frac{\tau^2}{2}\left(\frac{1}{\infty^2} - \frac{1}{\left(\frac{\tau}{2}\right)^2}\right)\right)\\ &= \frac{\tau^2}{2}\left(\ln\infty + 2\tau\left(0 - \frac{2}{\tau}\right) - \frac{\tau^2}{2}\left(0 - \frac{4}{\tau^2}\right)\right) = \frac{\tau^2}{2}\left(\infty - 2\tau\frac{2}{\tau} + \frac{\tau^2}{2}\frac{4}{\tau^2}\right)\\ &= \frac{\tau^2}{2}(\infty - 4 + 2) = \tau^2\frac{\infty - 2}{2} = \tau^2(\infty - 1) = \infty;\end{aligned} \tag{3.590}$$

this result unfolds an infinite dispersion for the residence time distribution, on account of the finite amount of fluid next to the inner wall of the vessel, where the non-sliding condition imposes an infinite residence time. Comparative inspection of Eq. (3.590) with the asymptotic behavior of Eq. (3.529) expressed by

$$\lim_{N\to 0}\sigma^2 = \frac{\tau^2}{0} = \infty, \tag{3.591}$$

complemented by Eq. (3.530), suggests that laminar flow stands opposite to plug flow viewed as asymptotic pattern when a cascade of equal-sized stirred flow vessels is used as a model for reasoning; this fully justifies detailed study of the former, in attempts to accurately model/design sterilization of liquid media.

3.4.2 Filter Processing

A filter is basically a porous membrane, bearing a pore size distribution designed to quantitatively remove all suspended particles with (equivalent) diameter above some given threshold. Filtration accordingly applies to phases containing particles in suspension and flown through; the filter matrix acts as a mechanical barrier, usually based on size (sometimes combined with charge) and meant to retain solid particles, thus proving technically feasible provided that a liquid or a gas serves as carrier to assure flowability. As expected, the larger the viscosity of the fluid, the larger the pumping power required to make it flow through a filter; this is why filtration of gases is a routine practice, whereas filtration of liquids is usually left only to media containing components labile to the high-temperature treatments typically performed in sterilization. Filtration is easier to bring about, and certainly more predictable than other nonthermal techniques of sterilization, for example, irradiation by γ-photons produced by such radioactive isotopes as ^{60}Co.

Note that a suspension is a biphasic mixture since suspended particles do not behave as true solutes, even though they might be nanosized; whereas a solution materializes a single phase, where molecular solvation plays a role. Solvation requires indeed extensive physicochemical interaction between solvent and solute molecules, whereas only mechanical forces are germane in suspension. A consequence of this realization is that a suspension represents a meta-equilibrium state; it might keep its morphology for long periods, but only when the terminal velocity of suspended particles is small enough to be essentially balanced by (natural or forced) upward convection streams. On the contrary, a solution represents a true chemical equilibrium, associated to nil change in Gibbs' energy when evolving from each molecular entity to another one (as product) or another set (as agglomerate). Therefore, intermolecular forces at chemical level are to be handled when filtering a solution through a medium impermeable to its solute, whereas only mechanical forces have to be dealt with should a plain suspension be at stake.

Problem 3.11

Although macroscopic filtration has been popular in efforts to get rid of suspended solid material in liquid media, too narrow pores will also raise a barrier to passage of dissolved (macro)molecules, such as proteins; these have to be taken into account further to resistance offered by the filter itself and the cake formed, when estimating the pumping power necessary to overcome the (overall) pressure difference between the two sides of the filter. Develop an expression for said estimate of the minimum pressure difference required; assume ideal solution behavior, for simplicity.

Solution

Consider a vessel possessing a partition, permeable only to solvent, between two sides, I and II, the former containing pure solvent and thus described by $x_{S,I,0} = 1$ for

(continued)

Problem 3.11 (continued)

mole fraction of solvent, and the other containing a solution of some solute characterized by initial mole fraction of solvent $x_{S,II,0} < 1$. This device is sketched in Fig. 3.21, at initial conditions (on the left) and after equilibrium has been attained (on the right); the solutions on either side depart from the same initial heights $h_{I,0} = h_{II,0}$, whereas initial and final equilibrium guarantee homogeneity of composition within either phase. Due to the initial difference in solvent concentration between sides, diffusion of solvent will take place from I to II through the membrane in general agreement with Fick's law, since migration of solute is not permitted by hypothesis owing to membrane pore diameter lying below solute molecular diameter. Bulk motion of solvent will accordingly take place until the heights of the two solutions reach $h_{I,eq} < h_{II,eq}$ under equilibrium; at this stage, the difference in hydrostatic pressures will exactly match the difference in osmotic pressure associated to the distinct concentrations of solvent in the adjacent solutions, that is, $x_{S,I,eq} = x_{S,I,0}$ and $1 > x_{S,II,eq} > x_{S,II,0}$, along with $h_{I,eq} < h_{I,0}$ and $h_{II,eq} > h_{II,0}$. The condition for chemical equilibrium of solvent between solutions, at the level of the free surface of solution I, reads

$$\mu_{S,I,eq}\{P_0, T\} = \mu_{S,II,eq}\{P_0 + \Pi, T\}, \tag{3.592}$$

where P_0 denotes atmospheric pressure and $\Pi \equiv \Pi\{x_{S,II,eq}\}$ denotes osmotic pressure associated to mole fraction of solvent $x_{S,II,eq}$; in view of being ideal solutions by hypothesis, Gibbs' fundamental relationship supports conversion of Eq. (3.592) to

(continued)

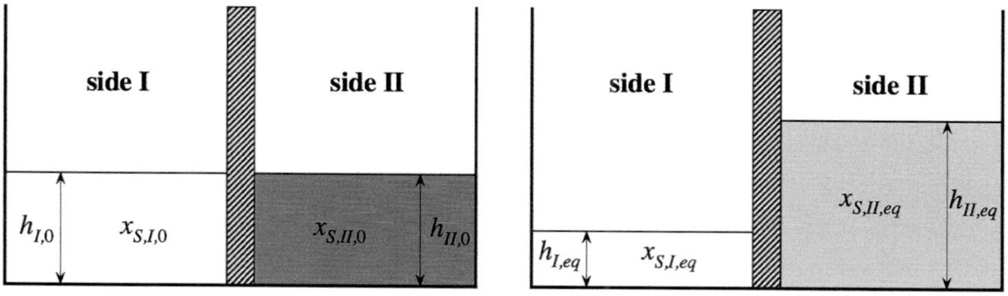

Fig. 3.21 Schematic representation of two-sided vessel, with a solution on side I, with height h_I and solvent molar fraction $x_{S,I}$, separated from a solution on side II, with height h_{II} and solvent molar fraction $x_{S,II}$, by a membrane permeable to only solvent, at startup (left) and after equilibrium has been reached (right)

Problem 3.11 (continued)

$$\mu_S^\theta\{P_0\} + \mathcal{R}T \ln x_{S,I,eq} = \mu_S^\theta\{P_0\} + \int\limits_{P_0}^{P_0+\Pi} \left(\frac{\partial \mu_S^\theta}{\partial P}\right)_{T,x} dP + \mathcal{R}T \ln x_{S,II,eq} \quad (3.593)$$

with superscript $^\theta$ referring to reference conditions (i.e., pure solvent at temperature T). Note that $x_{S,I,eq} = x_{S,I,0} = 1$, and thus $\ln x_{S,I,eq} = 0$ permit simplification of Eq. (3.593) to

$$0 = \int\limits_{P_0}^{P_0+\Pi} \left(\frac{\partial \mu_S^\theta}{\partial P}\right)_{T,x} dP + \mathcal{R}T \ln x_{S,II,eq}, \quad (3.594)$$

where $\mu_S^\theta\{P_0\}$ was meanwhile cancelled between sides. After recalling that

$$\left(\frac{\partial \mu_S^\theta}{\partial P}\right)_{T,x} = v_S^\theta \quad (3.595)$$

with v_S^θ denoting molar volume of pure solvent also at temperature T, Eq. (3.594) becomes

$$\int\limits_{P_0}^{P_0+\Pi} v_S^\theta \, dP + \mathcal{R}T \ln x_{S,II,eq} = 0; \quad (3.596)$$

in view of the incompressibility of liquids, v_S^θ remains essentially constant despite changes in (hydrostatic) pressure, so it will be taken off the kernel as

$$v_S^\theta \int\limits_{P_0}^{P_0+\Pi} dP = -\mathcal{R}T \ln x_{S,II,eq}. \quad (3.597)$$

Trivial application of the fundamental theorem of integral calculus converts Eq. (3.597) to

(continued)

Problem 3.11 (continued)

$$v_S^\theta \left. P \right|_{P_0}^{P_0+\Pi} = -\mathcal{R}T \ln x_{S,II,eq}, \tag{3.598}$$

or else

$$v_S^\theta (P_0 + \Pi - P_0) = -\mathcal{R}T \ln x_{S,II,eq}; \tag{3.599}$$

after dropping symmetrical terms in the left-hand side, one obtains

$$v_S^\theta \Pi = -\mathcal{R}T \ln x_{S,II,eq}, \tag{3.600}$$

where isolation of Π leaves

$$\Pi = -\mathcal{R}\frac{T}{v_S^\theta} \ln x_{S,II,eq} \tag{3.601}$$

with Π being accessible via hydrostatic arguments, as product of $h_{II,eq} - h_{I,eq}$ by acceleration of gravity, g, and by mass density of liquid, ρ. If the liquid medium to be filtered contains a solute at mole fraction x_s, besides suspended particles, then the minimum pressure difference across the filtering membrane, ΔP_{min}, should read

$$\Delta P_{min} = \Delta P_{filter} + \Delta P_{cake} - \mathcal{R}\frac{T}{v_S^\theta} \ln\{1 - x_s\} \tag{3.602}$$

since $x_s = 1 - x_S$ applies to a binary solution, which obviously exceeds the pressure drop on account of the filter itself, ΔP_{filter}, plus the pressure drop caused by the cake formed by suspended material already removed from the filtrate, ΔP_{cake}. As expected, no osmotic pressure develops when pure solvent is at stake, because $x_s = 0$, and thus $-\mathcal{R}T \ln \{1 - 0\}/v_S^\theta = -\mathcal{R}T(\ln 1)/v_S^\theta = 0$ in Eq. (3.602); the last term will instead override when x_s approaches unity because $-\ln \{1 - x_s\} \rightarrow -\ln \{1 - 1^-\} = -\ln 0^+ = \infty$, while both ΔP_{filter} and ΔP_{cake} remain finite.

3.4.2.1 Liquid Media

When the liquid medium of interest contains heat-labile ingredients, for example, enzymes, growth factors, or other fragile proteins (as happens with serum and other complex media, meant for growth of animal cells), denaturation thereof brought about by heat discards the possibility of performing thermal sterilization; plain microfiltration should be utilized in this case, using membranes and containing devices as illustrated in Fig. 3.22. Typical membranes are made of cellulose esters or other polymers, with pore diameters within 0.22–0.45 μm; as medium permeates the filter, bacteria (with diameter of the order of 1 μm), as well as larger microorganisms and particles are screened out, and accordingly collect on the outer surface of the membrane where they form a filtration cake. Viruses and mycoplasms will get through those filters (in series with cake), though; passage of mycoplasms can be prevented by using filters of 0.1 μm for nominal pore size—a practice that has become standard routine in animal cell culture.

When modelling (micro)filtration of liquid media, buildup of the aforementioned cake on its surface should be considered; therefore, the overall pressure drop, ΔP_{tot}, undergone by the liquid as it passes through the filter will have, in general, to take two components into account, namely,

$$\Delta P_{tot} = \Delta P_{filter} + \Delta P_{cake}, \tag{3.603}$$

where ΔP_{filter} denotes pressure drop due to the (porous) filter itself and ΔP_{cake} denotes pressure drop brought about by the cake. According to Hagen and Poiseuille's law (to be derived elsewhere),

$$\Delta P = \frac{8L}{R^2} \mu v; \tag{3.604}$$

here, L and R denote length and radius, respectively, of (cylindrical) pipe, μ denotes viscosity of liquid, and v denotes (average) nominal velocity thereof. Equation (3.604) is applicable under laminar regime, as typically encountered in filtration; this is due chiefly to the minute radii of the interstices available for liquid flow and large filtration area, both of which contribute to a small Reynolds' number. Since $8L/R^2$ is but a geometric factor characteristic of the pipe, one may resort to Eq. (3.604) as template to write

$$\Delta P = \mathcal{R} \mu v, \tag{3.605}$$

applicable under laminar flow in the form of percolation through a particulate bed; here, $\mathcal{R}$ denotes hydraulic resistance offered by the medium, dependent on its morphology, that is, porosity and tortuosity, besides obviously geometry. Nominal velocity can be defined as

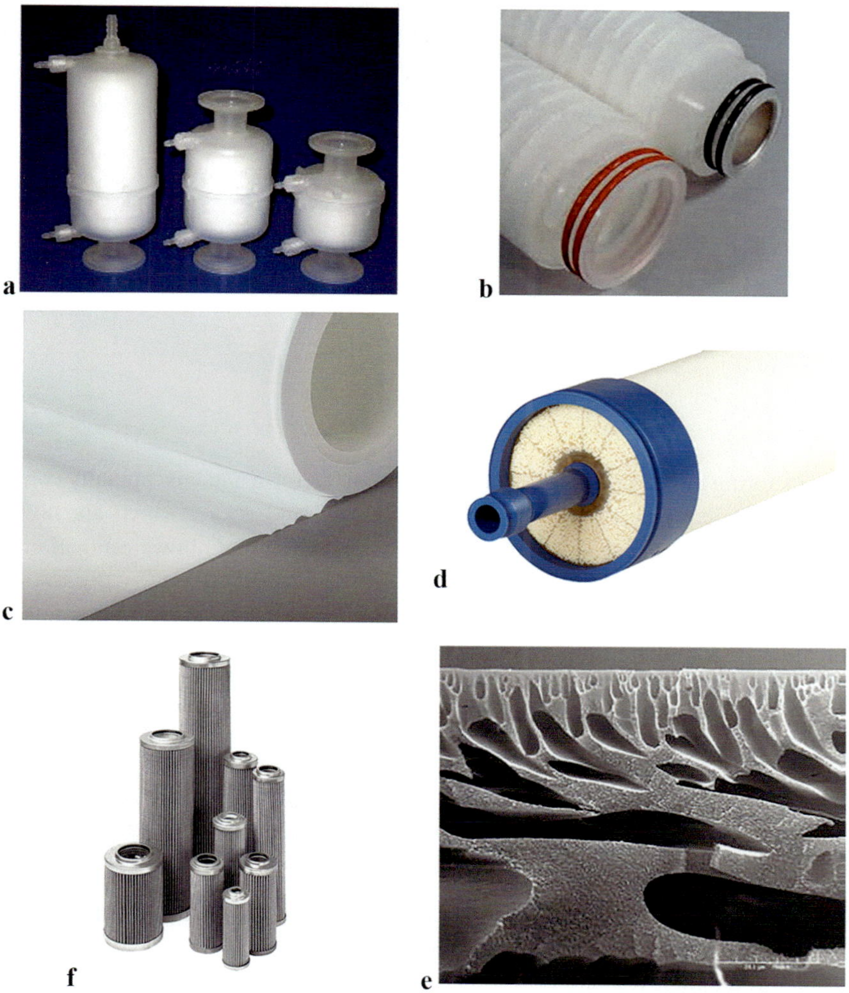

Fig. 3.22 Examples of liquid filtration media: (**a**) 0.2 μm hydrophobic capsules (courtesy of Serene, Mumbai, India) and 0.45 μm PTFE membrane filter in (**b**) cartridge form (courtesy of Critical Process Filtration, Nashua NH, USA) and (**c**) sheet form (courtesy of W. L. Gore & Associates, Newark DE, USA); (**d**) polyethersulfone/polyvynilpropylene microfiltration module (courtesy of Lenntech, Delfgauw, The Netherlands), with (**e**) electron scanning micrography of cross section (courtesy of PerfuseCell, Holte, Denmark); and (**f**) metal membrane filter in pleated form (courtesy of Parker Hannifin France SAS, France)

$$v \equiv \frac{1}{A}\frac{dV}{dt}, \tag{3.606}$$

where V denotes volume of filtrate, A denotes cross-sectional area of filter, and t denotes time; insertion of Eq. (3.606) turns Eq. (3.605) to

$$\Delta P = \mathcal{R}\mu \frac{1}{A}\frac{dV}{dt}, \tag{3.607}$$

which prompts reformulation of Eq. (3.603) to

$$\Delta P_{tot} = \mathcal{R}_{filter}\mu \frac{1}{A}\frac{dV}{dt} + \mathcal{R}_{cake}\mu \frac{1}{A}\frac{dV}{dt}, \tag{3.608}$$

or else

$$\Delta P_{tot} = \frac{\mu}{A}\left(\mathcal{R}_{filter} + \mathcal{R}_{cake}\right)\frac{dV}{dt} \tag{3.609}$$

after factoring out μ/A and dV/dt. Although simple in form and conceptually correct, use of Eq. (3.609) requires a priori knowledge of $\mathcal{R}_{filter}$ and $\mathcal{R}_{cake}$, which are not readily available in practice; hence, some model will be necessary that allows estimation thereof from easily measured physical parameters.

Toward this goal, remember that dependence of pressure drop on nominal liquid velocity, under laminar flow through a bed constituted by small spherical particles of (uniform) radius r_P, reads

$$\Delta P_{cake} = \frac{37.5}{r_P{}^2}\frac{(1-\varepsilon)^2}{\varepsilon^3}\mu L v, \tag{3.610}$$

widely known as Karman and Kozeny's equation (and to be derived in due course); here, ε and L denote porosity and thickness of bed, respectively, while μ and v denote viscosity and nominal velocity of liquid, respectively. Following insertion of Eq. (3.606), one gets

$$\Delta P_{cake} = \frac{37.5}{r_P{}^2}\frac{(1-\varepsilon)^2}{\varepsilon^3}L\mu \frac{1}{A}\frac{dV}{dt} \tag{3.611}$$

from Eq. (3.610); comparative inspection with Eqs. (3.607) and (3.611) indicates that

$$\mathcal{R}_{cake} = \frac{37.5}{r_P{}^2}\frac{(1-\varepsilon)^2}{\varepsilon^3}L, \tag{3.612}$$

so Eq. (3.609) will look like

$$\Delta P_{tot} = \frac{\mu}{A}\left(\mathcal{R}_{filter} + \frac{37.5}{r_P{}^2}\frac{(1-\varepsilon)^2}{\varepsilon^3}L\right)\frac{dV}{dt}, \tag{3.613}$$

with $\mathcal{R}_{cake} \equiv \mathcal{R}_{cake}\{L\}$ unfolding a linear function as per Eq. (3.612).

As the filtration cake builds up with time, L in Eq. (3.613) will gradually increase; in attempts to relate cake thickness and filtrate volume, one should set up a mass balance to solids as

$$LA(1-\varepsilon)\rho_P = c_P(V + LA\varepsilon), \tag{3.614}$$

where ρ_P denotes mass density of solid particles and c_P denotes starting concentration of solids in the crude liquid medium. Following elimination of parenthesis in its right-hand side, Eq. (3.614) becomes

$$LA(1-\varepsilon)\rho_P = Vc_P + LA\varepsilon c_P; \tag{3.615}$$

the left-hand side of Eq. (3.615) represents the mass of solids that constitute the cake, whereas Vc_P in the right-hand side represents the mass of solids removed from volume V of filtrate and $LA\varepsilon c_P$ represents the mass of solids still suspended in the volume of liquid filling up the pores within the cake. Factoring out of LA converts Eq. (3.615) to

$$LA\big((1-\varepsilon)\rho_P - \varepsilon c_P\big) = Vc_P; \tag{3.616}$$

isolation of L then unfolds

$$L = \frac{c_P}{A\big((1-\varepsilon)\rho_P - \varepsilon c_P\big)}V. \tag{3.617}$$

Elimination of L between Eqs. (3.613) and (3.617) gives rise to

$$\Delta P_{tot} = \frac{\mu}{A}\left(\mathcal{R}_{filter} + \frac{37.5}{r_P{}^2}\frac{(1-\varepsilon)^2}{\varepsilon^3}\frac{c_P}{A\big((1-\varepsilon)\rho_P - \varepsilon c_P\big)}V\right)\frac{dV}{dt}, \tag{3.618}$$

while division of both sides by μ/A produces

$$\left(\mathcal{R}_{filter} + \frac{37.5(1-\varepsilon)^2 c_P}{Ar_P{}^2\varepsilon^3\big((1-\varepsilon)\rho_P - \varepsilon c_P\big)}V\right)\frac{dV}{dt} = \frac{A\,\Delta P_{tot}}{\mu}; \tag{3.619}$$

Equation (3.619) conveys a differential dependence of V on t, function of a set of germane operating parameters, subjected to

$$V|_{t=0} = 0 \tag{3.620}$$

as initial condition.

When ΔP_{tot} is kept essentially constant, as often happens when a centrifugal pump is employed, then separation of variables in Eq. (3.619) is in order as

$$\int_0^V \left(\mathcal{R}_{filter} + \frac{37.5(1-\varepsilon)^2 c_P}{A r_P{}^2 \varepsilon^3 \left((1-\varepsilon)\rho_P - \varepsilon c_P\right)} \widetilde{V} \right) d\widetilde{V} = \frac{A\,\Delta P_{tot}}{\mu} \int_0^t d\widetilde{t} \tag{3.621}$$

with the aid of Eq. (3.620); the fundamental theorem of integral calculus has it that

$$\frac{1}{2} \left. \frac{\left(\mathcal{R}_{filter} + \frac{37.5(1-\varepsilon)^2 c_P}{A r_P{}^2 \varepsilon^3 \left((1-\varepsilon)\rho_P - \varepsilon c_P\right)} \widetilde{V} \right)^2}{\frac{37.5(1-\varepsilon)^2 c_P}{A r_P{}^2 \varepsilon^3 \left((1-\varepsilon)\rho_P - \varepsilon c_P\right)}} \right|_0^V = \frac{A\,\Delta P_{tot}}{\mu} \left. \widetilde{t} \right|_0^t, \tag{3.622}$$

where elimination of denominators unfolds

$$\left(\mathcal{R}_{filter} + \frac{37.5(1-\varepsilon)^2 c_P}{A r_P{}^2 \varepsilon^3 \left((1-\varepsilon)\rho_P - \varepsilon c_P\right)} V \right)^2 - \mathcal{R}_{filter}{}^2 = \frac{75(1-\varepsilon)^2 c_P}{A r_P{}^2 \varepsilon^3 \left((1-\varepsilon)\rho_P - \varepsilon c_P\right)} \frac{A\,\Delta P_{tot}}{\mu} t, \tag{3.623}$$

along with replacement of integrated functions by upper and lower limits of integration. After dropping A between numerator and denominator in the right-hand side, and moving $\mathcal{R}_{filter}{}^2$ to the right-hand side as well, Eq. (3.623) gives rise to

$$\left(\mathcal{R}_{filter} + \frac{37.5(1-\varepsilon)^2 c_P}{A r_P{}^2 \varepsilon^3 \left((1-\varepsilon)\rho_P - \varepsilon c_P\right)} V \right)^2 = \mathcal{R}_{filter}{}^2 + \frac{75(1-\varepsilon)^2 c_P\,\Delta P_{tot}}{\mu r_P{}^2 \varepsilon^3 \left((1-\varepsilon)\rho_P - \varepsilon c_P\right)} t, \tag{3.624}$$

so square roots may be taken of both sides to get

$$\mathcal{R}_{filter} + \frac{37.5(1-\varepsilon)^2 c_P}{Ar_P{}^2\varepsilon^3\left((1-\varepsilon)\rho_P - \varepsilon c_P\right)} V = \sqrt{\mathcal{R}_{filter}{}^2 + \frac{75(1-\varepsilon)^2 c_P\,\Delta P_{tot}}{\mu r_P{}^2\varepsilon^3\left((1-\varepsilon)\rho_P - \varepsilon c_P\right)} t}; \quad (3.625)$$

isolation of V finally unfolds

$$V = \frac{Ar_P{}^2\varepsilon^3\left((1-\varepsilon)\rho_P - \varepsilon c_P\right)}{37.5(1-\varepsilon)^2 c_P}\left(\sqrt{\mathcal{R}_{filter}{}^2 + \frac{75(1-\varepsilon)^2 c_P\,\Delta P_{tot}}{\mu r_P{}^2\varepsilon^3\left((1-\varepsilon)\rho_P - \varepsilon c_P\right)} t} - \mathcal{R}_{filter}\right), \quad (3.626)$$

which entails the functional form sought for $V \equiv V\{t\}$.

In most cases of practical interest, the filtration cake is not allowed to build up excessively, so the amount of suspended solids remaining in the liquid retained inside the cake interstices is small; this means that $V + LA\varepsilon \approx V$ in Eq. (3.614), and consequently $(1-\varepsilon)\rho_P - \varepsilon c_P \approx (1-\varepsilon)\rho_P$ from Eq. (3.616) on. Under these circumstances, Eq. (3.626) simplifies to

$$V = \frac{Ar_P{}^2\varepsilon^3(1-\varepsilon)\rho_P}{37.5(1-\varepsilon)^2 c_P}\left(\sqrt{\mathcal{R}_{filter}{}^2 + 75\frac{(1-\varepsilon)^2 c_P\,\Delta P_{tot}}{\mu r_P{}^2\varepsilon^3(1-\varepsilon)\rho_P} t} - \mathcal{R}_{filter}\right), \quad (3.627)$$

where $1-\varepsilon$ will cancel between numerator and denominator, and R_{filter} will factor out as

$$V = \frac{Ar_P{}^2\varepsilon^3\rho_P\mathcal{R}_{filter}}{37.5(1-\varepsilon)c_P}\left(\sqrt{1 + \frac{75(1-\varepsilon)c_P\,\Delta P_{tot}}{\mu r_P{}^2\varepsilon^3\rho_P\mathcal{R}_{filter}{}^2} t} - 1\right). \quad (3.628)$$

A further division of both sides of Eq. (3.628) by $Ar_P{}^2\varepsilon^3\rho_P\mathcal{R}_{filter}/(1-\varepsilon)c_P$ gives rise to

$$\frac{(1-\varepsilon)c_P}{Ar_P{}^2\varepsilon^3\rho_P\mathcal{R}_{filter}} V = \frac{\sqrt{1 + 75\frac{(1-\varepsilon)c_P\,\Delta P_{tot}}{\mu r_P{}^2\varepsilon^3\rho_P\mathcal{R}_{filter}{}^2} t} - 1}{37.5}; \quad (3.629)$$

Equation (3.629) is depicted in Fig. 3.23 as dimensionless volume of filtrate, $(1-\varepsilon)c_P V/Ar_P{}^2\varepsilon^3\rho_P\mathcal{R}_{filter}$, versus dimensionless time, $(1-\varepsilon)c_P\,\Delta P_{tot}t/\mu r_P{}^2\varepsilon^3\rho_P\mathcal{R}_{filter}{}^2$. Note the square-root shape of the curve, encompassing a steep departure from zero, which gradually slows down as time elapses from the maximum at startup (as indicated); the pressure drop remains essentially constant.

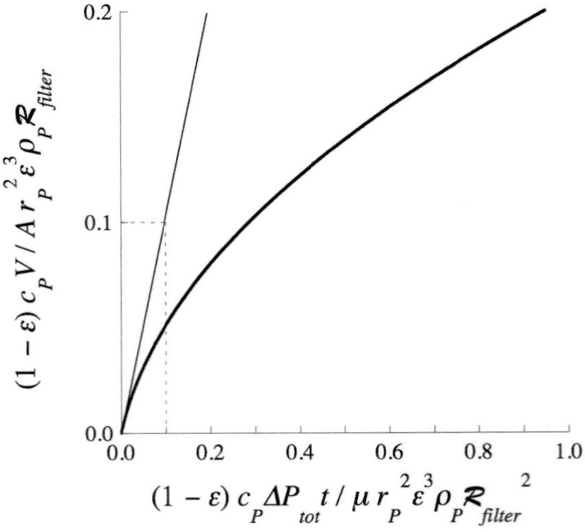

Fig. 3.23 Variation of dimensionless volume of filtrate, $(1-\varepsilon)c_P V/A{r_P}^2\varepsilon^3\rho_P\mathcal{R}_{filter}$, versus dimensionless time, $(1-\varepsilon)c_P\,\Delta P_{tot}t/\mu {r_P}^2\varepsilon^3\rho_P{\mathcal{R}_{filter}}^2$ (——), with indication of maximum rate of filtration (___)

If the volumetric filtration rate, Q, is kept essentially constant, as happens when a positive displacement pump is employed, then Eq. (3.619) becomes

$$\left(\mathcal{R}_{filter}+\frac{37.5(1-\varepsilon)^2c_P}{A{r_P}^2\varepsilon^3\left((1-\varepsilon)\rho_P-\varepsilon c_P\right)}V\right)Q=\frac{A\,\Delta P_{tot}}{\mu}, \tag{3.630}$$

on account of

$$\frac{dV}{dt}=Q; \tag{3.631}$$

integration of Eq. (3.631), again with the aid of Eq. (3.620) for initial condition, leads to

$$\int_0^V d\tilde{V}=Q\int_0^t d\tilde{t}, \tag{3.632}$$

which promptly becomes

$$\tilde{V}\Big|_0^V=Q\,\tilde{t}\Big|_0^t \tag{3.633}$$

or, equivalently,

$$V = Qt. \tag{3.634}$$

Insertion of Eq. (3.634) converts Eq. (3.630) to

$$\left(\mathcal{R}_{filter} + \frac{37.5(1-\varepsilon)^2 c_P}{A r_P{}^2 \varepsilon^3 \left((1-\varepsilon)\rho_P - \varepsilon c_P \right)} Qt \right) Q = \frac{A \Delta P_{tot}}{\mu}, \tag{3.635}$$

where isolation of $\Delta P_{tot}\{t\}$ unfolds

$$\Delta P_{tot} = \frac{\mu Q}{A} \left(\mathcal{R}_{filter} + \frac{37.5(1-\varepsilon)^2 c_P Q}{A r_P{}^2 \varepsilon^3 \left((1-\varepsilon)\rho_P - \varepsilon c_P \right)} t \right). \tag{3.636}$$

If the approximation $(1 - \varepsilon)\rho_P - \varepsilon c_P \approx (1 - \varepsilon)\rho_P$ holds (as explored before), then Eq. (3.636) simplifies to

$$\Delta P_{tot} = \frac{\mu Q}{A} \left(\mathcal{R}_{filter} + \frac{37.5(1-\varepsilon)^2 c_P Q}{A r_P{}^2 \varepsilon^3 (1-\varepsilon)\rho_P} t \right), \tag{3.637}$$

which simplifies further to

$$\Delta P_{tot} = \frac{\mu Q}{A} \left(\mathcal{R}_{filter} + \frac{37.5(1-\varepsilon) c_P Q}{A r_P{}^2 \varepsilon^3 \rho_P} t \right) \tag{3.638}$$

once $1 - \varepsilon$ is dropped between numerator and denominator; to reduce overparameterization, both sides may to advantage be divided by $\mu Q R_{filter}/A$ to obtain

$$\frac{A}{\mu Q \mathcal{R}_{filter}} \Delta P_{tot} = 1 + 37.5 \frac{(1-\varepsilon) c_P Q}{A r_P{}^2 \varepsilon^3 \rho_P \mathcal{R}_{filter}} t, \tag{3.639}$$

as depicted in Fig. 3.24 using $A \Delta P_{tot} / \mu Q \mathcal{R}_{filter}$ as dimensionless pressure drop and $(1-\varepsilon) c_P Q t / A r_P{}^2 \varepsilon^3 \rho_P \mathcal{R}_{filter}$ as dimensionless time. In this case, the volumetric flow rate through the filter plus cake remains constant, but the associated pressure drop increases linearly with time after departing from a non-nil vertical intercept; this a consequence of cake thickness growing linearly with filtrate volume as per Eq. (3.617), and thus also with time in view of Eq. (3.610).

Fig. 3.24 Variation of dimensionless volume of filtrate, $A\Delta P_{tot}/\mu Q\mathcal{R}_{filter}$, versus dimensionless time, $(1-\varepsilon)c_P Qt/Ar_P{}^2\varepsilon^3\rho_P\mathcal{R}_{filter}$

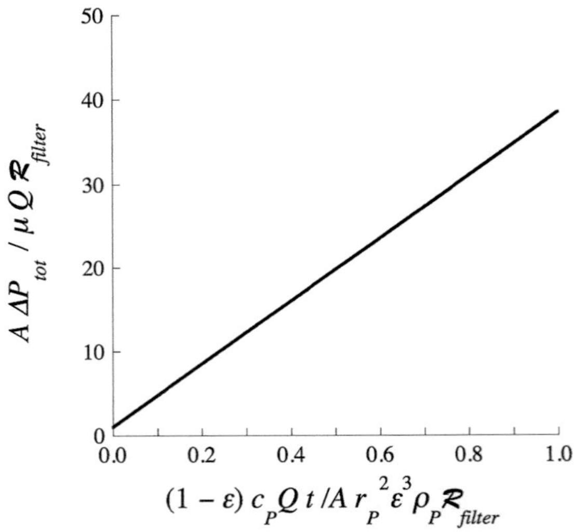

Problem 3.12
Consider the general case of surface filtration, concomitant with cake buildup, resorting to either centrifugal or positive displacement pumping devices.

(a) Find the linear asymptote of the $(1-\varepsilon)c_P V/Ar_P{}^2\varepsilon^3\rho_P\mathcal{R}_{filter}$-vs.-$(1-\varepsilon)c_P\,\Delta P_{tot}t/\mu r_P{}^2\varepsilon^3\rho_P\mathcal{R}_{filter}{}^2$ plot in the vicinity of startup, should filtration be driven by a centrifugal pump.
(b) For a given filter, investigate the effect of each operating parameter, in attempts to minimize (only) the rate of increase in pressure due to cake buildup, should a positive displacement pump be employed.

Solution

(a) After simplifying notation in Eq. (3.629) to

$$\alpha V = \frac{\sqrt{1+75\beta t}-1}{37.5}, \tag{3.640}$$

coupled with

$$\alpha \equiv \frac{(1-\varepsilon)c_P}{Ar_P{}^2\varepsilon^3\rho_P\mathcal{R}_{filter}} \tag{3.641}$$

and

(continued)

Problem 3.12 (continued)

$$\beta \equiv \frac{(1-\varepsilon)c_P \Delta P_{tot}}{\mu r_P{}^2 \varepsilon^3 \rho_P \mathcal{R}_{filter}{}^2} \tag{3.642}$$

as definition of auxiliary parameters α and β, respectively, one may proceed to multiplication of both numerator and denominator of Eq. (3.640) by the conjugate binomial of the latter, that is,

$$\alpha V = \frac{(\sqrt{1+75\beta t}-1)(\sqrt{1+75\beta t}+1)}{37.5(\sqrt{1+75\beta t}+1)}; \tag{3.643}$$

Equation (3.643) breaks down to

$$\alpha V = \frac{(\sqrt{1+75\beta t})^2 - 1^2}{37.5(\sqrt{1+75\beta t}+1)} = \frac{1+75\beta t-1}{37.5(\sqrt{1+75\beta t}+1)} = \frac{75\beta t}{37.5(\sqrt{1+75\beta t}+1)} = \frac{2\beta t}{1+\sqrt{1+75\beta t}} \tag{3.644}$$

since the numerator is but a difference of squares, and after cancelling symmetrical terms in numerator, and finally dropping common factors between numerator and denominator. At early times, $1 + 75\beta t \approx 1$, so Eq. (3.644) reduces to

$$\lim_{t \to 0} \alpha V = \lim_{t \to 0} \frac{2\beta t}{1+\sqrt{1}} = \frac{2\beta t}{2} = \beta t; \tag{3.645}$$

one therefore concludes on the unit slope of the tangent to the aforementioned plot of αV versus βt, as outlined via the thin straight line in Fig. 3.23.

(b) Minimizing the rate of increase in ΔP_{tot} as given by Eq. (3.639), without changing its departing value, is mathematically equivalent to decreasing its slope with regard to t. This implies decreasing Qc_P in its numerator, thus reducing the load of suspended particles meant for removal per unit time; or else increasing A, $r_P{}^2$, or ρ_P in denominator, thus increasing filtration area or particle radius that facilitate permeation of cake by liquid medium, or increasing density of cake that reduces liquid path. The outstanding effect of cake porosity deserves extra study, since ε appears explicitly in both numerator (as $1-\varepsilon$) and denominator (as ε^3) of the aforementioned slope; one should thus resort to

(continued)

Problem 3.12 (continued)

$$\frac{\partial}{\partial \varepsilon}\left(\frac{37.5(1-\varepsilon)c_P Q}{A r_P{}^2 \varepsilon^3 \rho_P \mathcal{R}_{filter}}\right) = \frac{37.5 c_P Q}{A r_P{}^2 \rho_P \mathcal{R}_{filter}} \frac{d}{d\varepsilon}\left(\frac{1-\varepsilon}{\varepsilon^3}\right), \tag{3.646}$$

which stands for (partial) derivative of slope of the straight line in Eq. (3.639) with regard to ε. Application of the rule of differentiation of a quotient converts Eq. (3.646) to

$$\begin{aligned}\frac{\partial}{\partial \varepsilon}\left(\frac{37.5(1-\varepsilon)c_P Q}{A r_P{}^2 \varepsilon^3 \rho_P \mathcal{R}_{filter}}\right) &= \frac{37.5 c_P Q}{A r_P{}^2 \rho_P \mathcal{R}_{filter}} \frac{(-1)\varepsilon^3 - (1-\varepsilon)3\varepsilon^2}{(\varepsilon^3)^2} \\ &= -\frac{37.5 c_P Q}{A r_P{}^2 \rho_P \mathcal{R}_{filter}} \frac{\varepsilon\Big(\varepsilon + 3(1-\varepsilon)\Big)}{\varepsilon^2 \varepsilon^4},\end{aligned} \tag{3.647}$$

where dropping of ε^2 between numerator and denominator, followed by condensation of terms alike unfold

$$\begin{aligned}\frac{\partial}{\partial \varepsilon}\left(\frac{37.5(1-\varepsilon)c_P Q}{A r_P{}^2 \varepsilon^3 \rho_P \mathcal{R}_{filter}}\right) &= -\frac{37.5 c_P Q}{A r_P{}^2 \rho_P \mathcal{R}_{filter}} \frac{\varepsilon + 3(1-\varepsilon)}{\varepsilon^4} \\ = -\frac{37.5 c_P Q}{A r_P{}^2 \rho_P \mathcal{R}_{filter}} \frac{\varepsilon + 3 - 3\varepsilon}{\varepsilon^4} &= -\frac{37.5 c_P Q}{A r_P{}^2 \rho_P \mathcal{R}_{filter}} \frac{3-2\varepsilon}{\varepsilon^4};\end{aligned} \tag{3.648}$$

in view of $0 < \varepsilon < 1$ by definition, and consequently $0 < 1 < 3 - 2\varepsilon < 3$, one finds that

$$\frac{\partial}{\partial \varepsilon}\left(\frac{37.5(1-\varepsilon)c_P Q}{A r_P{}^2 \varepsilon^3 \rho_P \mathcal{R}_{filter}}\right) < 0 \tag{3.649}$$

as $37.5 c_P Q / A r_P{}^2 \rho_P \mathcal{R}_{filter} > 0$ and $\varepsilon^4 > 0$. One therefore concludes that an increase in porosity favors a slower increase in pressure buildup as filtration progresses, because more room is available as interstices for permeation by fluid.

3.4.2.2 Gaseous Media

When cultures are processed aerobically, irrespective of batch or continuous mode of operation, air must be supplied on a permanent basis due to the poor solubility of oxygen in aqueous media, along with its extensive use for respiratory metabolism. Loads of 10^3–10^4 microorganisms m_{air}^{-3} are typical, usually carried by solid or liquid aerosols, so sterilization of air in advance is a must; at large-scale operation, heat sterilization of air would be impractical, so filtration has become a routine tool.

Compacted beds, or pads of fibrous materials (e.g., glass wood) known as depth filters, have been widely employed (see Fig. 3.25). The interfiber distance ranges within 2–10 μm, that is, ca. tenfold the diameter of target bacteria and spores; hence, they penetrate the filter to varying depth before being eventually arrested, thus gradually decreasing membrane porosity—unlike happens with surface filtering of liquid media, where particles accumulate on the surface to form a cake. Combination of various layers of interwoven or cross-laid material is efficient toward circumventing fast clogging, while pressure drop is kept below a reasonable threshold; for a prespecified quality of the emerging air, the depth of the filter required depends on both volumetric flow rate and contamination load of the source air.

A combination of impaction, interception, electrostatic interaction, and even diffusion accounts for the phenomenological performance of air filters. Special care is to be exercised

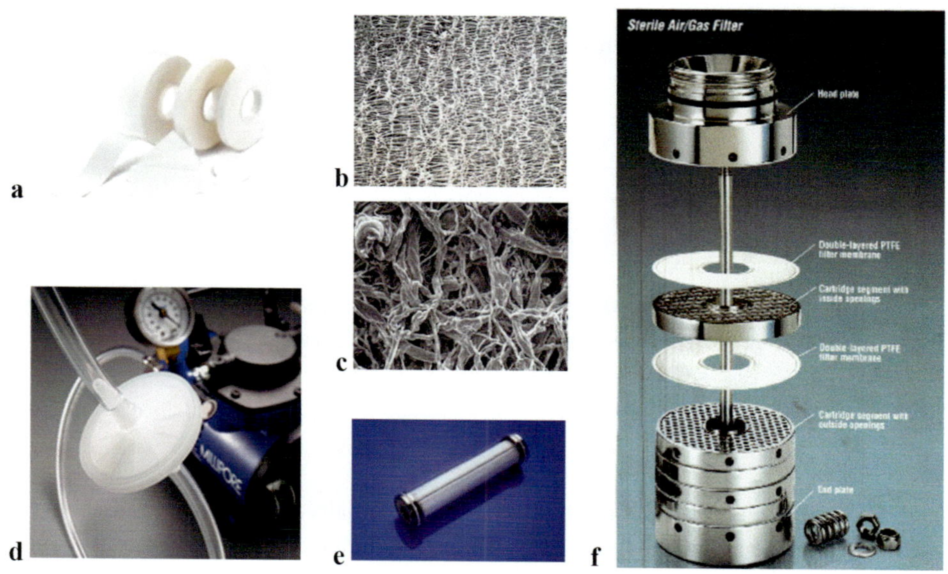

Fig. 3.25 Examples of air filtration media: (**a**) hydrophobic PTFE membranes for venting (courtesy of Sartorius, Goettingen, Germany), with (**b**, **c**) micrographic details of porous matrices (courtesy of Advantech MFS, Dublin CA, USA); (**d**) in-line gas filter (courtesy of Merck, Burlington MA, USA); and filtration compact module in (**e**) cartridge form (courtesy of Applied Membrane Technology, Minnetonka MN, USA) or (**f**) segmented design (courtesy of Micropure, Cleveland OH, USA)

to avoid increase in air flow rate beyond specification or intake of very moist air. The former carries suspended particles with kinetic energy that may be sufficiently high to span the whole depth of the filter; whereas the latter may lead to water condensation inside the filter, which not only increases pressure drop but also causes channeling of gas flow and provides conditions suitable for growth of microorganisms meanwhile collected within the porous matrix.

To simulate performance of a depth filter, one should revisit Eq. (3.609) as merely

$$\Delta P = \frac{\mu}{A} \mathcal{R}_{filter} \frac{dV}{dt}, \tag{3.650}$$

because no cake builds on its surface (and consequently $R_{cake} \approx 0$); hence, pressure drop arises solely from passage through the filter material, with Eq. (3.650) satisfying again

$$V|_{t=0} = 0 \tag{3.651}$$

as initial condition (see Eq. (3.620)). Since the particles suspended in air will actually penetrate the porous matrix of the filter and accumulate therein, its fraction of voids, ε_f, will gradually decrease in time. The resistance raised by the filter to fluid flow will accordingly abide to

$$\mathcal{R}_{filter} = \kappa \frac{1-\varepsilon_f}{\varepsilon_f} \tag{3.652}$$

in its simplest form, where κ denotes a proportionality constant specific for each material and manufacture process utilized; insertion in Eq. (3.607) leads to

$$\Delta P = \frac{\mu\kappa}{A} \frac{1-\varepsilon_f}{\varepsilon_f} \frac{dV}{dt}, \tag{3.653}$$

to be used in lieu of Eq. (3.610). Despite its essentially empiric nature, it should be emphasized that the functional form of Eq. (3.652) accommodates two obvious asymptotic behaviors, namely,

$$\lim_{\varepsilon_f \to 0} \mathcal{R}_{filter} = \kappa \frac{1-0}{0} = \frac{\kappa}{0} = \infty \tag{3.654}$$

and

$$\lim_{\varepsilon_f \to 1} \mathcal{R}_{filter} = \kappa \frac{1-1}{1} = \kappa 0 = 0; \tag{3.655}$$

Equation (3.654) implies an infinite resistance, as expected from a compact matrix with no void spaces available for fluid flow, whereas Eq. (3.655) accommodates a nil resistance, as anticipated if no solid medium existed at all.

A mass balance to solids reads, in this case,

$$V_f\Big(\big(1-\varepsilon_f\big) - \big(1-\varepsilon_{f,0}\big)\Big)\rho_P = c_P\big(V + V_f\varepsilon_f\big) \tag{3.656}$$

as alias to Eq. (3.614), which breaks down to

$$V_f\big(\varepsilon_{f,0} - \varepsilon_f\big)\rho_P = c_P\big(V + V_f\varepsilon_f\big) \tag{3.657}$$

upon removing inner parentheses and cancelling unity with its negative afterward; here, V_f denotes filter volume, $\varepsilon_{f,0}$ denotes initial filter porosity, ρ_P denotes mass density of particles retained, c_P denotes mass concentration of particles in air, and V denotes volume of air filtrate. After getting rid of parentheses, Eq. (3.657) becomes

$$\varepsilon_{f,0}\rho_P V_f - \varepsilon_f \rho_P V_f = c_P V + \varepsilon_f c_P V_f, \tag{3.658}$$

where ε_f and V_f will factor out as

$$V_f(\rho_P + c_P)\varepsilon_f = \varepsilon_{f,0}\rho_P V_f - c_P V; \tag{3.659}$$

isolation of ε_f prompts

$$\varepsilon_f = \frac{\varepsilon_{f,0}\rho_P V_f - c_P V}{(\rho_P + c_P)V_f} = \frac{\varepsilon_{f,0}\rho_P}{\rho_P + c_P} - \frac{c_P}{(\rho_P + c_P)V_f} V \tag{3.660}$$

along with splitting of the resulting fraction, or else

$$\varepsilon_f = \frac{\varepsilon_{f,0}}{1 + \frac{c_P}{\rho_P}} - \frac{\frac{c_P}{\rho_P}}{\left(1 + \frac{c_P}{\rho_P}\right)V_f} V \tag{3.661}$$

after dividing both numerator and denominator of both terms by ρ_P. Once Eq. (3.653) is revisited as

$$\Delta P = \frac{\mu\kappa}{A}\left(\frac{1}{\varepsilon_f} - 1\right)\frac{dV}{dt} \tag{3.662}$$

upon splitting of the second factor in its right-hand side, insertion of Eq. (3.661), as $\varepsilon_f \equiv \varepsilon_f\{V\}$, ensues to get

$$\Delta P = \frac{\mu\kappa}{A}\left(\frac{1}{\frac{\varepsilon_{f,0}}{1+\frac{c_P}{\rho_P}} - \frac{\frac{c_P}{\rho_P}}{\left(1+\frac{c_P}{\rho_P}\right)V_f}V} - 1\right)\frac{dV}{dt}; \tag{3.663}$$

Equation (3.663) gives rise to

$$\Delta P = \frac{\mu\kappa}{A}\left(\frac{1+\frac{c_P}{\rho_P}}{\varepsilon_{f,0} - \frac{\frac{c_P}{\rho_P}}{V_f}V} - 1\right)\frac{dV}{dt}, \tag{3.664}$$

upon multiplication of both numerator and denominator of the first term in parenthesis by $1 + c_P/\rho_P$.

When centrifugal pumping is at stake, both V and dV/dt vary in time since an essentially constant ΔP is delivered; hence, Eq. (3.664) integrates, via separation of variables, to

$$\int_0^V \left(\frac{1+\frac{c_P}{\rho_P}}{\varepsilon_{f,0} - \frac{\frac{c_P}{\rho_P}}{V_f}\tilde{V}} - 1\right)d\tilde{V} = \frac{A\,\Delta P}{\mu\kappa}\int_0^t d\tilde{t}, \tag{3.665}$$

where Eq. (3.651) was conveniently brought on board and the constancy of $A\,\Delta P/\mu\kappa$ taken advantage of. The fundamental theorem of integral calculus may now be invoked to write

$$\left.\left(\left(1+\frac{c_P}{\rho_P}\right)\frac{\ln\left\{\varepsilon_{f,0} - \frac{\frac{c_P}{\rho_P}}{V_f}\tilde{V}\right\}}{-\frac{\frac{c_P}{\rho_P}}{V_f}} - \tilde{V}\right)\right|_0^V = \frac{A\,\Delta P}{\mu\kappa}\,\tilde{t}\,\Big|_0^t \tag{3.666}$$

instead of Eq. (3.665), which reduces to

$$-\left(1+\frac{c_P}{\rho_P}\right)\frac{\ln\left\{\varepsilon_{f,0}-\frac{\frac{c_P}{\rho_P}}{V_f}V\right\}}{\frac{\frac{c_P}{\rho_P}}{V_f}}-V+\left(1+\frac{c_P}{\rho_P}\right)\frac{\ln\left\{\varepsilon_{f,0}-0\right\}}{\frac{\frac{c_P}{\rho_P}}{V_f}}=\frac{A\,\Delta P}{\mu\kappa}(t-0) \quad (3.667)$$

upon replacement of the indicated functions by upper and lower limits of integration. Condensation of the two logarithmic terms converts Eq. (3.667) to

$$\frac{1+\frac{c_P}{\rho_P}}{\frac{\frac{c_P}{\rho_P}}{V_f}}\ln\frac{\varepsilon_{f,0}}{\varepsilon_{f,0}-\frac{\frac{c_P}{\rho_P}}{V_f}V}-V=\frac{A\,\Delta P}{\mu\kappa}t, \quad (3.668)$$

which will be solved for t as

$$t=\frac{\mu\kappa}{A\,\Delta P}\left(\frac{1+\frac{c_P}{\rho_P}}{\frac{\frac{c_P}{\rho_P}}{V_f}}\ln\frac{\varepsilon_{f,0}}{\varepsilon_{f,0}-\frac{\frac{c_P}{\rho_P}}{V_f}V}-V\right) \quad (3.669)$$

or, after multiplying both numerator and denominator of the argument of the logarithm by V_f, and reducing the number of fraction levels,

$$t=\frac{\mu\kappa}{A\,\Delta P}\left(\frac{1+\frac{c_P}{\rho_P}}{\frac{c_P}{\rho_P}}V_f\ln\frac{\varepsilon_{f,0}V_f}{\varepsilon_{f,0}V_f-\frac{c_P}{\rho_P}V}-V\right); \quad (3.670)$$

therefore, a functional relationship has been reached for $V \equiv V\{t\}$, although in implicit form.

As discussed previously, the inventory of otherwise suspended material meanwhile accumulated within the porous matrix is normally of a lesser importance—this realization being especially true when the carrier thereof is a gas; in other words, $V_f\varepsilon_f$ becomes small compared to V in Eq. (3.657) after a short time upon startup, and thus $(\rho_P + c_P)V_f \approx \rho_P V_f$ in Eq. (3.660), or else $1 + c_P/\rho_P \approx 1$ in Eq. (3.661) and following ones. Under these circumstances, Eq. (3.670) reduces to just

$$t=\frac{\mu\kappa}{A\,\Delta P}\left(\frac{V_f}{\frac{c_P}{\rho_P}}\ln\frac{\varepsilon_{f,0}V_f}{\varepsilon_{f,0}V_f-\frac{c_P}{\rho_P}V}-V\right), \quad (3.671)$$

where multiplication of both sides by $A\,\Delta P/\mu\kappa$ unfolds

$$\frac{A\,\Delta P}{\mu\kappa}t = \frac{V_f}{\frac{c_P}{\rho_P}} \ln \frac{\varepsilon_{f,0} V_f}{\varepsilon_{f,0} V_f - \frac{c_P}{\rho_P} V} - V; \tag{3.672}$$

a further multiplication of both sides by $c_P/\rho_P V_f$ gives rise to

$$\frac{A c_P \Delta P}{\mu\kappa\rho_P V_f}t = \ln \frac{\varepsilon_{f,0}}{\varepsilon_{f,0} - \frac{c_P}{\rho_P V_f} V} - \frac{c_P}{\rho_P V_f} V, \tag{3.673}$$

together with division of both numerator and denominator of the argument of the logarithm by V_f. A plot of dimensionless volume, $c_P V/\rho_P V_f$, as a function of dimensionless time, $A c_P\, \Delta P\, t/\mu\kappa\rho_P V_f$, is provided in Fig. 3.26 for selected values of (germane) parameter $\varepsilon_{f,0}$. The volume of filtrate increases rapidly with filtration time at startup, and then at a slower and slower pace as the filter gets clogged with otherwise suspended material in air. The rate of filtration, or slope of the straight-line tangent to the curve at each point, is highest when the largest fraction of voids are initially available, as measured by $\varepsilon_{f,0}$; such slope mathematically satisfies, in general,

$$\frac{dV}{dt} = \frac{\frac{A\,\Delta P}{\mu\kappa}}{\frac{1}{\varepsilon_{f,0} - \frac{c_P}{\rho_P V_f} V} - 1}, \tag{3.674}$$

stemming from Eq. (3.664) after setting $1 + c_P/\rho_P \approx 1$.

Since the solid particles will accumulate inside the finite volume of voids provided by the porous filter, an upper limit exists for its loading capacity—unlike happened with

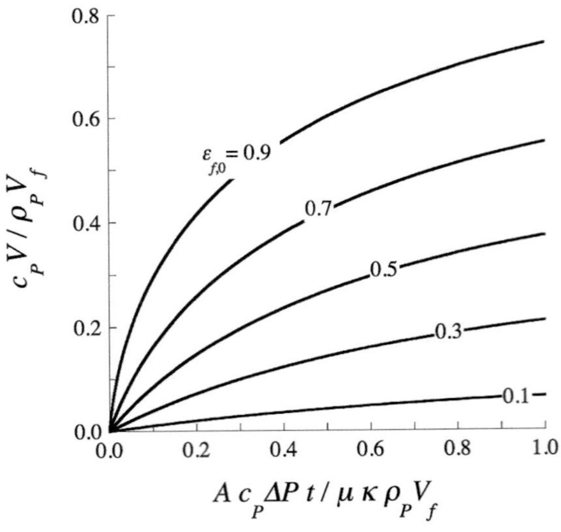

Fig. 3.26 Variation of dimensionless filtrate volume, $c_P V/\rho_P V_f$, versus dimensionless filtration time, $A c_P\, \Delta P\, t/\mu\kappa\rho_P V_f$, for selected values of dimensionless parameter $\varepsilon_{f,0}$

surface filtration, where the solid particles accumulate as a cake allowed to grow outward; such asymptotic behavior can be ascertained from Eq. (3.673) via

$$\lim_{\frac{c_P}{\rho_P V_f}V \to \frac{c_P}{\rho_P V_f}V_{cr}} \frac{A c_P \Delta P}{\mu \kappa \rho_P V_f} t = \infty \;, \tag{3.675}$$

with V_{cr} denoting critical (or maximum) volume of filtrate possible before full clogging of filter. Equation (3.675) should reflect an infinite value for the argument of the logarithmic function in Eq. (3.673), since $c_P V/\rho_P V_f$ as such must remain finite; this condition, associated to

$$\varepsilon_{f,0} - \frac{c_P}{\rho_P V_f} V_{cr} = 0 \tag{3.676}$$

since $\varepsilon_{f,0}/0 = \infty$, may be algebraically rearranged to yield

$$\frac{c_P}{\rho_P V_f} V_{cr} = \varepsilon_{f,0}. \tag{3.677}$$

After retrieving Eq. (3.657) in its approximate version as

$$V_f \left(\varepsilon_{f,0} - \varepsilon_f \right) \rho_P = c_P V \tag{3.678}$$

consistent with postulated $V_f \varepsilon_f \ll V$, elimination of $\varepsilon_{f,0}$ becomes possible at the expense of Eq. (3.677), namely,

$$V_f \left(\frac{c_P}{\rho_P V_f} V_{cr} - \varepsilon_f \big|_{V=V_{cr}} \right) \rho_P = c_P V_{cr}; \tag{3.679}$$

removal of parenthesis then unfolds

$$c_P V_{cr} - \varepsilon_f \big|_{V=V_{cr}} \rho_P V_f = c_P V_{cr}, \tag{3.680}$$

where common terms will be dropped from both sides to get

$$\varepsilon_f|_{V=V_{cr}}\, \rho_P V_f = 0 \tag{3.681}$$

or, equivalently,

$$\varepsilon_f|_{V=V_{crt}} = 0 \tag{3.682}$$

as per $\rho_P V_f \neq 0$. Equation (3.682) indicates that, after filtering a volume V_{cr} of air, the porosity of the filter turns nil, so no more filtration will be feasible; therefore, Eq. (3.677) describes the vertical intercept of the horizontal asymptotes that drive the behavior of the curves in Fig. 3.26, after a sufficiently long time undergoing filtration.

In case the air is pumped by a positive displacement device, its volumetric flow rate, Q, will be constant, and so will the derivative of filtrate volume with regard to time, that is,

$$\frac{dV}{dt} = Q \tag{3.683}$$

to be hereby used instead of Eq. (3.674); note that Eq. (3.683) supports a linear relationship between V and t, according to

$$\int dV = Q \int dt \tag{3.684}$$

that degenerates to

$$V = k + Qt \tag{3.685}$$

with k denoting an arbitrary integration constant. Compatibility of Eq. (3.685) with Eq. (3.651) as initial condition enforces

$$0 = k + Q0 = k, \tag{3.686}$$

which may be inserted back in Eq. (3.685) to produce merely

$$V = Qt; \tag{3.687}$$

Equation (3.687) is identical to Eq. (3.634), as expected. In view of Eqs. (3.683) and (3.687), it becomes possible to redo Eq. (3.664) as

$$\Delta P = \frac{\mu \kappa}{A} \left(\frac{1 + \frac{c_P}{\rho_P}}{\varepsilon_{f,0} - \frac{\frac{c_P}{\rho_P}}{V_f} Qt} - 1 \right) Q; \tag{3.688}$$

a mathematical descriptor for the increase in pressure drop with time has accordingly been produced. Recalling again the simplifying hypothesis $1 + c_P/\rho_P \approx 1$, reasonable as far as a suspension of solids in a gas is under scrutiny, Eq. (3.688) reduces to

$$\Delta P = \frac{\mu \kappa}{A} \left(\frac{1}{\varepsilon_{f,0} - \frac{\frac{c_P}{\rho_P}}{V_f} Qt} - 1 \right) Q \tag{3.689}$$

that conveys $\Delta P \equiv \Delta P\{t\}$ through a hyperbolic functionality. Multiplication of both sides by $A/\mu\kappa Q$, coupled with reduction of fraction levels reduce overparameterization of Eq. (3.689) to

$$\frac{A}{\mu \kappa Q} \Delta P = \frac{1}{\varepsilon_{f,0} - \frac{c_P Q}{\rho_P V_f} t} - 1 \tag{3.690}$$

as depicted in Fig. 3.27. The pressure drop develops as more air undergoes filtration, departing from

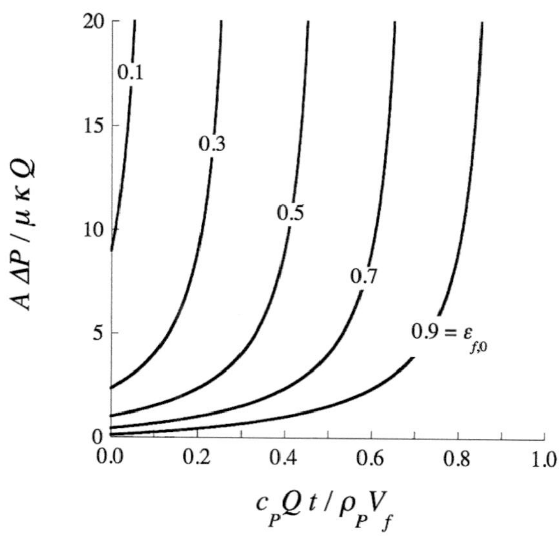

Fig. 3.27 Variation of dimensionless pressure drop, $A\,\Delta P/\mu\kappa Q$, versus dimensionless time, $c_P Qt/\rho_P V_f$, for selected values of dimensionless parameter $\varepsilon_{f,0}$

$$\lim_{\frac{c_P Q}{\rho_P V_f} t \to 0} \frac{A}{\mu \kappa Q} \Delta P = \frac{1}{\varepsilon_{f,0} - 0} - 1 = \frac{1}{\varepsilon_{f,0}} - 1 \tag{3.691}$$

based on Eq. (3.690) as descriptor of the vertical intercept of the curves; such increase in ΔP becomes faster and faster as time goes by, to be eventually driven by a vertical asymptote satisfying

$$\lim_{\frac{c_P Q}{\rho_P V_f} t \to \varepsilon_{f,0}} \frac{A}{\mu \kappa Q} \Delta P = \frac{1}{\varepsilon_{f,0} - \varepsilon_{f,0}} - 1 = \frac{1}{0} - 1 = \infty - 1 = \infty \tag{3.692}$$

under which conditions the pump would obviously break down. Therefore, when more void space is initially available, filtration can be carried out for a longer period of time before the increase in pressure reaches some predefined upper threshold.

Depth filters have been increasingly replaced by membrane cartridge filters, especially in industrial applications; these are constituted by heat-sterilizable polymeric membranes, and act instead as surface filters when trapping contaminants (as in a sieve). To maximize filtering area per unit volume, such cartridges come with a pleated filter, containing small and uniformly sized pores of diameter below 0.45 μm; the material is usually hydrophobic, so as to minimize chance of wetting. To reduce rapid fouling of the membrane caused by larger particles (e.g., oil or water droplets, or even foam) suspended in the incoming air, prefilters are normally provided upstream or as a part of the cartridge itself; on the other hand, membrane filters themselves must be steam sterilized in advance, or subjected to γ- or X-radiation prior to use.

Filters are also used to sterilize gases leaving the bioreactor, thus avoiding release of any microorganisms entrained in aerossols, which form in its overhead portion as a consequence of bubble collapse or just intense stirring. Since bioreactors are regularly operated under high cell densities, the concentration of cells in those aerosols will likely exceed by orders of magnitude those found in typical air. Containment is thus a must, when said microorganisms are potentially harmful to plant personnel or the environment—as happens with pathogens or recombinant strains.

Problem 3.13

Although an upper limit exists for the useful life of an air filter, corresponding to its complete clogging, replacement thereof is supposed to take place at an earlier stage—well before V reaches V_{cr}.

(continued)

Problem 3.13 (continued)

(a) If the overall (dead) time taken to disassemble, clean/replace, and reassemble the filter is denoted by t_{cln}, devise a graphical strategy to find the optimum time, t_{opt}, for replacement of the filter, using maximum volumetric productivity of filtration as objective function, in the presence of centrifugal pumping.

(b) Plot t_{opt}-vs.-t_{cln}, as obtained in *a*), but in dimensionless form, for selected values of $\varepsilon_{f,0}$, and comment.

(c) Inspection of the functional form of Eq. (3.673) unfolds two terms in the right-hand side with opposing effects upon its left hand side. Prove that *t* is, nevertheless, a monotonic function of *V*, within the range of physical interest.

Solution

(a) The volumetric productivity, η, should hereby be defined by

$$\eta \equiv \frac{V\{t\}}{t + t_{cln}}, \tag{3.693}$$

that is, volume filtered divided by underlying filtration time, *t*, additively corrected by cleaning/replacement time, t_{cln}. The necessary condition for maximum productivity will accordingly read

$$\left.\frac{d\eta}{dt}\right|_{t=t_{opt}} = 0, \tag{3.694}$$

with t_{opt} denoting best filtration time; Eq. (3.694) becomes

$$\left.\frac{d}{dt}\left(\frac{V\{t\}}{t + t_{cln}}\right)\right|_{t_{opt}} = 0 \tag{3.695}$$

upon insertion of Eq. (3.693). Application of the rule of differentiation of a quotient transforms Eq. (3.695) to

$$\left.\frac{\frac{dV}{dt}(t + t_{cln}) - V}{(t + t_{cln})^2}\right|_{t_{opt}} = 0, \tag{3.696}$$

which degenerates to

(continued)

Problem 3.13 (continued)

$$\left.\frac{dV}{dt}(t+t_{cln})\right|_{t_{opt}}=V\big|_{t_{opt}} \tag{3.697}$$

after multiplying both sides by $(t + t_{cln})^2 \neq 0$; a final division of both sides by $t + t_{cln} \neq 0$ unfolds

$$\left.\frac{dV}{dt}\right|_{t_{opt}}=\left.\frac{V}{t+t_{cln}}\right|_{t_{opt}}. \tag{3.698}$$

A graphical interpretation of Eq. (3.698) is provided by Fig. 3.28. According to Eq. (3.698), t_{opt} is the time when the straight line, defined by points of coordinates $(-t_{cln}, 0)$ and $(t_{opt}, V\{t_{opt}\})$, is parallel to (and actually coincident with) the straight-line tangent to the $V\{t\}$ curve at point of coordinates $(t_{opt}, V\{t_{opt}\})$. To investigate the nature of this stationary point, one should calculate the second-order derivative (in its vicinity), namely,

$$\frac{d^2}{dt^2}\left(\frac{V\{t\}}{t+t_{cln}}\right)\equiv\frac{d}{dt}\left(\frac{d}{dt}\left(\frac{V\{t\}}{t+t_{cln}}\right)\right), \tag{3.699}$$

where Eq. (3.696) can be recalled to obtain

(continued)

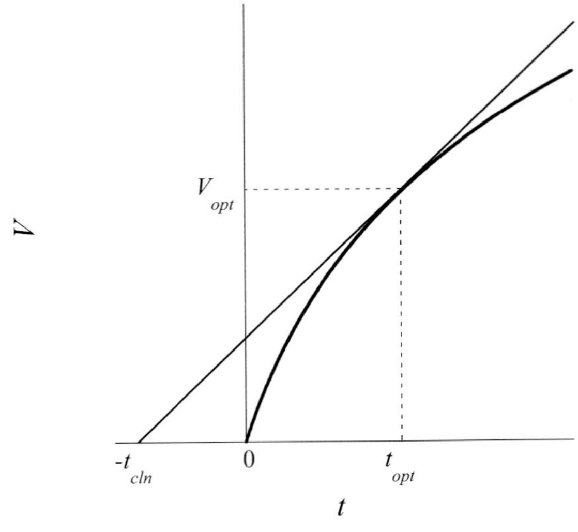

Fig. 3.28 Variation of filtrate volume, V, versus filtration time, t, and indication of optimum filtration time, t_{opt}, and corresponding volume of filtrate, V_{opt}

Problem 3.13 (continued)

$$\frac{d^2}{dt^2}\left(\frac{V\{t\}}{t+t_{cln}}\right)=\frac{d}{dt}\left(\frac{\frac{dV}{dt}(t+t_{cln})-V}{(t+t_{cln})^2}\right); \tag{3.700}$$

the classical rules of differentiation may again be invoked to write

$$\frac{d^2}{dt^2}\left(\frac{V\{t\}}{t+t_{cln}}\right)=\frac{\left(\frac{d^2V}{dt^2}(t+t_{cln})+\frac{dV}{dt}-\frac{dV}{dt}\right)(t+t_{cln})^2-\left(\frac{dV}{dt}(t+t_{cln})-V\right)2(t+t_{cln})}{\left((t+t_{cln})^2\right)^2}, \tag{3.701}$$

where cancellation of symmetrical terms, followed by dropping of $t + t_{cln}$ from both numerator and denominator leave

$$\frac{d^2}{dt^2}\left(\frac{V\{t\}}{t+t_{cln}}\right)=\frac{\frac{d^2V}{dt^2}(t+t_{cln})-2\frac{dV}{dt}(t+t_{cln})+2V}{(t+t_{cln})^3}. \tag{3.702}$$

At $t = t_{opt}$, Eq. (3.698) applies and Eq. (3.702) accordingly yields

$$\left.\frac{d^2}{dt^2}\left(\frac{V\{t\}}{t+t_{cln}}\right)\right|_{t_{opt}}=\left.\frac{\frac{d^2V}{dt^2}(t+t_{cln})-2\frac{V}{t+t_{cln}}(t+t_{cln})+2V}{(t+t_{cln})^3}\right|_{t_{opt}}, \tag{3.703}$$

which reduces to

$$\left.\frac{d^2}{dt^2}\left(\frac{V\{t\}}{t+t_{cln}}\right)\right|_{t_{opt}}=\left.\frac{\frac{d^2V}{dt^2}(t+t_{cln})-2V+2V}{(t+t_{cln})^3}\right|_{t_{opt}}=\left.\frac{\frac{d^2V}{dt^2}(t+t_{cln})}{(t+t_{cln})^3}\right|_{t_{opt}} \tag{3.704}$$

upon dropping factors from numerator and denominator, and then cancelling out symmetrical terms in numerator; one finally obtains

$$\left.\frac{d^2}{dt^2}\left(\frac{V\{t\}}{t+t_{cln}}\right)\right|_{t_{opt}}=\left.\frac{\frac{d^2V}{dt^2}}{(t+t_{cln})^2}\right|_{t_{opt}} \tag{3.705}$$

(continued)

Problem 3.13 (continued)

upon cancellation of $t + t_{cln}$ between numerator and denominator. Inspection of Fig. 3.26 unfolds a concavity for $V\{t\}$ facing down, that is, mathematically described by

$$\frac{d^2V}{dt^2} < 0; \tag{3.706}$$

insertion of Eq. (3.706) converts Eq. (3.705) to

$$\left.\frac{d^2}{dt^2}\left(\frac{V\{t\}}{t + t_{cln}}\right)\right|_{t_{opt}} < 0, \tag{3.707}$$

in view of $(t + t_{cln})^2 > 0$. Equations (3.695) and (3.707) are sufficient to guarantee a maximum for $V\{t\}$, when $t = t_{opt}$ as per Eq. (3.698).

(b) After defining auxiliary parameters a and b as

$$a \equiv \frac{Ac_P \Delta P}{\mu \kappa \rho_P V_f} \tag{3.708}$$

and

$$b \equiv \frac{c_P}{\rho_P V_f}, \tag{3.709}$$

respectively, it is possible to rewrite Eq. (3.673) as

$$at = \ln \frac{\varepsilon_{f,0}}{\varepsilon_{f,0} - bV} - bV, \tag{3.710}$$

and consequently calculate

$$\frac{d(at)}{d(bV)} = \frac{\frac{\varepsilon_{f,0}}{(\varepsilon_{f,0} - bV)^2}}{\frac{\varepsilon_{f,0}}{\varepsilon_{f,0} - bV}} - 1 = \frac{1}{\varepsilon_{f,0} - bV} - 1 \tag{3.711}$$

upon applying the rules of differentiation, and dropping common factors between numerator and denominator afterward; by the same token, it is possible to reformulate Eq. (3.698) as

(continued)

Problem 3.13 (continued)

$$\left.\left(\frac{d(bV)}{d(at)} = \frac{bV}{at + at_{cln}}\right)\right|_{at_{opt}}, \tag{3.712}$$

following multiplication of both sides by $b/a \neq 0$, and placement of a and b under the differential operators in view of their constancy. After reciprocals are taken of both sides, Eq. (3.712) turns to

$$\left.\left(\frac{1}{\frac{d(bV)}{d(at)}} = \frac{at + at_{cln}}{bV}\right)\right|_{at_{opt}}, \tag{3.713}$$

where the rule of calculation of the derivative of an inverse function has it that

$$\left.\left(\frac{d(at)}{d(bV)} = \frac{at + at_{cln}}{bV}\right)\right|_{at_{opt}}; \tag{3.714}$$

insertion of Eqs. (3.710) and (3.711) in (3.714) leads then to

$$\left.\left(\frac{1}{\varepsilon_{f,0} - bV} - 1 = \frac{\ln \frac{\varepsilon_{f,0}}{\varepsilon_{f,0} - bV} - bV + at_{cln}}{bV}\right)\right|_{at_{opt}}, \tag{3.715}$$

with splitting of the right-hand side unfolding

$$\left.\left(\frac{1}{\varepsilon_{f,0} - bV} - 1 = \frac{1}{bV} \ln \frac{\varepsilon_{f,0}}{\varepsilon_{f,0} - bV} - 1 + \frac{at_{cln}}{bV}\right)\right|_{at_{opt}} \tag{3.716}$$

or, after cancelling out -1 between sides,

$$\left.\left(\frac{1}{\varepsilon_{f,0} - bV} = \frac{1}{bV} \ln \frac{\varepsilon_{f,0}}{\varepsilon_{f,0} - bV} + \frac{at_{cln}}{bV}\right)\right|_{at_{opt}}. \tag{3.717}$$

Upon multiplying both sides by bV, Eq. (3.717) becomes

(continued)

Problem 3.13 (continued)

$$\left(\frac{bV}{\varepsilon_{f,0}-bV}=\ln\frac{\varepsilon_{f,0}}{\varepsilon_{f,0}-bV}+at_{cln}\right)\Bigg|_{at_{opt}}, \tag{3.718}$$

from which at_{cln} can be isolated as

$$\left(at_{cln}=\frac{bV}{\varepsilon_{f,0}-bV}-\ln\frac{\varepsilon_{f,0}}{\varepsilon_{f,0}-bV}\right)\Bigg|_{at_{opt}}; \tag{3.719}$$

a plot of $bV\big|_{at_{opt}}$ as a function of at_{cln}, for selected values of $\varepsilon_{f,0}$, is provided in Fig. 3.29. Note the resemblance of this plot with Fig. 3.26, although with left- and downward distortion of the curves, since the tangency point precedes the region of more notorious leveling off; once again, a larger fraction of voids initially available permits the filtration process be carried over for a longer period under optimal operation, since there is more room to accumulate suspended particles.

(c) Any change in monotony of a continuous function requires existence of a stationary point, which, in this case, should be characterized by

$$\frac{d\left(\frac{Ac_P\,\Delta P}{\mu\kappa\rho_P V_f}t\right)}{d\left(\frac{c_P}{\rho_P V_f}V\right)}=0, \tag{3.720}$$

(continued)

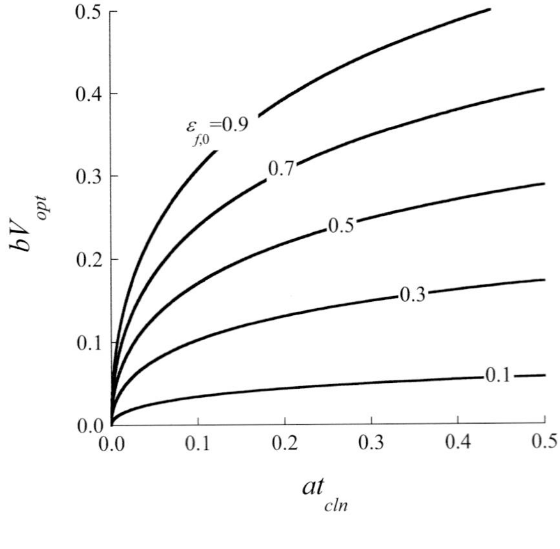

Fig. 3.29 Variation of optimum dimensionless filtrate volume, bV_{opt}, versus dimensionless filter cleaning time, at_{cln}, for selected values of initial filter porosity, $\varepsilon_{f,0}$

Problem 3.13 (continued)

stemming from the functional form of the lumped variables in Eq. (3.673); this is equivalent to writing

$$-\frac{\dfrac{\varepsilon_{f,0}(-1)}{\left(\varepsilon_{f,0}-\frac{c_P}{\rho_P V_f}V\right)^2}}{\dfrac{\varepsilon_{f,0}}{\varepsilon_{f,0}-\frac{c_P}{\rho_P V_f}V}}-1=0, \tag{3.721}$$

in view of the classical rules of differentiation. Upon reduction in number of fraction levels and collapse of similar factors, Eq. (3.721) simplifies to

$$\frac{1}{\varepsilon_{f,0}-\frac{c_P}{\rho_P V_f}V}=1, \tag{3.722}$$

which may instead appear as

$$\varepsilon_{f,0}-\frac{c_P}{\rho_P V_f}V=1 \tag{3.723}$$

after getting rid of denominators. Isolation of $c_P V/\rho_P V_f$ in Eq. (3.723) leads to

$$\frac{c_P}{\rho_P V_f}V=\varepsilon_{f,0}-1<0, \tag{3.724}$$

because $0<\varepsilon_{f,0}<1$ by definition; therefore, no change in monotony can occur when dimensionless volume spans the interval $[0, \infty[$, as originally claimed.

3.5 Materials and Devices

Besides special materials and devices for structural support/containment and direct contact, a number of design features are necessary to assure in situ sterilization, inoculation, sampling, cleaning, and maintenance/repair of industrial bioreactors—further to guaranteeing aseptic conditions under regular operation. Such features are meant for long-term utilization of such bioreactors, which constitute the vast majority of industrial-scale apparatuses.

Stainless steel is the standard construction material when sterile operation is aimed at (see Fig. 3.30a); besides its inherent corrosion-proof features, it is able to withstand from

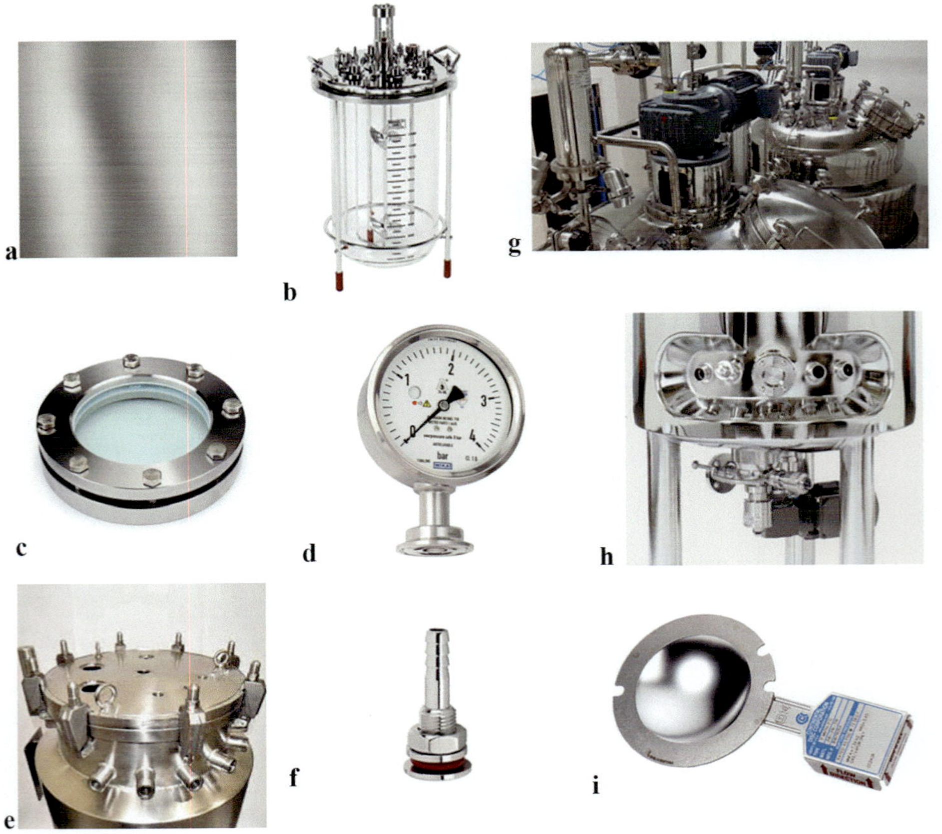

Fig. 3.30 Examples of bioreactor materials: (**a**) grade 304 stainless steel (courtesy of Pittella, Ronkonkoma NY, USA), (**b**) glass vessel (courtesy of Broadley-James Ltd., Bradford, UK), (**c**) sight glass (courtesy of Visilume Ltd., Glossop, UK), (**d**) pressure gauge (courtesy of WIKA, Klingenberg, Germany), (**e**) inlet/outlet ports (courtesy of Apparate-Maschine-Systeme Technology GmbH, Limbach-Oberfrohna, Germany), (**f**) air inlet/outlet adapter (courtesy of BCE Scientific, Buzzard, UK), (**g**) top entrance hole (courtesy from Zhenjiang East Biotech Equipment & Technology, Jiangsu, China), (**h**) harvest nozzle (courtesy of GEA, Dusseldorf, Germany), and (**i**) rupture disk (courtesy of Continental Disk Co., Zoeterwoude-Rijndijk, Netherlands)

high-level vacuum to processing pressures up to 3 atm, and up to temperatures of 150–180 °C, thus safely undergoing (repeated) sterilization and cleaning cycles. Grade 304 stainless steel is the most common austenitic stainless steel, and known indeed for its corrosion resistance; it contains Ni within 8–10.5%(w/w) and Cr within 18–20%(w/w), with Mn, Si, and C as minor alloying elements.

Mild steel with stainless steel cladding has also been an option, whereas less expensive grades of stainless steel can be employed for the jacket and other surfaces that do not contact the broth. Interior steel surfaces are polished to a bright mirror-finish to facilitate cleaning and sterilization, and wells on the interior of the vessel are ground flush before polishing; electropolishing is preferred over its mechanical counterpart, because the latter leaves tiny ridges and grooves in the metal susceptible of accumulating dirt and microorganisms. Copper and copper-containing alloys are to be avoided in all parts of the fermenter in contact with the culture, because even trace copper ions prove toxic to most cells. When the vessel capacity does not exceed ca. 30 L, glass is preferred as construction material for being smooth, nontoxic, corrosion-proof, and transparent (for easy inspection of contents); an example is provided in Fig. 3.30b. The lining in contact with the fermentation medium and broth should be nonreactive, nonabsorptive, and nonadsorptive.

A hole is normally provided at the top of large apparatuses to permit entrance of workers for cleaning, preventive maintenance, and repair (see Fig. 3.30g); the whole top is instead removable, in the case of smaller units. Furthermore, a harvest nozzle located at the lowest point of the bioreactor is meant for draining in full (see Fig. 3.30h). In the case of glass vessels, the ports and pipe fittings are usually equipped with stainless steel plates containing a number of screws/bolts (see also Fig. 3.30b). A (lighted) vertical sight glass (see Fig. 3.30c) is commonly installed in large-scale fermenters for routine inspection of their contents; nozzles are also present for addition of inoculum, liquid medium, acid/alkali, and antifoam agent, as well as a top- or bottom-entering stirrer shaft in the case of mechanical stirring (see Fig. 3.30e). Inlet and outlet tubing is provided for charge and discharge, under either batch or continuous operation, besides supply and exhaust pipes for air in aerobic cultures, nitrogen (in anaerobic cultures), or carbon dioxide (in phototrophic cultures; see Fig. 3.30f). Side ports for pH, temperature, and dissolved oxygen sensors are a minimum requirement (see again Fig. 3.30e), further to a pressure gauge for pressure monitoring (see Fig. 3.30d) and a rupture disk for emergency release of overpressure (see Fig. 3.30i), which are frequently mounted on the head plate. A steam-sterilizable port for inoculum and another port for sample collection should also be provided.

If the goal of some biotechnological company is to launch a new product, then optimization of bioreactor design and culture performance holds a much lower priority than releasing the product to the market quickly. This is so not only to assure safety of use by the prospective consumer, but also economic feasibility of the associated production process; recall, in this regard, the heuristic rule that the first to get to the market with a novel product will earn 80% of the profits associated with it and the like. Under these circumstances, ready-made, off-the-shelf, disposable bioreactors have become an attractive option; examples are provided in Fig. 3.31. None of the time-consuming cleaning and sterilization facilities and procedures are hereby required; in conventional fermentation, they are known to take a major proportion of capital and operating costs; for instance, boilers, clean-steam generators, steam-in-place, and cleaning-in-place systems are no longer needed. If single-use components are applied everywhere in a manufacturing plant, a chief fraction of infrastructure normally associated with bioprocessing can be eliminated altogether; often this more than compensates for the costs involved in having to replace disposable items, along with significant savings of energy, water, and cleaning chemicals, besides savings of time and labor.

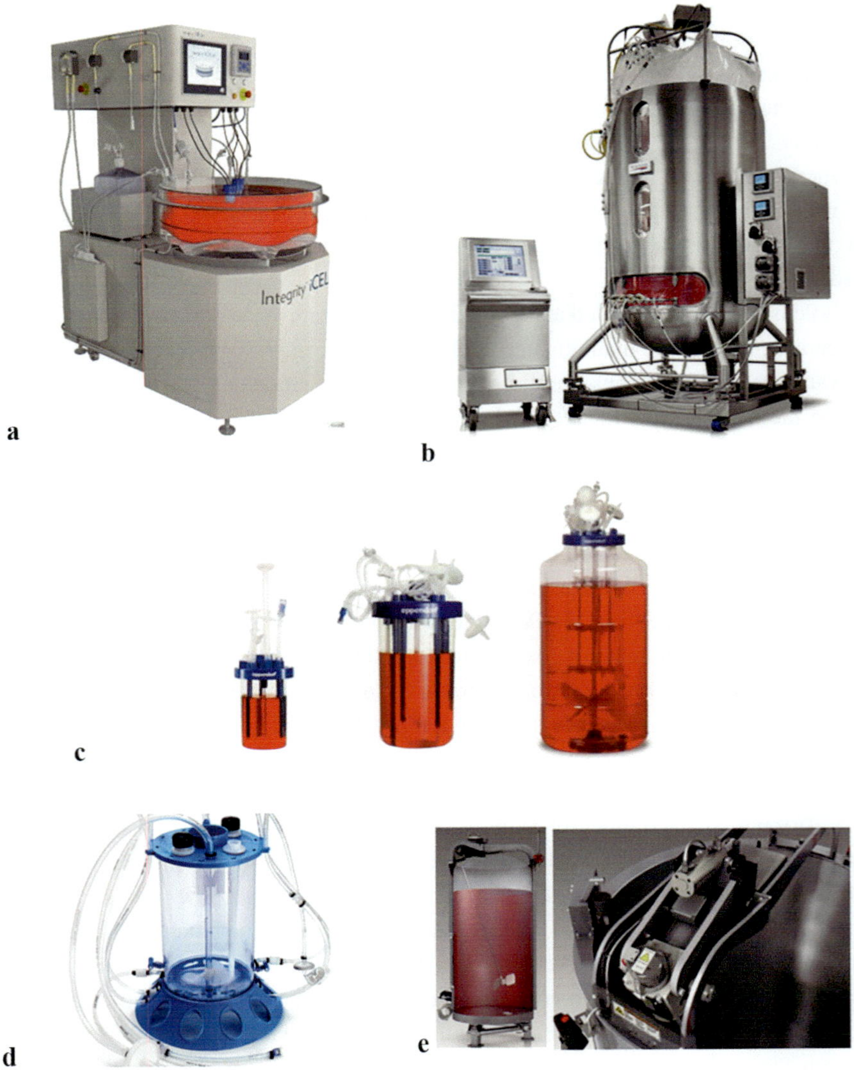

Fig. 3.31 Examples of disposable bioreactors: (**a**) industrial fixed bed bioreactor (courtesy of Pall Co., Port Washington NY, USA), (**b**) industrial built-in impeller stirred bioreactor (courtesy of Enigma—saintia solusindo, Bantem, Indonesia), bench scale stirred bioreactors (**c**) of different sizes (courtesy of Eppendorf, Hamburg, Germany) and (**d**) a single size (courtesy of Vekamaf Services, Rotterdam, Netherlands), and (**e**) cross-sectional view of stirred bioreactor (left) with detail of stirrer top support (right) (courtesy of Thermo Fisher Scientific, Waltham MA, USA)

Filter membranes and cartridges, depth filter pads, and aseptic connectors are among the single-use items employed routinely in large-scale bioprocessing. Furthermore, plastic process bags (with volumes up to 2–3 m^3 level) can replace stainless steel vessels in liquid storage, and disposable bioreactor or cell culture bags, with capacities up to 1 m^3, are also available for some applications. These upper sizes are constrained by the intrinsic strength of the bag material, coupled with the narrow range of options provided for mixing of culture contents; since inside stirring is difficult when collapsible construction materials are used, external agitation is preferred, namely, via placement of bioreactors bags on a rocking/tilting platform. In some cases, single-use bioreactors come with a disposable internal paddle mixing and sparger system. Performance in either case is obviously poorer than if using conventional bioreactors, in terms of mixing, heat transfer, and oxygen transfer, because efficient stirring cannot be produced in the first place; this is why their application is usually restricted to animal cell cultures, since oxygen demands and heat loads are low, and rapid mixing is not advised owing to the shear sensitivity of the cells. On the other hand, the increasing amount of new biotechnological products coming from animal cell culture has provided an impetus for use of disposable bioreactors, in alternative to repeatedly preparing small-to-medium amounts of media and hygienizing perennial equipment, namely, for production of recombinant and therapeutic proteins, antibodies, viruses, and vaccines.

An important concern derived from disposable item-based bioprocessing is the environmental impact of single-use plasticware, which must be incinerated a posteriori for being biohazardous material, or disposed of as landfill after autoclaving or chemical decontamination. A factor of 100-fold often applies when a large-scale bioreactor is replaced by single-use, small-scale counterparts, and the possibility to reuse or recycle most disposables is limited by the underlying risk of contamination or biological carryover—further to the need to separate the multiple types of plastic typically present in each item prior to recycling. The environmental burden associated to generation of plasticware, versus the environmental effects of steel production in terms of prior generation of electricity and high-temperature heat and waste disposal, may still point at the former option as preferred; however, a careful life cycle analysis is to be carried out on a case-by-case basis for a more educated decision.

Problem 3.14

Bioreactors are normally cylindrical in shape, and built from metal sheet duly welded at its extremities—thick enough to withstand pressure gauge during regular operation. Find the best geometric configuration of one such cylinder for an intended volume V, with (internal) radius R and height L, using minimum area of sheet, A, as objective function.

Solution

The total area of metal sheet necessary to construct a (hollow) cylinder is given by

$$A = 2\pi R^2 + 2\pi RL, \tag{3.725}$$

(continued)

Problem 3.14 (continued)

corresponding to top and bottom bases each of area πR^2, and lateral surface of area $2\pi RL$. The corresponding (inner) volume reads

$$V = \pi R^2 L, \tag{3.726}$$

which may be solved for L as

$$L = \frac{V}{\pi R^2}; \tag{3.727}$$

Equation (3.727) allows conversion of Eq. (3.725) to

$$A = 2\pi R^2 + 2\pi R \frac{V}{\pi R^2}, \tag{3.728}$$

or merely

$$A = 2\pi R^2 + \frac{2V}{R} \tag{3.729}$$

upon dropping πR from both numerator and denominator, with $A \equiv A\{R\}$ since V has been set in advance. The necessary condition for a minimum of area with regard to radius, that is,

$$\frac{dA}{dR} = 0, \tag{3.730}$$

becomes in this case

$$\frac{d}{dR}\left(2\pi R^2 + \frac{2V}{R}\right) = 4\pi R - \frac{2V}{R^2} = 0 \tag{3.731}$$

at the expense of Eq. (3.729), coupled with the classical theorems on differentiation; multiplication of both sides by $R^2/2$ leads to

$$2\pi R^3 - V = 0, \tag{3.732}$$

which prompts

(continued)

Problem 3.14 (continued)

$$R_{opt} = \left(\frac{V}{2\pi}\right)^{\frac{1}{3}} \tag{3.733}$$

following isolation of R, and referred to by subscript $_{opt}$ for abiding to Eq. (3.730) as optimality condition. Insertion of Eq. (3.733) converts Eq. (3.727) to

$$L_{opt} \equiv L\big|_{R=R_{opt}} = \left.\frac{V}{\pi R^2}\right|_{R=R_{opt}} = \frac{V}{\pi\left(\left(\frac{V}{2\pi}\right)^{\frac{1}{3}}\right)^2} = \frac{2^{\frac{2}{3}}\pi^{\frac{2}{3}}V^{1-\frac{2}{3}}}{\pi} = \frac{2^{\frac{2}{3}}}{\pi^{\frac{1}{3}}}V^{\frac{1}{3}}, \tag{3.734}$$

while ordered division of Eq. (3.734) by Eq. (3.733) unfolds

$$\frac{L_{opt}}{R_{opt}} = \frac{\frac{2^{\frac{2}{3}}}{\pi^{\frac{1}{3}}}V^{\frac{1}{3}}}{\frac{V^{\frac{1}{3}}}{2^{\frac{1}{3}}\pi^{\frac{1}{3}}}} = 2^{\frac{2}{3}+\frac{1}{3}}\pi^{\frac{1}{3}-\frac{1}{3}}V^{\frac{1}{3}-\frac{1}{3}}; \tag{3.735}$$

lumping of exponents gives finally rise to

$$\frac{L_{opt}}{R_{opt}} = 2 \tag{3.736}$$

for optimum geometric configuration, which is the heuristic rule utilized in cylindrical bioreactor design, that is, its height should equal twice its radius.

3.6 Aseptic Operation

Aside from the food sector, most fermentations are carried out with pure (or axenic) microbial cultures that require operation under aseptic conditions—to be maintained for periods extending to several days or even a couple of months; this requirement proves critical in the case of slow-growing cultures (e.g., microalgae, animal cells), otherwise they would be quickly overrun by contaminant microorganisms (e.g., bacteria).

Sterilization is the key step to assure asepsy; in fact, 3–5% of all industrial-scale fermentations are lost due to faulty sterilization, even though the frequency of unwanted contamination varies considerably between processes. For instance, no more than 2% fermentations aimed at antibiotic production are lost since said type of metabolite(s) confers an intrinsic protection against bacterial and phage contamination; conversely, contamination rates up to 15% have been reported in animal cell-mediated, large-scale production of specific proteins, owing to their inherent slow growth and need of nutrient-rich media.

Owing to the large capacities at stake, industrial fermenters are designed to carry out sterilization procedures in situ using pressurized steam; examples are conveyed in Fig. 3.32a–c, at distinct size scales. Such devices as internal structures, ports, nozzles, connections, and other attachments are thus to be kept to a minimum, so as to ensure that steam reaches all parts of the equipment in a more or less uniform fashion. Furthermore, the bioreactor should be exempt of crevices and stagnant areas since liquids or solids tend to accumulate there; this is why polished welded joints are preferred to alternative coupling methods. Small cracks or gaps in joints, as well as fine fissures in welds are likely to harbor microbial contaminants, so they should be avoided at all cost in fermenter construction.

For effective sterilization, air must be quantitatively displaced by steam from the inside of vessels, pipes, and connections, and all nutrient medium and air supplied afterward must be sterile. Immediately after the flow of steam through the fermenter is discontinued,

Fig. 3.32 Examples of in situ sterilization systems for (**a**) bench scale bioreactor (courtesy of BioEquip, Shanghai, China), (**b**) pilot scale bioreactor (courtesy of Beijing Sentinain Technology Inc., Beijing, China), and (**c**) industrial-scale bioreactor (courtesy of Solaris Biotechnology, Mantovano, Italy), with (**d**) detail of steam inlet and outlet ports to jacket (courtesy of bbi-biotech GmbH, Berlin, Germany)

introduction of air (previously sterilized) via filtration helps maintain a slight overpressure relative to the surroundings, thus discouraging entry of airborne contaminants. Filters that prevent passage of microorganisms are as well fitted at outlet air/gas exhaust lines; they not only help contain the culture inside the vessel and avoid airborne release of process microorganisms to the environment, but also prevent contamination outside-in should an accidental drop occur in operating pressure.

Since liquid flow to and from a fermenter is controlled by valves, construction of these devices should be compatible with aseptic operation; a few examples are conveyed in Fig. 3.33. Such common design as gate valves tends to leak around their stem, and consequently accumulate broth solids in the closing mechanism; hence, they prove incompatible with high levels of sterility, so pinch and diaphragm valves (see Fig. 3.33a) are instead recommended. Their underlying design resorts to a flexible sleeve or weir, so the closing mechanism remains isolated from the contents of the pipe without dead spaces within the valve structure. Rubber or neoprene, able to withstand repeated sterilization cycles, has been widely employed to fashion the valve closure; the chief disadvantage is their need to be checked for wear at regular times, in attempts to prevent valve failure. Wherever strict asepsy is not required, ball and plug valves have been successfully used to reduce costs associated with bioreactor construction; examples are included in Fig. 3.33b, c, respectively.

Another point in stirred bioreactors susceptible to contamination is where the stirrer shaft enters the vessel body; the gap in between is to be duly sealed, yet operation for long periods may cause wearing of the seal and pave the way for airborne contaminants. Mechanical seals are normally utilized in large fermenters, in which one part of the assembly is stationary while the other rotates with the shaft; the precision-machined surfaces of both components are pressed together by springs or expanding bellows, and cooled and lubricated by water. Mechanical seals with running surfaces made of silicon carbide, paired with tungsten carbide are frequently recommended for fermenter application. Stirrer seals are especially critical if the bioreactor possesses a bottom-entering stirrer—in which case double mechanical seals are typically installed to prevent leakage of liquid medium by gravity. In smaller apparatuses (i.e., up to ca. 800 L), meant to handle

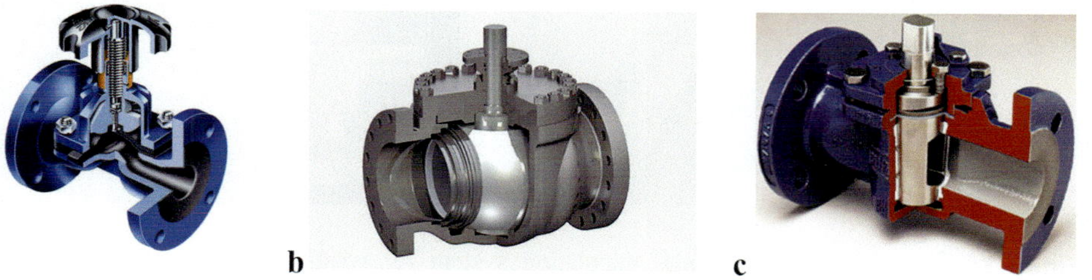

a b c

Fig. 3.33 Examples of process valves as cross-sectional views of (**a**) weir-type diaphragm valve (courtesy of Crane Process Flow Technologies, Gwent, UK), (**b**) ball valve (courtesy from Wenzhou Goldtop Valve, Wenzhou, China), and (**c**) plug valve (courtesy of Flexachem, Cork, Ireland) relatively nonviscous broths under aeration, magnetic drives can be employed to couple the stirrer shaft with the motor and accordingly bring about sufficient mixing; the fermenter body is not pierced by the stirrer shaft in this configuration. A magnet in a housing, on the outside of the fermenter, is driven by the stirrer motor; another magnet is placed inside and attached to the end of the stirrer shaft, and held in place by bearings.

Of particular concern is assurance of aseptic conditions throughout inoculation and sampling of liquid medium in a bioreactor; examples of devices utilized to perform these operations are provided in Fig. 3.34. Besides their being typically built in stainless steel to avoid corrosion and facilitate sterilization, no contact with the outer environment is permitted; hence, any open ends shown are actually meant for connection via/to suitable piping.

In large-scale fermentations, inocula are transferred from a smaller (or secondary) vessel to the larger (or primary) vessel; both are kept under (slight) air overpressure to avoid contamination from the environment. A typical layout for pipes and valves in this case is

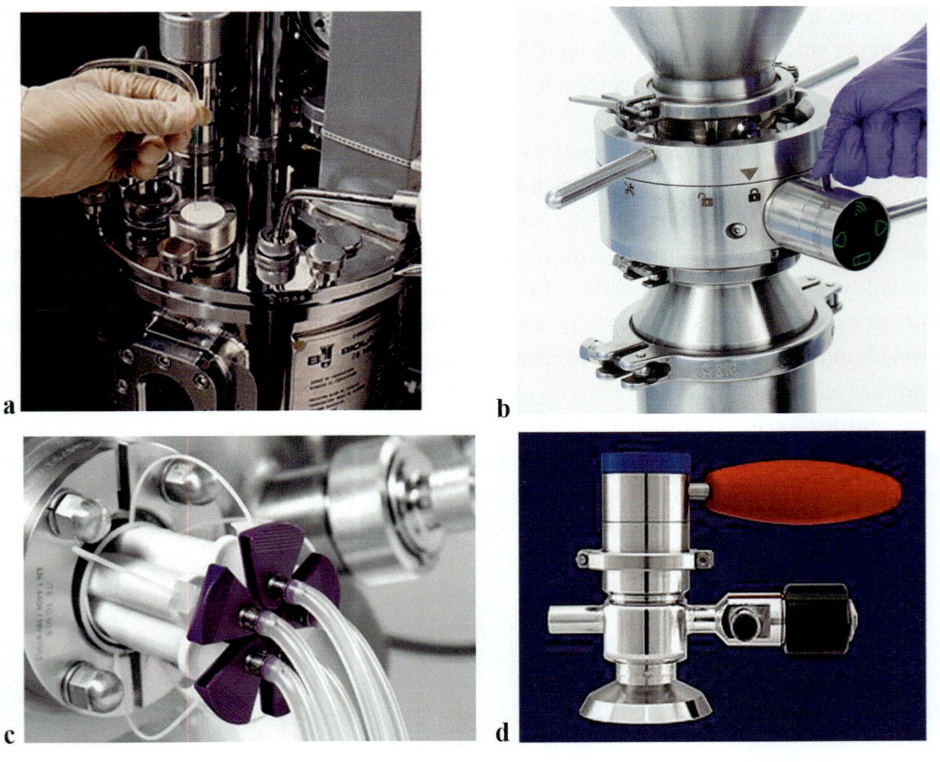

Fig. 3.34 Examples of (**a**) aseptic inoculation system for small-scale bioreactor (courtesy of QualiTru, Oakdale MN, USA) and (**b**) split butterfly valve for large-scale inoculation (courtesy of Chargepoint Technology, Liverpool, UK), (**c**) aseptic sampling system for large-scale bioreactor (courtesy of Sartorius Stedim Biotech, Bohemia NY, USA), and (**d**) three-way valve for periodic sampling (courtesy of Kieselmann GmbH, Knittlingen, Germany)

provided in Fig. 3.35a. Besides connecting pipes and regular valves, a steam trap and two connection/bleeding valves are also included (see illustrations in Fig. 3.35b, c, respectively); an example of inoculum vessel, as auxiliary mobile apparatus, is provided in Fig. 3.35d.

The sequence of steps involved in preliminary sterilization as per Fig. 3.35a, followed by inoculation proper, are tabulated in Table 3.1. The primary and secondary vessels, together with all piping and valves, are sterilized in advance and separately before the cells are added thereto or grown therein. The first step is connection of the secondary vessel to the network, through connectors A and B. Since these connectors were previously open, they must be sterilized in advance; steam is thus flown chiefly through H, F, and I before leaving at the steam trap, as well as through H and slowly bled at A, and through H and F and slowly bled at B. After several minutes, the connection between secondary and primary vessels has become sterile; it is then cooled down by passing sterile air, under positive pressure. The next step is supply of inoculum from the secondary vessel into the main one, resorting to overpressured sterile air, for a sufficient time to quantitatively remove the inoculum and assure emptying of the line between vessels afterward; this constitutes indeed the simplest method for aseptic transfer, via blowing the culture into the bioreactor. Isolation of the fermenter ensues, thus allowing disconnection of the secondary vessel prior to engaging in regular operation.

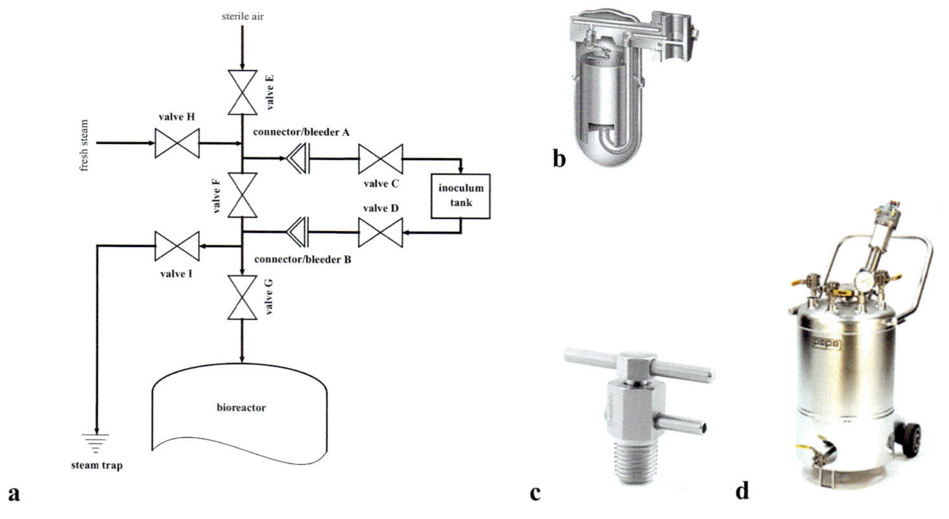

Fig. 3.35 Typical layout for bioreactor inoculation, as (**a**) schematic view of pipe and valve connections, with specification of fresh steam supply (*s*), sterile air supply (*a*), inoculum tank (*i*), steam trap disposal (*t*), and bioreactor (*b*), (**b**) cross-sectional view of steam trap (courtesy of Armstrong International, Three Rivers MI, USA), (**c**) bleed valve (courtesy of Swagelok, Solon OH, USA), and (**d**) mobile inoculation vessel (courtesy of Pope Scientific, Saukville WI, USA)

Table 3.1 Sequence of operations involved in bioreactor inoculation, with specification of valves open (□), partially open/closed (▒), and closed (■), as well as fresh steam supply (s), sterile air supply (a), inoculum tank (i), steam trap disposal (t), and bioreactor (b)

Step	Valves									Path
	A	B	C	D	E	F	G	H	I	
Connection of inoculum tank	□	□	■	■	■	■	■	■	■	$A \leftarrow C - i - D \rightarrow B$
Sterilization of connecting lines	▒	▒	■	■	■	□	■	□	□	$s \rightarrow H - F - I \rightarrow t$ $s \rightarrow H \rightarrow A$ $s \rightarrow H - F \rightarrow B$
Cooling of connecting lines	■	■	■	■	□	□	□	■	■	$a \rightarrow E - F - G \rightarrow b$
Inoculation	□	□	□	□	□	■	□	■	■	$a \rightarrow E - A - C - i - D - B - G \rightarrow b$
Disconnection of inoculum tank	□	□	■	■	■	■	■	■	■	$A \leftarrow H - i - I \rightarrow B$
Regular operation	■	■	■	■	□	□	□	■	■	$a \rightarrow E - F - G \rightarrow b$

Sampling ports are fitted to ducts through the wall of a bioreactor to permit continuous or intermittent removal of liquid medium for analysis and control; a typical layout compatible with aseptic handling is shown in Fig. 3.36a. Note the common presence of a three-way valve (see Fig. 3.36b, c), which stops flow momentarily in the regular direction while diverging it to the sampling vessel where an aliquot is taken before being returned to the initial position again. If the frequency of sampling so justifies, aliquots may be collected automatically (see Fig. 3.36d).

The sequence of steps involved in previous sterilization, followed by inoculation proper, is tabulated in Table 3.2. A steam barrier is maintained between reactor and outside environment under normal operation to preserve asepsy. The pipe connecting the fermenter to the sampling port is then sterilized, and bleeding permitted at B. Right before sampling, A is open briefly to release a portion of medium to help cool the pipe while carrying away any condensate that would otherwise dilute the sample; this liquid is discarded. Sampling proper takes place through B. Once complete, another stage of sterilization of sample line takes place, including bleeding of residual medium; this prevents any putative contaminants, which might have gained access during sampling, from travelling into the fermenter before resuming regular operation.

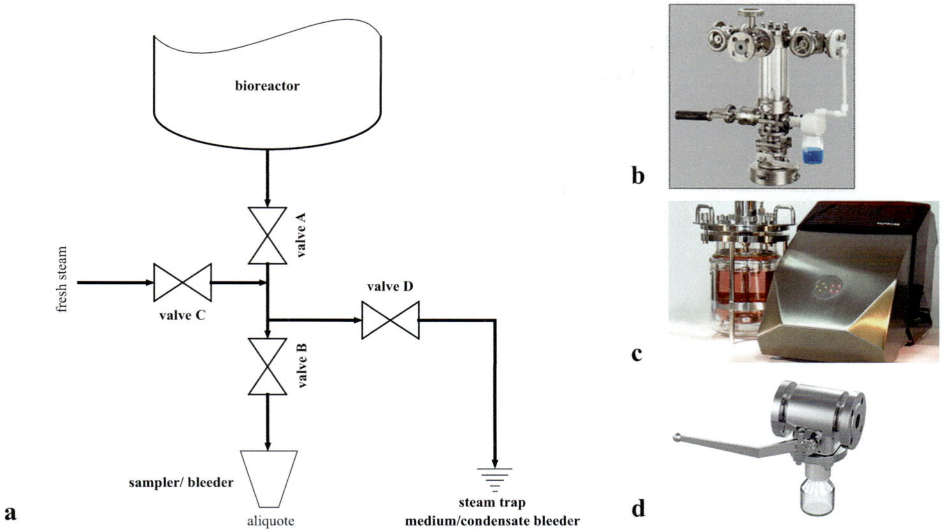

Fig. 3.36 Typical layout for bioreactor sampling as (**a**) schematic view of pipe and valve connections, with specification of fresh steam supply (*s*), aliquot collection (*a*), steam trap disposal (*t*), and bioreactor (*b*), (**b**) manual sampling system (courtesy of Belfast Controls, Dubai, UAE) and (**c**) automatic sampling apparatus (courtesy of bbi-biotech, Berlin, Germany), and (**d**) inline sampling valve (courtesy of AMG PEsch, Cologne, Germany)

Table 3.2 Sequence of operations involved in bioreactor sampling, with specification of valves open ( ), partially open/closed ( ), and closed ( ), as well as fresh steam supply (*s*), steam disposal (*t*), aliquot (*a*), and bioreactor (*b*)

Step	Valves				Path
	A	B	C	D	
Regular operation	closed	closed	open	open	$s \rightarrow C - D \rightarrow t$
Sterilization of connecting lines	closed	partially open/closed	partially open/closed	closed	$s \rightarrow C - B \rightarrow a$
Cooling of connecting lines	open	closed	closed	open	$b \rightarrow A - D \rightarrow t$
Sampling	open	open	closed	closed	$b \rightarrow A - B \rightarrow a$
Sterilization of connecting lines	closed	open	open	closed	$s \rightarrow C - B \rightarrow a$

Problem 3.15

A chemostat will be fed liquid medium, previously sterilized continuouswise with condensing steam in a concentric tube-and-shell heat exchanger; assume plug flow of both fluid phases.

(continued)

Problem 3.15 (continued)

(a) Determine the temperature profile of medium along the sterilizer.
(b) Express the amount of heat added to the medium, as a function of the logarithmic mean of the driving force for heat transfer.
(c) Derive a mathematical expression relating degree of reduction in viable microorganisms to area of heat transfer spanned by medium.

Solution

(a) Given the double-tube configuration and the plug flow pattern, with condensing steam supplied via the outer tube and medium flown through the inner one, an enthalpic balance to the sterilizer will take the form

$$A_i v_m \rho_m c_{P,m}\, dT_m = \lambda_s\, dm_s, \tag{3.737}$$

pertaining to the two phases from a thermodynamic point of view; complemented by

$$A_i v_m \rho_m c_{P,m}\, dT_m = U(T_s - T_m)dA_t, \tag{3.738}$$

encompassing the two phases from a kinetic point of view. Here, A_i and A_t denote inner cross-sectional area and outer heat transfer area, respectively; v_m, ρ_m, and $c_{P,m}$ denote velocity, mass density, and thermal heat capacity of medium, respectively; T_m and T_s denote temperature of medium and steam, respectively; λ_s denotes latent heat of condensation of steam; m_s denotes mass flow rate of steam; and U denotes overall heat transfer coefficient between phases. Integration of Eq. (3.738) may proceed, via separation of variables, as

$$\int_{T_{m,in}}^{T_m} \frac{-d\widetilde{T}_m}{T_s - \widetilde{T}_m} = -\frac{U}{A_i v_m \rho_m c_{P,m}} \int_0^{A_t} d\widetilde{A}_t \tag{3.739}$$

at the expense of

$$T_m|_{A_t=0} = T_{m,in} \tag{3.740}$$

for boundary condition, with $T_{m,in}$ denoting (given) inlet temperature of medium, and U, ρ_m, and $c_{P,m}$ taken as independent of temperature; the fundamental theorem of integral calculus will now be invoked to write

(continued)

Problem 3.15 (continued)

$$\ln\{T_s - \widetilde{T}_m\}\Bigg|_{T_{m,in}}^{T_m} = -\frac{U}{A_i v_m \rho_m c_{P,m}}\,\widetilde{A}_t\Bigg|_0^{A_t} \tag{3.741}$$

in lieu of Eq. (3.739), which degenerates to

$$\ln\frac{T_s - T_m}{T_s - T_{m,in}} = -\frac{U}{A_i v_m \rho_m c_{P,m}} A_t \tag{3.742}$$

once upper and lower limits of integration are brought on board. After taking exponentials of both sides, Eq. (3.742) gives rise to

$$\frac{T_s - T_m}{T_s - T_{m,in}} = \exp\left\{-\frac{U}{A_i v_m \rho_m c_{P,m}} A_t\right\}, \tag{3.743}$$

where isolation of T_m conveys

$$T_m = T_s - (T_s - T_{m,in})\exp\left\{-\frac{U}{A_i v_m \rho_m c_{P,m}} A_t\right\}, \tag{3.744}$$

thus providing a descriptor for the temperature profile of medium along the heat exchanger as the transfer area in the sterilizer is spanned thereby.

(b) Integration of Eq. (3.737), by separation of variables, leads to

$$A_i v_m \rho_m c_{P,m} \int_{T_{m,in}}^{T_m} d\widetilde{T}_m = \lambda_s \int_0^{m_s} d\widetilde{m}_s \tag{3.745}$$

under a constant λ_s, with the aid of

$$T_m\big|_{m_s=0} = T_{m,in} \tag{3.746}$$

as boundary condition compatible with Eq. (3.740); after invoking again the fundamental theorem of integral calculus, Eq. (3.745) becomes

$$A_i v_m \rho_m c_{P,m}\,\widetilde{T}_m\Bigg|_{T_{m,in}}^{T_m} = \lambda_s\,\widetilde{m}_s\Bigg|_0^{m_s}, \tag{3.747}$$

which readily reduces to

(continued)

Problem 3.15 (continued)

$$A_i v_m \rho_m c_{P,m}(T_m - T_{m,in}) = \lambda_s m_s. \tag{3.748}$$

Isolation of $A_i v_m \rho_m c_{P,m}$ from Eq. (3.748) unfolds

$$A_i v_m \rho_m c_{P,m} = \frac{\lambda_s m_s}{T_m - T_{m,in}}, \tag{3.749}$$

which will be inserted in Eq. (3.742) to obtain

$$\ln \frac{T_s - T_m}{T_s - T_{m,in}} = -\frac{U}{\frac{\lambda_s m_s}{T_m - T_{m,in}}} A_t; \tag{3.750}$$

upon removal of the minus sign in the right-hand side on account of taking the reciprocal of the argument of the logarithmic function, Eq. (3.750) reappears as

$$\ln \frac{T_s - T_{m,in}}{T_s - T_m} = \frac{U}{\frac{\lambda_s m_s}{T_m - T_{m,in}}} A_t, \tag{3.751}$$

while solution for $\lambda_s m_s$ ensues as

$$\lambda_s m_s = UA_t \frac{T_m - T_{m,in}}{\ln \frac{T_s - T_{m,in}}{T_s - T_m}} = UA_t \frac{T_m - T_s + T_s - T_{m,in}}{\ln \frac{T_s - T_{m,in}}{T_s - T_m}} = UA_t \frac{T_s - T_{m,in} - (T_s - T_m)}{\ln \frac{T_s - T_{m,in}}{T_s - T_m}} \tag{3.752}$$

with T_s meanwhile added and subtracted to the numerator in the right-hand side, followed by algebraic reorganization thereof. Note that the left-hand side of Eq. (3.752) represents the total heat transferred per unit time between phases, say Q, according to

$$Q = \lambda_s m_s; \tag{3.753}$$

elimination of $\lambda_s m_s$ between Eqs. (3.752) and (3.753) leads finally to

(continued)

Problem 3.15 (continued)

$$Q = UA_t \Delta T_{lm}, \tag{3.754}$$

should ΔT_{lm} abide to

$$\Delta T_{lm} = \frac{T_s - T_{m,in} - (T_s - T_m)}{\ln \frac{T_s - T_{m,in}}{T_s - T_m}} \tag{3.755}$$

known as logarithmic mean, between entrance and any point along the heat exchanger, of the temperature driving force for heat transfer.

(c) The rate of death of microorganisms has classically been described by first-order kinetic constant k_d, namely,

$$\frac{dC_X}{dt} = -k_d C_X \tag{3.756}$$

that mimics Eq. (3.18), knowing that

$$C_X|_{t=0} = C_{X,0} \tag{3.757}$$

prevails at inlet (see Eq. (3.19)). On the other hand, the relationship between time elapsed upon entrance into the sterilizer and transfer area tracked so far by the medium looks like

$$dA_t = 2\pi R_i v_m \, dt, \tag{3.758}$$

as long as R_i denotes radius of inner tube and plug flow prevails; this permits transformation of Eq. (3.756) to

$$\frac{dC_X}{\frac{dA_t}{2\pi R_i v_m}} = -k_d C_X, \tag{3.759}$$

or else

$$-\frac{dC_X}{dA_t} = \frac{k_d}{2\pi R_i v_m} C_X \tag{3.760}$$

(continued)

Problem 3.15 (continued)

following division of both sides by $2\pi R_i v_m$. Integration by separation of variables converts Eq. (3.760) to

$$\int_{C_{X,0}}^{C_X} \frac{d\widetilde{C}_X}{\widetilde{C}_X} = -\frac{1}{2\pi R_i v_m} \int_0^{A_t} k_d\left\{T_m\left\{\widetilde{A}_t\right\}\right\} d\widetilde{A}_t, \tag{3.761}$$

where Eq. (3.757) was already brought on board and the ultimate dependence of k_d on A_t through T_m was made explicit. Combination with Eq. (3.744) produces

$$\int_{C_{X,0}}^{C_X} \frac{d\widetilde{C}_X}{\widetilde{C}_X} = -\frac{1}{2\pi R_i v_m} \int_0^{A_t} k_d\left\{T_s - (T_s - T_{m,in})\exp\left\{-\frac{U}{A_i v_m \rho_m c_{P,m}}\widetilde{A}_t\right\}\right\} d\widetilde{A}_t \tag{3.762}$$

as alias to Eq. (3.761); application of the fundamental theorem of integral calculus to the left-hand side justifies, in turn,

$$\ln \widetilde{C}_X\Big|_{C_{X,0}}^{C_X} = -\frac{1}{2\pi R_i v_m} \int_0^{A_t} k_d\left\{T_s - (T_s - T_{m,in})\exp\left\{-\frac{U}{A_i v_m \rho_m c_{P,m}}\widetilde{A}_t\right\}\right\} d\widetilde{A}_t. \tag{3.763}$$

Equation (3.763) readily yields

$$\ln \frac{C_X}{C_{X,0}} = -\frac{1}{2\pi R_i v_m} \int_0^{A_t} k_d\left\{T_s - (T_s - T_{m,in})\exp\left\{-\frac{U}{A_i v_m \rho_m c_{P,m}}\widetilde{A}_t\right\}\right\} d\widetilde{A}_t \tag{3.764}$$

or, after isolating $C_X/C_{X,0}$,

(continued)

Problem 3.15 (continued)

$$\frac{C_X}{C_{X,0}} = \exp\left\{ -\frac{1}{2\pi R_i v_m} \int_0^{A_t} k_d \left\{ T_s - (T_s - T_{m,in}) \exp\left\{ -\frac{U}{A_i v_m \rho_m c_{P,m}} \widetilde{A}_t \right\} \right\} d\widetilde{A}_t \right\}. \tag{3.765}$$

All is left now is expressing k_d as a function of temperature, namely,

$$k_d = k_{d,0} \exp\left\{ -\frac{E_{act}}{\mathscr{R}T} \right\} \tag{3.766}$$

as per Arrhenius' law (see Eq. (3.20)), with preexponential factor $k_{d,0}$ and activation energy E_{act} to be made available, besides ideal gas constant $\mathscr{R}$; insertion of Eq. (3.766) in Eq. (3.765) finally supports

$$\frac{C_X}{C_{X,0}} = \exp\left\{ -\frac{k_{d,0}}{2\pi R_i v_m} \int_0^{A_t} \exp\left\{ -\frac{E_{act}}{\mathscr{R}\left(T_s - (T_s - T_{m,in}) \exp\left\{ -\frac{U}{A_i v_m \rho_m c_{P,m}} \widetilde{A}_t \right\} \right)} \right\} d\widetilde{A}_t \right\}, \tag{3.767}$$

which conveys degree of reduction in viable microorganisms, $C_X/C_{X,0}$, as a function of area of heat transfer, A_t, already spanned by medium, with numerical integration needed to calculate the outstanding integral.

Further Reading

Aiba, S., Humphrey, N. F., & Millis, N. F. (1965). *Biochemical engineering*. Academic Press.

Amaha, M., & Sakaguchi, K. (1957). The mode and kinetics of death of the bacterial spores by moist heat. *Journal of General and Applied Microbiology, 3*, 163–192.

Battley, E. H. (1979). Alternative methods of calculating free energy change accompanying the growth of *S. cerevisiae* (Hansen) on three substrates. *Biotechnology and Bioengineering, 21*, 1929–1961.

Cameron, B. (1987). Mechanical seals for bioreactors. *Chemical Engineering, 442*, 41–42.

Chisti, Y. (1992). Assure bioreactor sterility. *Chemical Engineering Progress, 88*, 41–42.

Chopra, C. L., Qasi, G. N., Chaturvedi, S. K., Gaind, C. N., & Atal, C. K. (1981). Production of citric acid by submerged fermentation. Effect of medium sterilisation at pilot-plant level. *Journal of Chemical Technology and Biotechnology, 31*, 122–126.

Deindoerfer, F. H., & Humphrey, A. E. (1959). Analytical method for calculating heat sterilization times. *Applied Microbiology, 7*, 256–274.

Dupré, A. (1867). Sixième memoire sur la théorie mécanique de la chaleur. *Annales de Chimie Physique, 11*, 194–220.

Karmanev, D. G. (1991). Model of the biofilm structure of *Thiobacillus ferrooxidans*. *Journal of Biotechnology, 20*, 51–64.

Khalili, P., & Vasiliu, D. (2010). An extension of the mean value theorem for integrals. *International Journal of Mathematical Education in Science and Technology, 41*, 707–710.

Parker, A. (1958). Sterilization of equipment, air and media. In R. Steel (Ed.), *Biochemical engineering* (pp. 97–121). Heywood.

Richards, J. W. (1968). *Introduction to industrial sterilization*. Academic Press.

Schlegel, H. G., & Zaborosch, C. (1993). *General microbiology*. Cambridge University Press.

Wang, D. I.-C., Scharer, J., & Humphrey, A. E. (1964). Kinetics of death of bacterial spores at elevated temperatures. *Applied Microbiology, 12*, 451–454.

Index

F. X. Malcata, *Operation Fundamentals in Bioreactor Engineering—Medium Handling*, https://doi.org/10.1007/978-3-032-03172-3

D

E

F

K

L

M

N